T0190140

Communications
in Computer and Information Science   1458

More information about this series at http://www.springer.com/series/7899

Zohreh Molamohamadi ·
Erfan Babaee Tirkolaee · Abolfazl Mirzazadeh ·
Gerhard-Wilhelm Weber (Eds.)

Logistics and Supply Chain Management

7th International Conference, LSCM 2020
Tehran, Iran, December 23–24, 2020
Revised Selected Papers

 Springer

Editors
Zohreh Molamohamadi ⓘ
Kharazmi University
Tehran, Iran

Abolfazl Mirzazadeh ⓘ
Kharazmi University
Tehran, Iran

Erfan Babaee Tirkolaee ⓘ
Istinye University
Istanbul, Turkey

Gerhard-Wilhelm Weber ⓘ
Poznań University of Technology
Poznań, Poland

ISSN 1865-0929   ISSN 1865-0937 (electronic)
Communications in Computer and Information Science
ISBN 978-3-030-89742-0   ISBN 978-3-030-89743-7 (eBook)
https://doi.org/10.1007/978-3-030-89743-7

This Springer imprint is published by the registered company Springer Nature Switzerland AG
The registered company address is: Gewerbestrasse 11, 6330 Cham, Switzerland

Preface

The Seventh International Conference on Logistics and Supply Chain Management (LSCM 2020) was held virtually in Tehran, Iran, during December 23–24, 2020. LSCM 2020 was organized by Kharazmi University with the collaboration of the Iran Logistics Society in six different panels. It attracted the attention of faculty members, scholars, educators, industry experts, and university students from all around the world.

LSCM 2020 aimed at providing a platform to transfer information, experience, and scientific and research findings of the recent theoretical and practical achievements in logistics and supply chain in both research and industry. The conference was an extraordinary opportunity to encourage experts of logistics and supply chain management to respond to current national and international supply chain issues and to enhance university-industry collaboration. In addition to paper presentations, the conference included several invited lectures and keynote speeches.

The subjects covered by LSCM 2020 included, but not were not limited to, the following:

- Strategic supply chain management
- Sustainable supply chain management
- Green and sustainable supply chain management
- Global supply chain management
- Lean/flexible/agile supply chain
- Pricing in supply chain
- Expert systems and information technology in supply chain
- Value creation in supply chain
- Optimization and decision making: methods and algorithms
- Marketing and supply/demand management
- Inventory control, production planning, and scheduling
- Transportation/distribution management
- Human resource management in supply chain
- Supply chain risk management
- Supply chain management in an uncertain environment

The conference involved the participation of people from 40 countries, including international scientific committee members and reviewers, and the submission of papers from more than 60 internationally-affiliated authors.

The conference team received 420 English and Persian papers, which were reviewed by at least three international reviewers (single blind reviews). Taking into consideration the invaluable reviewers' comments in the first round of review, the papers were reviewed once more (with much stricter criteria) to select the most appropriate ones for publication in this Springer volume. Finally, the two-round review process resulted in the selection of 20 papers (18 full papers and 2 short ones) for this book in the Communications in Computer and Information Science series.

LSCM 2020 provided an enjoyable and energetic knowledge-transferring atmosphere for its participants (as some feedback revealed), and the conference team members worked hard to foster a friendly and professional environment. We are grateful to everyone involved in the organization of the conference and to all the participants.

August 2021

Zohreh Molamohamadi
Erfan Babaee Tirkolaee
A. Mirzazadeh
Gerhard-Wilhelm Weber

Organization

General Chair

Azizollah Habibi   Kharazmi University, Iran

Scientific Committee Manager

Abolfazl Mirzazadeh   Kharazmi University, Iran

Scientific Committee

Zohreh Molamohamadi (Co-manager)	Kharazmi University, Iran
Gerhard-Wilhem Weber (Editor)	Poznan University of Technology, Poland
Erfan Babaee Tirkolaee (Editor)	Istinye University, Turkey
Ruben Ruiz	Valencia Polytechnic University, Spain
Alexander Dolgui	IMT Atlantique, France
Turan Paksoy	Kenya Technical University, Turkey
Sankar Kumar Roy	Vidyasagar University, India
Sadia Samar Ali	King Abdul Aziz University, Saudi Arabia
Tatiana Tchemisova Cordeiro	University of Aveiro, Portugal
Michael Pecht	University of Maryland, USA
Kathryn Stecke	University of Texas at Dallas, USA
Maher Agi	Rennes School of Business, France
Richard Allmendinger	University of Manchester, UK
Bernardo Almada-Lobo	University of Porto, Portugal
Zeynep Alparslan Gok	Süleyman Demirel University, Turkey
Gursel A. Suer	Ohio University, USA
Fayçal Belkaid	Abou Bekr Belkaid University of Tlemcen, Algeria
Marilisa Botte	University of Naples Federico II, Italy
Jaouad Boukachour	University du Havre, France
Leopoldo E. Cárdenas-Barrón	Tecnológico de Monterrey, Mexico
Kevin Cullinane	University of Gothenburg, Sweden
Eren Özceylan	Gaziantep University, Turkey
Oliveria Fabiana	Amazonas State University, Brazil
Michael G. Kay	North Carolina State University, USA
Michel Gendreau	Polytechnique Montreal, Canada
Josef Jablonsky	University of Economics, Czech Republic

Mohd Khairol Anuar Ariffin	University Putra Malaysia, Malaysia
Sachin Kumar	University of Plymouth, UK
Farouk Yalaoui	University of Technology of Troyes, France
Zulkiflle Leman	University Putra Malaysia, Malaysia
Mahantesh Nadakatti	Cogte Institute of Technology, India
Gholamreza Nakhaeizadeh	Karlsruhe Institute of Technology, Germany
Mustapha Oudani	International University of Rabat, Morocco
Chefi Triki	Sultan Qaboos University, Oman
Ahmet Özturk	Undag Univesity, Turkey
Jalal Safari	Siemens, Norway
Konstantina Skouri	University of Ioannina, Greece

Executive and Coordinating Committee

H. Ghaffari	Iran
M. V. Sebt	Iran
H. D. Ardakani	Iran
H. Selki	Iran
B. Moghtadai	Iran
L. Chehreghani	Iran
Z. Alihoseini	Iran
M. Musa	Iran
M. Ashtari	Iran
M. Haezi	Iran
S. Mirzaei	Iran
Sh. Izadi	Iran
M. R. Fazli	Iran
Z. Pourfarash	Iran
E. Akbarian	Iran
F. Abdi	Iran
E. Ebrahimi	Iran
S. Fatemi	Iran
E. Karami	Iran
F. Sasanian-Asl	Iran
S. H. Tayyar	Iran

Additional Reviewers

Tarak Alshatshat	University of Benghazi, Libya
Aidin Delgoshaei	Kharazmi University, Iran
Seyed Babak Ebrahimi	Khajeh Nasir Toosi University of Technology, Iran
Alireza Goli	University of Isfahan, Iran
Ramesh Lekurwale	K. J. Somaiya College of Engineering, India

Froilan Mobo   Philippine Merchant Marine Academy,
                               Philippines
Pınar Usta   Isparta University of Applied Sciences, Turkey
Hosein Seyyed Esfahani   Islamic Azad University-Tehran North, Iran
Rahul S. Mor   National Institute of Food Technology
                               Entrepreneurship and Management, India
Roya Soltani   Khatam University, Iran
Nurullah Yilmaz   Süleyman Demirel Üniversitesi, Turkey

Contents

Production/Scheduling and Transportation in Supply Chain Management

Sustainable and Resilient Supply Chain Management

Humanitarian Supply Chain Management

Information Technology in Supply Chain Management

A Robust Two-Echelon Periodic Multi-commodity RFID-Based Location Routing Problem to Design Petroleum Logistics Networks: A Case Study

Erfan Babaee Tirkolaee[1]([✉]) [iD], Alireza Goli[2] [iD], and Gerhard-Wilhem Weber[3,4] [iD]

[1] Department of Industrial Engineering, Istinye University, 34010 Istanbul, Turkey
erfan.babaee@istinye.edu.tr
[2] Department of Industrial Engineering and Future Studies Faculty of Engineering, University of Isfahan, Isfahan, Iran
goli.a@eng.ui.ac.ir
[3] Faculty of Engineering Management, Poznan University of Technology, Poznan, Poland
gerhard.weber@put.poznan.pl
[4] Institute of Applied Mathematics, Middle East Technical University, Ankara, Turkey

Abstract. This study proposes a robust two-echelon periodic multi-commodity Location Routing Problem (LRP) by the use of RFID which is one of the most useful utilities in the field of Internet of Things (IoT). Moreover, uncertain demands are considered as the main part to design multi-level petroleum logistics networks. The different levels of this chain contain plants, warehouse facilities, and customers, respectively. The locational and routing decisions are made on two echelons. To do so, a novel mixed-integer linear programming (MILP) model is presented to determine the best locations for the plants and warehouses and also to find the optimal routes between plant level and warehouse facilities level, for the vehicles and between warehouse facilities level and customers' level in order to satisfy all the uncertain demands. To validate the proposed model, the CPLEX solver/GAMS software is employed to solve several problem instances. These problems are analyzed with different uncertain conditions based on the applied robust optimization technique. Finally, a case study is evaluated in Farasakou Assaluyeh Company to demonstrate the applicability of our methodology and find the optimal policy.

Keywords: Location routing problem · Multi-commodity logistics · Robust optimization · IoT · RFID system

1 Introduction

In recent years, advances in information technology have made the virtual world more effective. The Internet of Things (IoT) is progressing day by day and its role in our lives is becoming more pronounced. Object intelligence in IoT is accomplished by using devices such as RFID, global positioning systems and other sensor equipment in networks connected to the Internet. Due to the importance of measuring energy and the

© Springer Nature Switzerland AG 2021
Z. Molamohamadi et al. (Eds.): LSCM 2020, CCIS 1458, pp. 3–23, 2021.
https://doi.org/10.1007/978-3-030-89743-7_1

sustainability of energy at any location, monitoring and control of resources have become increasingly important. To monitor devices, a large amount of information is handled daily and decisions are made based on the type of information. The data is automatically retrieved by the sensors. Internet technology provides data transfer objects without the need for manpower.

On the other hand, the competition between companies in charge of supplying goods and services has highly intensified. Nowadays, corporations and companies need integration and flexibility of all production activities, from raw material procurement to final delivery to the consumers in the supply chain [1–3]. Distribution and support systems contain this supply chain process. Planning, production, control of products, storage, providing services and their relevant information from the starting point to the consumption point must be addressed to fulfill customer requirements. Logistics is a significant element of Supply Chain Management (SCM). It is found out that logistic costs constitute a huge section of companies' budgets and their working capital; however, these costs can be significantly decreased by an attentive configuration of the supply chain. In fact, this distribution network is especially critical at the end of the chain due to its involvement in many small product flows towards end-customers or retailers [4, 5].

Nowadays, the global economy is highly dependent on the petroleum industry which needs a responsive and economically optimal supply chain design. Accordingly, Location-Routing Problem (LRP) is one of the most important and applicable issues in distribution management and logistics or in designing an efficient petroleum supply chain. LRP is the integration of Facility Location Problem (FLP) and Vehicle Routing Problem (VRP) within a network that needs to be optimized simultaneously. In other words, designing this network contains two hard combinatorial optimization problems (NP-hard problems) in order to optimally locate the central depots and determine the serving routes of the vehicle which supply the customers' demands from these depots [6, 38, 39]. As a great point, each of these problems has widespread applications separately.

In the literature review, LRP has been studied widely. The failure to consider both the location and routing problem simultaneously and its subsequent increase in supply chain support costs is why it attracted so much attention. As it was mentioned, both FLP and VRP belong to the NP-hard problems; consequently, LRP is also a problem with the complexity of NP-hard. Thus, solving the large-sized LRP using exact solution techniques is almost not easy and possible within a reasonable time [7, 8]. The idea of synchronizing depot location and vehicle routing decisions returns to approximately 50 years ago. On that occasion, the internal dependency of these different two kinds of decisions was already clarified but the optimization methodologies and computers application were not developed enough to reach an integrated treatment [9]. Gandy and Dohrn [10] are probably the first researchers who investigated customers' visits while locating depots by introducing a non-linear function of distance (sales function) in which sales have an indirect relationship with distance from the depot. The potential advantages obtained by including vehicles' routing decisions while locating depots were first determined by Salhi and Rand [11]. They demonstrated that the classical strategy of solving an FLP and a VRP separately often yields suboptimal solutions. Then, various versions of LRP have been studied during the last years. In the following, different important variants of the LRP with different developed solution methods are reviewed.

At the beginning of the introduction of LRP, authors investigated the LRP with unca-pacitated depots, for example, see Tuzun and Burke [12]. After the survey done by Nagy and Salhi [13], most researchers addressed the LRP considering capacity constraints for depots and vehicles, called Capacitated LRP (CLRP). According to the current research, multi-echelon distribution systems are regarded with widespread applications. For example, the given network can include three levels (plants, central warehouses, and end-customers), with locational decisions in the first level, the second level, or both [14, 15]. Prodhon and Prins [16] reviewed the recent research on LRPs in logistics network design. They covered the literature since the last survey done by Nagy and Salhi [13]. They analyzed the best well-known algorithms, benchmarks, and variants of the prob-lem. Martínez-Salazar et al. [17] suggested a novel mathematical model for a bi-objective LRP and solved it by two different metaheuristics.

In the Two-Echelon Location Routing Problem (LRP-2E), the supplying routes are determined to fulfill the demand of the depots from several main facilities or "plants" besides locating the depots optimally. These supplying routes represent the first distri-bution echelon. On the other hand, the routes originate from the aforementioned depots (called satellite depots or simply "satellites") towards customers constitute the second distribution level [18, 19]. The first work on LRP-2E after Madsen's contributions [15] was introduced by Lin and Lei [20]. Their proposed model consists of a set of pro-duction plants, a set of large customers, and a set of small customers. The objective of the model was to identify the location of the uncapacitated satellite depots (referred to distribution centers), to find an optimal subset of the large customers to be visited in the first routing level and to build the supplying routes for both levels. Nguyen et al. [21] addressed LRP-2E with a single central depot which is already located and a set of potential capacity-constrained satellites with establishing costs as an important loca-tion factor. Nekooghadirli et al. [22] employed a novel bi-objective Location Routing Inventory Problem (LRIP) to design a distribution network. The two objectives include minimizing the total cost and the maximum average time for delivering commodities to the customers. They tried different metaheuristics to solve the proposed problem. Rahmani et al. [23] offered two kinds of local search methods to tackle their proposed Multi-products LRP-2E with pickup and delivery. They evaluated the efficiency of the proposed solution methods by solving benchmarks. Vidović et al. [24] suggested a mixed-integer linear programming (MILP) mathematical model for an LRP-2E for designing recycling logistics networks considering profit. They developed efficient heuristics to treat large-sized problems. They tested their proposed heuristics for different parameters to get a better insight. Wang et al. [25] applied metaheuristics for an LRP-2E with time windows based on clustering the customers. They suggested a bi-objective model in order to minimize total cost and maximize service reliability.

On the other hand, there is few robust related research done in the field of study under uncertain conditions. Three of the most recent related research is reviewed in the following.

Schiffer and Walther [26] proposed a robust electric LRP with time windows consid-ering uncertain customer patterns according to the spatial customer distribution, service time windows and demand. They analyzed the benefit of a robust optimization tech-nique with operational feasibility and savings in total cost. Cheng et al. [27] presented a two-stage robust approach for designing a reliable logistics network. They adopted a two-stage robust optimization method where locational decisions are made before disruptions occur. They tackled the problem using a column-and-constraint-generation exact algorithm. Jabbarzadeh et al. [28] developed a robust approach to design a closed-loop supply chain network considering disruptions. Their proposed model was able to optimize facility locations and transshipments under a set of disruption scenarios using a Lagrangian Relaxation Algorithm (LRA). A maritime supply chain design for the transportation of wood pellets was proposed by Andersen et al. [3]. They integrated the locational, routing and scheduling decisions and proposed a novel MILP model to be employed as a decision support system. They demonstrated the applicability of their proposed methodology by solving a real case study problem in Norway.

In this research, a two-echelon periodic LRP considering the role of RFID and vehicle usage duration limitation is considered which is designed for strategic planning of naval petroleum SCM. After a deep examination of the related literature, the main contributions of this research are as follows:

- Developing a novel robust mathematical model for the periodic multi-commodity LRP-2E with demand uncertainty considering three levels of plants, warehouses, and customers in which two different fleets of vehicles are defined for each echelon,
- Considering RFID in LRP problem to accelerate information transaction,
- Investigating the applicability of the proposed model by evaluating a case study,
- Investigating the effects of different uncertain conditions on the problem,
- Applying a sensitivity analysis of the maximum available time for vehicles to evaluate different conditions in the case study.

The remaining sections of the study are organized as follows: Sect. 2 defines the problem and the developed robust mathematical model. In Sect. 3, a case study is discussed and in Sect. 4, the computational results including model validation, investigation of the case study and sensitivity analysis are given. Section 5 includes the concluding remarks and some suggestions for future works.

2   Model Development

In the proposed problem, there are a set of candidate locations for plants (level 1), a set of warehouses (level 2), and a set of customers (level 3) which plants and warehouses need to be established at the best-given candidate locations in a proposed graph network. The customers' locations are predefined in the network. Furthermore, the capacity of the candidate plants and warehouses are initially specified and considered as important input information in the problem besides the demands of the warehouses and customers. It is worth indicating that warehouses have proper demands that are not just adopted by customers' demands, but also due to the strategic plans of the management, it may be adopted by inventory policy, selling products in the free markets, etc.

There are two types of vehicles which have different limited capacities to transport goods and have a usage time limitation. As demand uncertainty has an important effect on the distribution, it is necessary that the information between vehicles and warehoused is being conveyed very fast. For this purpose, each vehicle has an RFID tag to be distinguished by supply chain and also has remote communication with warehouses. Using such communication system has fix cost that the supply chain should consider them. These two specifications of the vehicles determine the number of used vehicles. The first-type vehicle serves the warehouses in the first echelon, and the second-type vehicle serves the customers in the second echelon. On the other hand, the first-type vehicle begins its trip from a specific plant and after serving the determined warehouses which are assigned to it, returns to that specific plant. The second-type vehicle begins its trip from a specific warehouse and after serving the customers returns to that specific warehouse.

Each customer would receive its demand for products from one established warehouses and each established warehouse would be assigned to one established plant. As one of the important factors in strategic planning, a planning horizon is considered. In the case of this periodic problem, the demands of the warehouses/customers may be different in each time period, the facility locational decision is directly affected by these demands. The objective of the proposed problem is to minimize the total cost throughout the supply chain by determining the optimal locations for facilities and planning optimal tours with a minimum possible number of vehicles such that the total demand of warehouses and customers is met. Moreover, since the demand parameters in the first and second echelons are regarded to be uncertain within an interval, the robust model is developed based on the robust optimization technique introduced by Bertsimas and Sim [34, 35]. So the main assumptions in the problem are as follows:

i. There are three levels of facilities including plants in level 1, warehouses in level two, and customers in level 3,

ii. Each plant and each warehouse have its own establishment cost,

iii. A planning horizon is defined as related to the strategic planning of the proposed naval petroleum SCM,

iv. There different types of products to be distributed,

v. The first echelon includes plants, warehouses, and available first-type vehicles. Warehouses would receive their demands for different products from plants,

vi. The second echelon includes warehouses, customers, and available second-type vehicles. Customers would receive their demands for different products from warehouses,

vii. All the warehouses and customers have demand for all products in each period which fluctuates in a given uncertainty interval,

viii. Vehicles can use RFID with specific fixed costs. Each RFID system can increase the availability time of each vehicle.

ix. Each type of vehicles has its given capacity, usage cost, and maximum allowable usage time,

x. The capacity of plants/warehouses is fixed in each period for each type of products,

xi. All demands should be satisfied in each echelon,

xii. Split deliveries are not allowed.

Figure 1 depicts the problem space schematically for a period. As it is clear, plants 1 and 2 are established to supply all the established warehouses. So there are 2 established plants, 5 established warehouses, and 18 customers in the supply chain.

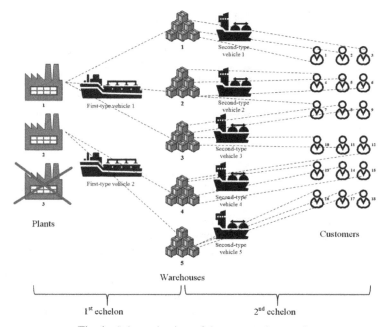

Fig. 1. Schematic view of the proposed network.

In order to describe the proposed MILP model, firstly the indices, sets, parameters, and variables are defined which are used in the mathematical model. Then, the mathematical model is given.

2.1 Indices and Sets

i, j, h: Index of nodes $(1, \ldots, n)$,
k: Index of vehicles,
k_1: Index of first-type vehicles,
k_2: Index of second-type vehicles,
r: Index of products,
p: Index of time periods,
f: Index of kind of RFID system,
TN: Set of total nodes,
TP: Set of total plants,
TW: Set of total warehouses,
TC: Set of total customers,
TR: Set of total products,
TT: Set of time periods,
K: Set of vehicles,
K_1: Set of first-type vehicles,
K_2: Set of second-type vehicles,
S: An arbitrary subset of nodes.

2.2 Parameters

C_{ij}: Distance between nodes i and j,
D_{irp}: Uncertain demand of warehouse i for r^{th} product in p^{th} period,
DD_{irp}: Uncertain demand of customer i for r^{th} product in p^{th} period,
$\overline{D}_{irp}$: Average demand of warehouse i for r^{th} product in p^{h} period,
$\overline{DD}_{irp}$: Average demand of customer i for r^{th} product in p^{th} period,
Q_{kr}: Capacity of kth vehicles for loading r^{th} product,
Cap_{ir}: Capacity of ith plant for supplying r^{th} product in each period,
Cap'_{ir}: Capacity of ith warehouse for supplying r^{th} product in each period,
te_{ij}: Traveling time between nodes i and j,
FC_i: Cost of establishing i^{th} plant,
FC'_i: Cost of establishing i^{th} warehouse,
CV_{kf}: Fixed cost of using k^{th} vehicles with f^{th} RFID system in each period,
$T1_{max,f}$: Maximum time availability of the first-type vehicles by using f^{th} RFID system,
$T2_{max,f}$: Maximum time of availability of the second-type vehicles by using f^{th} RFID system,
M: A large number,
α: Distance conversion factor per distance unit.

2.3 Variables

X_{ijkp}: Binary variable taking the value of 1 if vehicle k travels from node i to node j in p^{th} period, otherwise 0,
Y_{ijp}: Binary variable taking the value of 1 if warehouse j is assigned to plant i in p^{th} period, otherwise 0,
Y'_{ijp}: Binary variable taking the value of 1 if customer j is assigned to warehouse j in p^{th} period, otherwise 0,
O_i: Binary variable taking the value of 1 if plant i is established, otherwise 0,
O'_i: Binary variable taking the value of 1 if warehouse i is established, otherwise 0,
U_{kft}: Binary variable taking the value of 1 if k^{th} vehicle with f^{th} RFID system is used in p^{th} period, otherwise 0.

2.4 Robust Mathematical Model

Robust optimization is known as an efficient technique to cope with uncertainty according to real-life situations [31, 33]. This section addresses a robust mathematical model for the proposed two-echelon periodic multi-commodity L is modeled according to the approach proposed by Bertsimas and Sim [35]. Accordingly, the uncertain demand of warehouse (customer) D_{irp} (DD_{irp}) is formulated as a symmetric and bounded random variable which takes value in $\left[\overline{D}_{irp} - \hat{D}_{irp}, \overline{D}_{irp} + \hat{D}_{irp}\right]$ $\left(\left[\overline{DD}_{irp} - \widehat{DD}_{irp}, \overline{DD}_{irp} + \widehat{DD}_{irp}\right]\right)$, where $\hat{D}_{irp}$ ($\widehat{DD}_{irp}$) is the deviation of $\overline{D}_{irp}$ ($\overline{DD}_{irp}$)

[36]. These parameters lead to uncertainty in the capacity constraints of vehicles in their corresponding echelon. The corresponding uncertain constraints are:

$$\sum_{j \in TW} \sum_{i \in TN \setminus TC} D_{jrp} X_{jik_1p} \leq Q_{k_1r} \quad \forall p \in TT, k_1 \in K_1, r \in R, \tag{1}$$

$$\sum_{j \in TC} \sum_{i \in TN \setminus TP} DD_{jrp} X_{jik_2p} \leq Q_{k_2r} \quad \forall p \in TT, k_2 \in K_2, r \in R, \tag{2}$$

$$\sum_{j \in TC} D_{jrp} Y_{ijp} \leq Cap_{ir} O_i \quad \forall i \in TW, p \in TT, r \in R, \tag{3}$$

$$\sum_{j \in TW} D_{jrp} Y'_{ijp} \leq Cap_{ir} O'_i \quad \forall i \in TW, p \in TT, r \in R, \tag{4}$$

So, the parameters Γ_{k_1rp} and Γ'_{k_2rp} are defined for each vehicle k_1 and k_2 for product r in period p respectively such that $\Gamma_{k_1rp} \in [0, |J_{k_1rp}|]$ d $\Gamma'_{k_2rp} \in [0, |J_{k_2rp}|]$, where J_{k_1rp} (J_{k_2rp}) is the set of the served warehouses for product r by vehicle k_1 (k_2) in period p and represents a set uncertain coefficients in the corresponding capacity constraint and usage time constraint of vehicle k_1 (k_2) in each period. In other words, Γ_{k_1rp} and Γ'_{k_2rp} ensure the robustness adjustment against the conservatism level of the solution.

Bertsimas and Sim [41] proved that it is improbable that all of the coefficients change at the same time. Hence, it is supposed that up to $\lfloor \Gamma_{k_1rp} \rfloor$ number of these coefficients may change for warehouses and up to $\lfloor \Gamma'_{k_2rp} \rfloor$ number of these coefficients may change for customers. Furthermore, even if more than $\lfloor \Gamma_{k_1rp} \rfloor . (\lfloor \Gamma'_{k_2rp} \rfloor)$ parameters fluctuate, then the robust solution will remain feasible with a very high probability because of the symmetric distribution of variables. Therefore, $\lfloor \Gamma_{k_1rp} \rfloor (\lfloor \Gamma'_{k_2rp} \rfloor)$ is called the conservation level for vehicle k_1 (k_2) in period p for product r.

Now, the proposed robust mathematical model under the proposed assumptions is as follows:

$$\text{minimize } Total\ Cost = \alpha \left(\sum_{k \in K} \sum_{i \in TN} \sum_{j \in TN} \sum_{p \in TT} C_{ij} X_{ijkp} \right)$$
$$+ \sum_{p \in TT} \sum_{k \in K} \sum_{f \in F} CV_{kf} U_{kfp} \tag{5}$$
$$+ \sum_{i \in TP} FC_i O_i + \sum_{i \in TW} FC'_i O'_i$$

subject to

$$\sum_{i \in TP} X_{ijk_1p} = 1 \quad \forall j \in TW, p \in TT, k_1 \in K_1, \tag{6}$$

$$\sum_{i \in TW} X_{ijk_2p} = 1 \quad \forall j \in TC, p \in TT, k_2 \in K_2, \tag{7}$$

$$\sum_{\substack{j \in TN \\ j \neq i}} X_{jikp} = \sum_{\substack{j \in TN \\ j \neq i}} X_{jikp} \quad \forall i \in TN, p \in TT, k \in K, \tag{8}$$

$$\sum_{j\in TW}\sum_{i\in TN\backslash TC}\overline{D}_{jrp}X_{ijk_1p} + z_{k_1rp}\Gamma_{k_1rp} + \sum_{j\in TW}\sum_{i\in TN\backslash TC}v_{ijk_1p} \le Q_{k_1r}$$
$$\forall p \in TT, k_1 \in K_1; K_1 \subset K, r \in R, \tag{9}$$

$$\sum_{j\in TC}\sum_{i\in TN\backslash TP}\overline{DD}_{jrp}X_{ijk_2p} + z'_{k_2rp}\Gamma'_{k_2rp} + \sum_{j\in TC}\sum_{i\in TN\backslash TP}v'_{ijk_2p} \le Q_{k_2r}$$
$$\forall p \in TT, k_2 \in K_2; K_2 \subset K, r \in R, \tag{10}$$

$$\sum_{j\in TC}\overline{D}_{jrp}Y_{ijp} + \sum_{k_1\in K_1}z_{k_1rp}\Gamma_{k_1rp} + \sum_{k_1\in K_1}\sum_{j\in TW}\sum_{i\in TN\backslash TC}v_{jik_1p} \le Cap_{ir}O_is$$
$$\forall i \in TW, p \in TT, r \in R, \tag{11}$$

$$\sum_{j\in TW}\overline{DD}_{jrp}Y'_{ijp} + \sum_{k_2\in K_2}z'_{k_2rp}\Gamma'_{k_2rp} + \sum_{k_2\in K_2}\sum_{j\in TW}\sum_{i\in TN\backslash TC}v'_{ijk_2p} \le Cap'_{ir}O'_i$$
$$\forall i \in TP, p \in TT, r \in R, \tag{12}$$

$$z_{k_1rp} + v_{ijk_1p} \ge \hat{D}_{jrp}E_{ijk_1p} \quad \forall i \in TN\backslash TC, j \in TW, p \in TT, r \in R, k_1 \in K_1, \tag{13}$$

$$z'_{k_2rp} + v'_{ijk_2p} \ge \widehat{DD}_{jrp}E'_{ijk_2p} \quad \forall i \in TP, j \in TC, p \in TT, r \in R, k_2 \in K_2, \tag{14}$$

$$-E_{ijk_1p} \le X_{ijk_1p} \le E_{ijk_1p} \quad \forall i \in TN\backslash TC, j \in TW, p \in TT, k_1 \in K_1, \tag{15}$$

$$-E'_{ijk_2p} \le X_{ijk_2p} \le E'_{ijk_2p} \quad \forall i \in TP, j \in TC, p \in TT, k_2 \in K_2, \tag{16}$$

$$\sum_{p\in TT}\sum_{j\in TW}Y_{ijp} \le MO_i \quad \forall i \in TP, \tag{17}$$

$$\sum_{p\in TT}\sum_{j\in TC}Y'_{ijp} \le MO'_i \quad \forall i \in TW, \tag{18}$$

$$\sum_{i\in TN}\sum_{j\in TN}te_{ij}X_{ijk_1p} \le T1_{max,f}U_{kfp} \quad \forall p \in TT, k_1 \in K_1, f \in F, \tag{19}$$

$$\sum_{i\in TN}\sum_{j\in TN}te_{ij}X_{ijk_2p} \le T2_{max,f}U_{kfp} \quad \forall p \in TT, k_2 \in K_2, f \in F, \tag{20}$$

$$\sum_{i\in TN}\sum_{j\in TN}X_{ijkp} \le MU_{kp}, \quad \forall p \in TT, k \in K, \tag{21}$$

$$\sum_{i\in S}\sum_{j\in S}X_{ijk_1p} \le |S| - 1, \quad \forall S \subseteq TW, p \in TT, k_1 \in K_1, \tag{22}$$

$$\sum_{i\in S}\sum_{j\in S}X_{ijk_2p} \le |S| - 1, \quad \forall S \subseteq TC, p \in TT, k_2 \in K_2, \tag{23}$$

$$\sum_{h\in TN}X_{jhk_1p} + \sum_{h\in TN}X_{ihk_1p} \le 1 + Y_{ijp} \quad \forall i \in TP, j \in TW, p \in TT, k_1 \in K_1, \tag{24}$$

$$\sum_{h \in TN} X_{jhk_2p} + \sum_{h \in TN} X_{ihk_2p} \leq 1 + Y'_{ijp} \qquad \forall i \in TW, p \in TT, j \in TC, k_2 \in K_2, \qquad (25)$$

$$O_i, O'_i, X_{ijkp}, U_{kp}, Y_{ijp}, Y'_{ijp} \in \{0, 1\}, \quad \forall i \in TN, k \in K, \qquad (26)$$

$$E_{ijk_1p}, E'_{ijk_2p}, v_{ijk_1p}, v'_{ijk_2p}, z_{k_1rp}, z'_{k_2rp} \geq 0 \\ \forall i \in TN, j \in TN, p \in TT, r \in R, k_1 \in K_1, k_2 \in K_2. \qquad (27)$$

Equation (5) shows the objective function which includes four parts: the first part refers to the total traveling cost of the vehicles in the planning horizon that is calculated by multiplying total traveled distance by the conversion factor, the second part calculates the total usage cost of the vehicles by considering selected RFID system, and the third part refers to the fixed establishment costs of the plants and warehouses, respectively. Constraints (6) and (7) ensure that all demands of each warehouse and each customer should be satisfied in each period, respectively. Constraint (8) indicates that the inner flows are equal to the outer flows for all nodes within the network (flow balance). In other words, if a vehicle enters a node, it should exit that node. Constraints (9) and (10) are related to the capacity limitations of the first- and second-type vehicles, respectively. Constraints (11) and (12) indicate the capacity limitations of the plants and warehouses, respectively. Constraints (13)–(16) are the corresponding constraints to the robustness based on [35]. It should be noted that the robustness variables of $(z_{k_1rp}, E_{ijk_1p}, v_{ijk_1p})$ and $(z'_{k_2rp}, E'_{ijk_2p}, v'_{ijk_2p})$ are employed to adjust the robustness of the solutions and apply the protection levels to the model. Constraints (17) explain that each warehouse can be assigned to a plant if only the plant has been already established. Similarly, Constraints (18) indicate that each customer can be assigned to a warehouse if only the warehouse has been already established. Constraints (19) and (20) are related to the consideration of the maximum allowable usage time of the first-type and second-type vehicles in each period, respectively. In these constraints, the selected RFID system for each type of vehicle can affect the usage time of each vehicle. Constraints (21) express that each vehicle in each echelon would be used when its usage cost has been paid. In other words, if a vehicle is not determined to be used, it can't construct any route. Constraints (22) and (23) are related to the elimination of sub-tours for each type of vehicle, respectively. Constraints (24) and (25) connect assigning and routing components in the first and second echelons, respectively. These constraints link the related variables of assignment and routing together in each period. Constraints (26) and (27) specify the types of variables.

3 Case Study

Here, a case study in the field of the supply chain and logistics of Farasakou Assaluyeh Company in Iran (form more information, see [37]) is evaluated by our proposed model to examine the applicability and efficiency of the model in a real world case. Farasakou Assaluyeh Company is the owner of the specialized port terminal of petroleum items which provides port services in the field of transit, import and export and distribution in

Fig. 2. The schematic view of the second level of Farasakou Assaluyeh Company [36].

the local zone, barging bunkering in the Persian Gulf region. A schematic view of the warehouses of the plant is shown in Fig. 2.

In the case study, the planning horizon is considered for one year period which depends on the demands defined annually. This plant provides a variety of petroleum products. In this research, two of the most important products are considered including gasoline and mazut or fuel oil. The products are delivered from plants located in Erbil, Iraq (level 1), stored in these plants' warehouses (level 2) and finally are exported to Dubai and Sharjah of United Arab Emirates (level 3). Actually, the role of Farasakou Assaluyeh Company is to store and distribute the aforementioned products. Moreover, in this case, the first level is defined as suppliers in that supply chain and the goal is to determine the number of required suppliers.

For transportation in the first echelon, fuel trucks (k_1) are used, and also between the second and third level of the supply chain (second echelon), ships (k_2) are used. Each type of vehicle can use one of two available RFID systems. The first system is a basic remote communication system with 6 h of total use time per day. The second system increases the use time to 8 h per day. However, the implementation cost for each vehicle with the second RFID system has a 20% extra cost. The planning horizon is for one month. By considering the input parameters (e.g. costs, transportation (shipping and transit) time between the nodes), the problem is solved optimally in order to locate the first and the second level's facilities to minimize the establishment and transportation costs.

Currently, there are 7 suppliers, 6 warehouses and 2 customers in this supply chain, where the plant's goal is to meet customer's demand for the two products. So, the main questions are:

1) How would be the optimal policy of the supply chain if a new product is added?
2) How would be the optimal policy if some parameters change?

4   Experimental Results

In this section, model validation is done by generating and solving 10 test problems in different sizes. To validate the suggested model and solve the problems, CPLEX solver of GAMS software is employed as one of the most common exact methods [29–32]. Then, the obtained results are analyzed by presenting the objective values, run times, number of the established plants/warehouses, and the number of the used vehicles in each problem. As mentioned, the model is coded in GAMS software, and it is run for 10 randomly generated problems in small and medium sizes with a 3600 s run time limitation by a system which its configuration is shown in Table 1. The input parameters values are generated randomly near to the values in the real world with a uniform distribution which is presented in Table 2. The obtained results are given in Tables 3, 4, 5. Table 3 shows the input information of the test problems and the objectives and run time values of the deterministic problems after solving optimally. Table 4 presents a comparison between the number of established facilities and the number of used vehicles determined in different deterministic problems. The obtained results of the deterministic and robust model which were optimally solved by CPLEX solver are presented in Table 5 for different conservatism levels. Note that conservatism levels for both echelons are applied simultaneously.

Table 1.   Computer system configuration.

Intel® Core™ i7-4720HQ CPU @ 2.60 GHz	Processor
8.00 GB (3.24 GB)	Installed Memory (RAM)
62-bit Operating System	System type

Figure 3 depicts the impacts of various conservatism levels on CPLEX run time. According to Fig. 3, we can see that by increasing the problem scale, the obtained run time will increase such that P10 cannot be solved by GAMS in 3600 s time limitation Furthermore, Fig. 4 shows the assessment of the different conservatism levels in the problems based on their objective values. By increasing the conservatism levels, the objective value of test problems increases where the maximum value is obtained at conservatism levels of $\Gamma_{k_1 rp} = 5$ and $\Gamma'_{k_2 rp} = 10$ in each problem. Furthermore, the maximum increase of objective value occurs in P8 and P9, which are more remarkable. It can be concluded that the conservatism levels have a great role in the robust problem.

Based on input data which is shown in Table 6, the results obtained for the case study are presented in Table 7. Moreover, the schematic result is depicted in Fig. 5. It is worthy to say that the values of demands deviations and conservatism levels are determined by the experts.

Table 2. Input parameters' values.

Parameter	Value	Parameter	Value
C_{ij}	Uniform $(2, 10)$	FC_i	Uniform $(100000, 200000)$
$\overline{D}_{irp}$	Uniform $(30, 50)$	FC'_i	Uniform $(5000, 10000)$
$\overline{DD}_{irp}$	Uniform $(2, 10)$	$CV_{k_1,f}$	900 for $f = 1$, 1080 for $f = 2$
α	10	$CV_{k_2,f}$	600 for $f = 1$, 720 for $f = 2$
$Q_{k_1 r}$	Uniform $(40, 60)$	$Q_{k_2 r}$	Uniform $(20, 40)$
Cap_{ir}	Uniform $(1000, 2000)$	$T1_{max,f}$	6 for $f = 1$, 8 for $f = 2$
Cap'_{ir}	$R \times \max_p\{D_{irp}\}$	$T2_{max,f}$	6 for $f = 1$, 8 for $f = 2$
te_{ij}	Uniform $(1, 15)$	M	10^8
$\hat{D}_{irp}$	$0.2\,\overline{D}_{irp}$	$\widehat{DD}_{irp}$	$0.2\,\overline{DD}_{irp}$
$\Gamma_{k_1 rp}$	0, 1, 2, 5	$\Gamma'_{k_2 rp}$	0, 2, 5, 10

Table 3. Scale of the test problems.

Problem	PCL	WCL	NC	NAV1	NAV2	NP	NT
P1	2	2	2	2	2	2	1
P2	3	5	8	3	3	3	2
P3	4	6	10	3	4	3	4
P4	6	8	18	4	5	3	6
P5	7	10	25	5	6	3	8
P6	8	12	30	5	6	4	10
P7	8	12	35	5	7	4	12
P8	10	14	40	6	8	5	12
P9	11	14	45	7	9	6	18
P10	12	15	50	8	10	6	24

PCL: Number of plant candidate locations.
WCL: Number of warehouse candidate Locations.
NC: Number of customers.
NAV1: Number of available first-type vehicles.
NAV2: Number of available second-type vehicle.
NP: Number of products.
NT: Number of time periods.

Table 4. The comparison results of deterministic problems.

Problem	Optimal PCL	Optimal WCL	Optimal NAV1	Optimal NAV2
P1	1	2	1	1
P2	2	3	2	2
P3	3	5	2	3
P4	4	6	3	4
P5	5	8	4	5
P6	6	10	4	5
P7	6	10	4	6
P8	7	11	5	6
P9	8	13	5	7
P10	-	-	-	-

Table 5. The obtained results of robust problems with different conservatism levels.

Problem	$\Gamma_{k_1 rp} = 0$ $\Gamma'_{k_2 rp} = 0$		$\Gamma_{k_1 rp} = 1$ $\Gamma'_{k_2 rp} = 2$		$\Gamma_{k_1 rp} = 2$ $\Gamma'_{k_2 rp} = 5$		$\Gamma_{k_1 rp} = 5$ $\Gamma'_{k_2 rp} = 10$	
	Obj. Value	Run time (sec)	Obj. Value	Run time (sec)	Obj. Value	Run time (sec)	Obj. Value	Run time (sec)
P1	122931.15	1.04	124039.42	1.16	135487.34	2.28	138244.68	3.40
P2	281054.48	2.78	283418.15	3.16	309342.33	5.60	317876.95	8.95
P3	455701.64	14.52	459403.76	16.11	500565.47	27.49	512070.57	40.64
P4	610153.38	320.47	615592.60	366.10	672784.40	489.84	689770.46	648.26
P5	772182.91	989.3	779144.04	1120.42	854359.39	1307.19	1081066.88	1840.89
P6	925249.70	1634.08	932865.80	1846.09	1017269.58	2074.32	1228713.10	2678.58
P7	927728.97	2101.66	935276.42	2401.57	1020499.50	2587.67	1205806.45	3488.85
P8	1087061.85	2986.81	1095756.66	3299.53	1199187.28	3600	1421305.11	3600
P9	1254804.19	3600	1267169.03	3600	1381678.70	3600	1579422.03	3600
P10	-*	3600	-*	3600	-*	3600	-*	3600

* No solution found within 3600 s.

By analyzing Fig. 5 and Table 7, we find that besides the available warehouses which satisfy customers' demands for two products, 8 additional warehouses should be established as it is shown in Fig. 5 by red dash lines. Furthermore, a supplier should be added in order to satisfy the customers' demands for two products plus one added product that is marked with red dash lines (according to the main question of the problem).

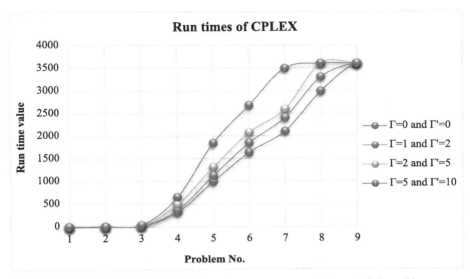

Fig. 3. CPLEX run times for different robust problems and deterministic problems.

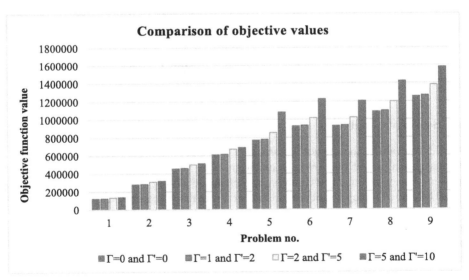

Fig. 4. CPLEX run times for different robust problems and deterministic problems.

Table 6. Input information of the case study.

PCL	WCL	NC	NAV1	NAV2	$\Gamma_{k_1 rp}$	$\Gamma'_{k_2 rp}$	$\hat{D}_{irp}$	$\widehat{DD}_{irp}$
9	16	2	10	10	1	2	$0.25\,\overline{D}_{irp}$	$0.25\,\overline{DD}_{irp}$

Table 7. Output information of the case study obtained by CPLEX.

PCL*	WCL*	NAV1*	NAV2*	Objective Value ($\times 10^7$)	Run time (sec)
8	14	7	10	79328.08	296.25

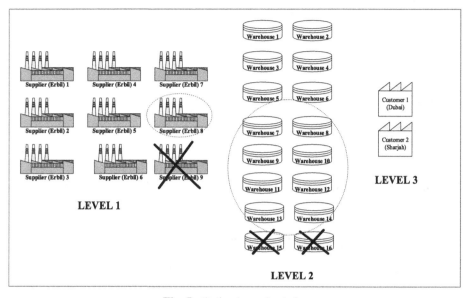

Fig. 5. Optimal supply chain.

4.1 Sensitivity Analysis

In order to evaluate the impact of parameters' fluctuations on the objective function, a sensitivity analysis of maximum available time for vehicles is studied in the real case problem. In fact, the managers of the supply chain are willing to know how much benefits would be earned if they consider more resources. In this research, the effects of different $T1_{max}$ and $T2_{max}$ on the objective are analyzed individually and simultaneously. Four decrease/increase intervals (i.e., -20%, -10%, 0%, $+10\%$ and $+20\%$) are taken into account for the parameters while the other parameters are fixed. The output results of the sensitivity analysis are shown in Table 8 and Fig. 6.

Table 8. Sensitivity analysis of the parameters of maximum available time for vehicles.

Parameter	Change interval				
	-20%	-10%	0	$+10\%$	$+20\%$
$T1_{max,f}$	80942.33	80279.02	79328.08	78409.46	73626.96
$T2_{max,f}$	87644.84	86036.86	79328.08	73241.24	70142.68
$T1_{max,f}$ and $T2_{max,f}$	**Infeasible**	91124.16	79328.08	70984.27	69241.01

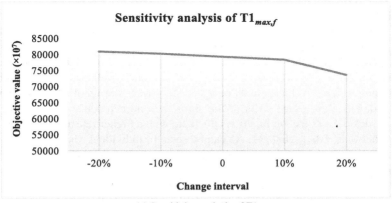

(a) Sensitivity analysis of $T1_{max,f}$

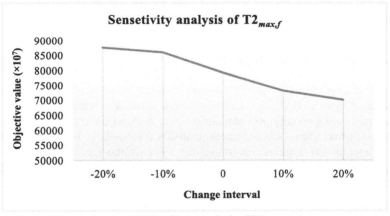

(b) Sensitivity analysis of $T2_{max}$

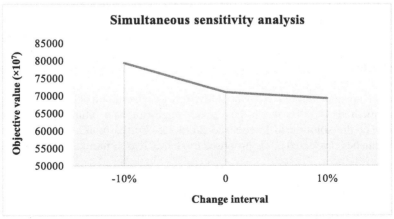

(c) Simultaneous sensitivity analysis of $T1_{max}$ and $T2_{max}$

Fig. 6. A comparison between the obtained results of sensitivity analysis performed on the maximum available time for vehicles.

As it is obvious in Fig. 6, the objective values will change significantly by changing of maximum available time for vehicles. These fluctuations of the objective function are more tangible for simultaneous parameters changing such that decreasing the parameters up to -10% simultaneously would lead to the maximum change (increase) in the objective function. In fact, by decreasing $T1_{max}$ or $T2_{max}$, more vehicles must be used, and by increasing it, fewer vehicles must be used. Moreover, routing of the vehicles would be affected. Since the usage costs of the ships (second echelon's vehicles) are more than fuel trucks, the objective functions fluctuate with a higher rate against changing of $T1_{max}$ rather than $T2_{max}$ changes. As can be seen, in individual sensitivity analysis of the parameters, the most change of the objective occurs for increasing the parameters. On the contrary, in the simultaneous sensitivity analysis, the most change occurs when the parameters decrease simultaneously.

These results show that the managers shall pay special attention to these possible results against changing the maximum available time for vehicles in each echelon in the supply chain. In fact, the optimal policy can be determined by analyzing different possible values of the parameters in order to consider the best levels of resources.

5   Conclusion and Future Work

Locating-routing problem is one of the main and useful issues in naval petroleum SCM and its related logistics decisions. In this research, a periodic robust multi-commodity two-echelon locating-routing problem considering demand uncertainty and maximum available time of vehicles is studied and evaluated in a real case study. Due to the increasing importance of IoT applications in logistics systems, the vehicles were considered to be equipped with RFID systems. The problem is formulated as an MILP model and the proposed model is validated by different sized test problems using CPLEX solver. Moreover, the test problems are analyzed with different conservatism levels defined for the proposed robust model. In order to test the applicability of the proposed model, a case study is investigated in Farasakou Assaluyeh Company. The obtained results determine the optimal policy for the company's development plan. It has been concluded that by considering a new additional product, 8 more warehouses and one more supplier (as a plant in level 1) should be established. Therefore, the optimal policy of supply chain consists of 8 suppliers, 14 warehouses, and 2 customers. Finally, a sensitivity analysis is implemented on the problem to study the behavior of the objective function against changing of maximum available time for vehicles.

Based on the main limitations of the study, the implementation of heuristics and metaheuristics methods [40, 41] would be a good suggestion as a future study to improve the quality of the solution in large-sized problems. Furthermore, the other objective functions can be considered in the proposed model such as minimization of total service time [32].

References

1. Ho, W., Lee, C.K.M., Ho, G.T.S.: Optimization of the facility location-allocation problem in a customer-driven supply chain. Oper. Manag. Res. **1**(1), 69–79 (2008)
2. Tirkolaee, E.B., Goli, A., Bakhsi, M., Mahdavi, I.: A robust multi-trip vehicle routing problem of perishable products with intermediate depots and time windows. Numer. Algebra Control Optim. **7**(4), 417–433 (2017)
3. Desport, P., Lardeux, F., Lesaint, D., Cairano-Gilfedder, C.D., Liret, A., Owusu, G.: A combinatorial optimisation approach for closed-loop supply chain inventory planning with deterministic demand. Eur. J. Ind. Eng. **11**(3), 303–327 (2017)
4. Martin, N., Verdonck, L., Caris, A., Depaire, B.: Horizontal collaboration in logistics: decision framework and typology. Oper. Manag. Res. **11**(1–2), 32–50 (2018). https://doi.org/10.1007/s12063-018-0131-1
5. Tan, C., Ji, S., Gui, Z., Shen, J., Fu, D.-S., Wang, J.: An effective data fusion-based routing algorithm with time synchronization support for vehicular wireless sensor networks. J. Supercomput. **74**(3), 1267–1282 (2017). https://doi.org/10.1007/s11227-017-2145-0
6. Hosseini, M.B., Dehghanian, F., Salari, M.: Selective capacitated location-routing problem with incentive-dependent returns in designing used products collection network. Eur. J. Oper. Res. **272**(2), 655–673 (2019)
7. Dai, Z., Aqlan, F., Gao, K., Zhou, Y.: A two-phase method for multi-echelon location-routing problems in supply chains. Expert Syst. Appl. **115**, 618–634 (2019)
8. Asadi, E., Habibi, F., Nickel, S., Sahebi, H.: A bi-objective stochastic location-inventory-routing model for microalgae-based biofuel supply chain. Appl. Energy **228**, 2235–2261 (2018)
9. Martinez, A.J.P., Stapleton, O., Van Wassenhove, L.N.: Field vehicle fleet management in humanitarian operations: a case-based approach. J. Oper. Manag. **29**(5), 404–421 (2011)
10. Watson-Gandy, C.D.T., Dohrn, P.J.: Depot location with van salesmen—a practical approach. Omega **1**(3), 321–329 (1973)
11. Salhi, S., Rand, G.K.: The effect of ignoring routes when locating depots. Eur. J. Oper. Res. **39**(2), 150–156 (1989)
12. Tuzun, D., Burke, L.I.: A two-phase tabu search approach to the location routing problem. Eur. J. Oper. Res. **116**(1), 87–99 (1999)
13. Nagy, G., Salhi, S.: Location-routing: Issues, models and methods. Eur. J. Oper. Res. **177**(2), 649–672 (2007)
14. Jacobsen, S.K., Madsen, O.B.: A comparative study of heuristics for a two-level routing-location problem. Eur. J. Oper. Res. **5**(6), 378–387 (1980)
15. Madsen, O.B.: Methods for solving combined two level location-routing problems of realistic dimensions. Eur. J. Oper. Res. **12**(3), 295–301 (1983)
16. Prodhon, C., Prins, C.: A survey of recent research on location-routing problems. Eur. J. Oper. Res. **238**(1), 1–17 (2014)
17. Martínez-Salazar, I.A., Molina, J., Ángel-Bello, F., Gómez, T., Caballero, R.: Solving a bi-objective transportation location routing problem by metaheuristic algorithms. Eur. J. Oper. Res. **234**(1), 25–36 (2014)
18. Crainic, T.G., Perboli, G., Mancini, S., Tadei, R.: Two-echelon vehicle routing problem: a satellite location analysis. Procedia Soc. Behav. Sci. **2**(3), 5944–5955 (2010)
19. Crainic, T.G., Mancini, S., Perboli, G., Tadei, R.: Impact of generalized travel costs on satellite location in the two-echelon vehicle routing problem. Procedia Soc. Behav. Sci. **39**, 195–204 (2012)
20. Lin, J.R., Lei, H.C.: Distribution systems design with two-level routing considerations. Ann. Oper. Res. **172**(1), 329 (2009)

21. Nguyen, V.P., Prins, C., Prodhon, C.: A multi-start iterated local search with tabu list and path relinking for the two-echelon location-routing problem. Eng. Appl. Artif. Intell. **25**(1), 56–71 (2012)
22. Nekooghadirli, N., Tavakkoli-Moghaddam, R., Ghezavati, V.R., Javanmard, S.: Solving a new bi-objective location-routing-inventory problem in a distribution network by meta-heuristics. Comput. Ind. Eng. **76**, 204–221 (2014)
23. Rahmani, Y., Cherif-Khettaf, W.R., Oulamara, A.: A local search approach for the two–echelon multi-products location–routing problem with pickup and delivery. IFAC-PapersOnLine **48**(3), 193–199 (2015)
24. Vidović, M., Ratković, B., Bjelić, N., Popović, D.: A two-echelon location-routing model for designing recycling logistics networks with profit: MILP and heuristic approach. Expert Syst. Appl. **51**, 34–48 (2016)
25. Wang, Y., Assogba, K., Liu, Y., Ma, X., Xu, M., Wang, Y.: Two-echelon location-routing optimization with time windows based on customer clustering. Expert Syst. Appl. **104**, 244–260 (2018)
26. Schiffer, M., Walther, G.: Strategic planning of electric logistics fleet networks: a robust location-routing approach. Omega **80**, 31–42 (2018)
27. Cheng, C., Qi, M., Zhang, Y., Rousseau, L.M.: A two-stage robust approach for the reliable logistics network design problem. Transp. Res. Part B Methodol. **111**, 185–202 (2018)
28. Jabbarzadeh, A., Haughton, M., Khosrojerdi, A.: Closed-loop supply chain network design under disruption risks: a robust approach with real world application. Comput. Ind. Eng. **116**, 178–191 (2018)
29. Tirkolaee, E.B., Aydın, N.S.: A sustainable medical waste collection and transportation model for pandemics. Waste Manage. Res. **211**, 734 (2021). https://doi.org/10.1177/0734242X2110 00437
30. Alinaghian, M., Tirkolaee, E.B., Dezaki, Z.K., Hejazi, S.R., Ding, W.: An augmented Tabu search algorithm for the green inventory-routing problem with time windows. Swarm Evol. Comput. **60**, 100802 (2021)
31. Tirkolaee, E.B., Mahmoodkhani, J., Bourani, M.R., Tavakkoli-Moghaddam, R.: A self-learning particle swarm optimization for robust multi-echelon capacitated location–allocation–inventory problem. J. Adv. Manuf. Syst. **18**(04), 677–694 (2019)
32. Tirkolaee, E.B., Abbasian, P., Weber, G.W.: Sustainable fuzzy multi-trip location-routing problem for medical waste management during the COVID-19 outbreak. Sci. Total Environ. **756**, 143607 (2021)
33. Yousefi Nejad Attari, M., Ebadi Torkayesh, A., Malmir, B., Neyshabouri Jami, E.: Robust possibilistic programming for joint order batching and picker routing problem in warehouse management. Int. J. Prod. Res. **59**, 4434 (2021). https://doi.org/10.1080/00207543.2020.176 6712
34. Bertsimas, D., Sim, M.: Robust discrete optimization and network flows. Math. Program. **98**(1), 49–71 (2003)
35. Bertsimas, D., Sim, M.: The price of robustness. Oper. Res. **52**(1), 35–53 (2004)
36. Ben-Tal, A., Nemirovski, A.: Robust solutions of linear programming problems contaminated with uncertain data. Math. Program. **88**(3), 411–424 (2000)
37. http://apc.co.ir/fa/. Accessed 23 May 2018
38. Yu, W., Liu, Z.: Vehicle routing problems with regular objective functions on a path. Naval Res. Logist. (NRL) **61**(1), 34–43 (2014)
39. Hennig, F., Nygreen, B., Lübbecke, M.E.: Nested column generation applied to the crude oil tanker routing and scheduling problem with split pickup and split delivery. Naval Res. Logist. (NRL) **59**(3–4), 298–310 (2012)

40. Ghaffarinasab, N., Zare Andaryan, A., Ebadi Torkayesh, A.: Robust single allocation p-hub median problem under hose and hybrid demand uncertainties: models and algorithms. Int. J. Manage. Sci. Eng. Manage. **15**(3), 184–195 (2020)
41. Goli, A., Tirkolaee, E.B., Aydin, N.S.: Fuzzy integrated cell formation and production scheduling considering automated guided vehicles and human factors. IEEE Trans. Fuzzy Syst. (2021). https://doi.org/10.1109/TFUZZ.2021.3053838

A Logistics Performance Index Review: A Glance at the World's Best and Worst Ten Performances

Cihan Çetinkaya[1] and Eren Özceylan[2]

[1] Department of Management Information Systems, Adana Alparslan Turkes Science and Technology University, Adana 01100, Turkey
ccetinkaya@atu.edu.tr
[2] Department of Industrial Engineering, Gaziantep University, Gaziantep 27027, Turkey
eozceylan@gantep.edu.tr

Abstract. Transport and logistics have a very significant place in international trade. Thus, the measurement of logistics performance is directly related with a country's trade income. The Logistics Performance Index (LPI) was created by the World Bank to measure the performance of countries in the field of logistics. The index is created after asking a number of categorized questions to employees and managers of several logistics companies located in each country and determining the results with respect to the question scores. Although it is not the only determiner in this sector, the LPI is very important because the supply chain members use this index as a decision making tool. The index is released biennially and there are 6 reports between 2007 and 2018. In this paper, 6 Logistics Performance Reports are analyzed; the top 10 and worst 10 performances are criticized. Also Turkey's overall performance is examined and discussed specifically. The intersection countries are highlighted and the reasons are discussed. It is determined that generally high income countries are performing well and the countries with internal disturbances are performing indifferently. It is believed that this research will be helpful to understand the global picture of logistics activities.

Keywords: Logistics performance index · Review · World Bank

1 Introduction

Logistics compasses the material and information flow of raw materials, semi products and finished goods from point of origin to point of customer. Also logistics activities include; transportation, materials handling, inventory management and packaging. Due to the large scope, small improvements will generate major effects within this field. In the developing world, logistics sector has gained acceleration and became one of the most important sectors. Also developments in information technologies make it difficult to compete in the world market. Strong infrastructure, effective planning, prudential investments, high qualified workers, technical developments are required to maintain a strong position at world competition. Because of the fierce competition, measuring the

© Springer Nature Switzerland AG 2021
Z. Molamohamadi et al. (Eds.): LSCM 2020, CCIS 1458, pp. 24–42, 2021.
https://doi.org/10.1007/978-3-030-89743-7_2

logistics performance became more important. According to the performance results, countries can figure out their situation in world market and they can identify their positions. These performance reports have also importance to guide countries to make their future plans [1].

The "Logistics Performance Index" was proposed firstly in 2007 by the World Bank to divine the logistics performances of nations. The LPI uses six scores to ascertain overall logistics performance [2]. In this way, it is crucial to understand and compare countries' logistical and transport systems considering other countries due to realize current bottlenecks from the governments' perspective [3].

In recent years, the number of published papers about LPI has increased. For instance, Marti et al. [4] analyzed the emphasis of LPI in international trade in their paper. Their results revealed that, the improvements in any fundamentals of the LPI can lead to compelling growth in international trade. Later, the mediator effect of LPI on gross domestic product and global competitiveness index was examined [5]. As a result, they report that the mediator effect of LPI considering relation between GCI and GDP was found as statistically significant. In addition, Çakır [6] made research on measuring the logistics performance of OECD (organization for economic co-operation and development) countries via fuzzy linear regression. It was reported that, the proposed methodology, considered as a practical choice for assessing logistics performance. Gani [7] explored the effect of logistics performance in international trade. The analysis draws on overall logistics performance as well as disaggregated measures of logistics specificities data for a large sample of countries. Yaprakli and Ünalan [1] analyzed the logistic performance index and last ten years' LPI for Turkey. Rezaei et al. [8] assigned weights to the six components of the LPI using the Best Worst Method (BWM), a multi-criteria decision-analysis method. Ekici et al. [2] propose an approach to set a course for policymakers in promoting the logistics performance of their countries. For this purpose, they analyzed the effect of the competitiveness pillars of the Global Competitiveness Index on the LPI. In addition, Turkey's logistics index is considering Netherlands, Belgium and Germany by Tabak and Yıldız [9]. Beysenbaev and Dus [10] proposed ways for improving the current LPI. They proposed an index with modification that both qualitatively and quantitatively demonstrates an main focus of 159 countries' logistics systems and subsystems, with respect to international statistical data, that can be utilized as a benchmarking tool for authorities. Göçer et al. [11] improved a methodological framework to recommend policies to augment the LPI score of pre-decided countries. The study applies qualitative and quantitative approaches to generate the recommendations for strategic decisions in an uncertain business environment. Finally, Sergi et al. [3] investigated the effect of strategic sub-components of the Global Competitiveness Index on the LPI. As a hypothesis, assumption is made that LPI and selected factors in GCI is unrelated, which were clustered as infrastructure, human factor, and institutions.

The aim of this paper is, firstly to examine LPI concept, and to analyze the top and bottom ten countries. Our country, Turkey's performance is also analyzed based on World Bank reports, covering 2007–2018. The paper consists of the following parts: *i*) introduction; *ii*) a brief description of the LPI; *iii*) analyzing of best and worst ten countries' LPI scores; *iv*) analyzing of Turkey's LPI performance; and *v*) conclusions.

2 Logistic Performance Index

The crucial work exhibiting the comparative environment of the world's logistics sector is the LPI [12]. The LPI is a benchmarking tool which is created for helping the countries to describe their opportunities and challenges in their logistics performance on trade. The LPI contains quantitative and qualitative measures and it assists to generate logistics friendly operations for the countries. It considers two approaches namely "domestic" and "international" and it measures the supply chain and logistics performances [13].

International LPI: Takes qualitative evaluations of a country in six areas into consideration by trading partners-logistics professionals that operate in the different country.

Domestic LPI: Considers qualitative and quantitative assessment of a country via logistics professionals that operate in the same country, including information on the logistics environment, core logistics process, institutions and performance time and cost data.

The LPI firstly was constituted by World Bank, in 2007. That year World Bank compared 150 countries based on seven dimensions. In 2010 World Bank compared 155 countries based on six dimensions. Because the "national logistics cost" was removed. In 2012, World Bank compared again 155 countries based on six dimensions. In 2014, 2016 and 2018 World Bank compared 160 counties based on six dimensions. The dimensions used in the International LPI were determined based on recent empirical and theoretical research and on the field experience of professionals involved in international freight forwarding. These are,

- Customs and border management clearance efficiency.
- The level of quality considering trade and transport.
- The ease of altering priced shipments from the competitive perspective.
- The competence and quality with respect to logistics services, trucking, forwarding and customs brokerage.
- The ability to track and trace consignments.
- The frequency with which shipments reach consignees within scheduled or expected delivery times [13].

These dimensions are divided into two basic categories:

- The policy regulation including the base inputs of the supply chain (custom, logistics services and quality infrastructure),
- Outputs of service offering performance (timeliness, international shipment, tracking and tracing).

2.1 Methodology of LPI

Logistics has different dimensions, so it is a challenging task to measure the performance across the countries. Analyzing the costs and time related with logistics operations—customs clearance, port processing, transport, and the like—is an outstanding beginning, and in some cases the associated information is available. But even when available, this information is not easy to confine into a single, consistent cross-country dataset, due to differences derives from structure. In addition, many crucial elements of logistics—such as service quality and process transparency, reliability and predictability—cannot be assessed by using only cost and time information.

To understand international and domestic LPI, there is a questionnaire which consists of 33 questions. The first part (questions 1–9) provides data about respondents. For example, respondent's position in company, organization level, number of employees in company, the country that respondent's currently working in. The second part of questionnaire (10–15) provides data to compute LPI. Every respondent give numbers to countries based on six core indicators (custom, infrastructure, and international shipment, quality regarding services in logistics, tracking and tracing, timeliness). Each respondent examines eight countries according to country which they currently work in. The methodology that is used for selecting country groups for survey respondents can be seen below in Table 1.

The LPI is an indicator of logistics performance that combines the data on six performance components into a single measure. In case some respondents do not give information on these six components, interpolation technique is utilized for gathering the missing values. The missing values are altered with the country mean response for each question, refilled by the respondent's average deviation from the country mean in the answered questions. The six LPI components are [14]:

1. The efficiency regarding customs and border clearance, labeled as "very low" (1) to "very high" (5) in survey question 10.
2. The quality with respect to trade and transport infrastructure, labeled as "very low" (1) to "very high" (5) in survey question 11.
3. The ease of arranging priced shipments from competitive perspective, labeled as "very difficult" (1) to "very easy" (5) in survey question 12.
4. The competence and quality for logistics services, labeled as "very low" (1) to "very high" (5) in survey question 13.
5. The ability to track and trace consignments, labeled as "very low" (1) to "very high" (5) in survey question 14.
6. The frequency with which shipments reach consignees within scheduled or expected delivery times, labeled as "hardly ever" (1) to "nearly always" (5) in survey question 15.

After respondent's ratings, normalization for question scores is done via subtraction of the sample mean and division with the standard deviation. To gather the international LPI, scores after normalization for each of the six original indicators are multiplied by their component loadings and then summed. Below in Table 2, component loadings for the international LPI can be seen.

Table 1. Methodology for selecting country groups for survey respondents [14].

	Respondents from low-income countries	Respondents from middle-income countries	Respondents from high-income countries
Respondents from coastal countries	Top fife significant export partner countries + Three crucial partner countries	Top three significant export partner countries + The most crucial partner country regarding import + Rondomly selected four countries, one from each country group: (a) Africa (b) East, South, and Central Asia (c) Latin America (d) Europe Iess Central Asla and OECD	Rondomly selected two countries in a list of five most crucial partner countries regarding export and five most significant partner countries considering import + Rondomly selected four countries, one from each country group: (a) Africa (b) East, South, and Central Asia (c) Latin America (d) Europe Iess Central Asia and OECD + Two countries randomly from the combined country groups a, b, c and d
Respondents from landlocked countries	Top four significant export partner countries + Two crucial import partner countries + Two land-bridge countries	Top three significant partner countries regarding export + The most crucial partner country regarding import + Two land-bridge countries + Rondomly selected two countries, one from each country group: (a) Africa, East, South and Central Asia, and Latin America (b) Europe Iess Central Aslla and OECD	

The component loadings are representing the weights which are given to each indicator. Since the loadings are very close to each other, the LPI is just about a simple average of the indicators. Although PCA is re-run for each LPI version, the weights remain steady from year to year. Thus, there is a high degree of comparability across the LPI versions.

Table 2. Component loadings for the international LPI [14].

Component	Weight
Customs	0.40
Infrastructure	0.42
International shipments	0.40
Logistics quality and competence	0.42
Tracking and tracing	0.41
Timeliness	0.40

Third part of the questionnaire (17–33) provides data for domestic LPI both qualitatively and quantitatively. If single value is indicated in a response, the answer is considered under the logarithm of that value. If range is indicated in a response, it is considered under the logarithm of the midpoint of that range. For instance, export distance can be pointed as less than 50 km, 50–100 km, 100–500 km, and so on. Therefore, 50–100 km response is regarded as log (75). Country scores are generated via exponentiation the average of responses in logarithms through all respondents for a given country. This approach corresponds to calculating a geometric average in levels. Scores for regions, income groups, and LPI quintiles are simple averages of the relevant country scores [14].

3 Analysis of Best and Worst Ten Countries' Logistic Performance Index

In this section the best and worst performing ten countries are analyzed regarding their LPI scores. The aggregated LPI scores are used between 2007 and 2018.

3.1 Best Ten Countries' LPI from 2007 to 2018

In 2007, we can see in Table 3 that Singapore which has ranked as first country with highest timeliness, tracking and tracing score. Singapore -an island state- is located in South-East Asia. This country has a close trade relation with Philippines, Indonesia, Australia, Taiwan and also USA, Japan, Hong Kong, Malaysia. Singapore is a substantial center in region trade cause of being transshipping port and having trough transit [16]. Also, qualified labor force, quality of infrastructure, justice of taxation and legal system are another reasons for having highest score. We can see that Netherlands has a close score to Singapore. Netherlands has three ports in the top twenty port list of Europe. Especially Rotterdam port is so important to the World. Netherlands has highest rank due to the advantages of wide-ranging highway, railway, pipe line, biggest inland water transport of the World. Also Germany, Sweden, Austria has good LPI score.

In 2010, we can see in Table 4 that Germany has ranked as the first with a score of 4.11. In Germany, the government policies aim to improve the combined transport.

Table 3. Top ten countries LPI rank and score in 2007 [15].

Country	LPI score	Customs	Infrastructure	International shipments	Logistics competence	Tracking & tracing	Timeliness
Singapore	4.19	3.90	4.27	4.04	4.21	4.25	4.53
Netherlands	4.18	3.99	4.29	4.05	4.25	4.14	4.38
Germany	4.10	3.88	4.19	3.91	4.21	4.12	4.33
Sweden	4.08	3.85	4.11	3.90	4.06	4.15	4.43
Austria	4.06	3.83	4.06	3.97	4.13	3.97	4.44
Japan	4.02	3.79	4.11	3.77	4.12	4.08	4.34
Switzerland	4.02	3.85	4.13	3.67	4.00	4.04	4.48
Hong Kong, China	4.00	3.84	4.06	3.78	3.99	4.06	4.33
United Kingdom	3.99	3.74	4.05	3.85	4.02	4.10	4.25
Canada	3.92	3.82	3.95	3.78	3.85	3.98	4.19

Table 4. Top ten countries LPI rank and score in 2010 [17].

Country	LPI score	Customs	Infrastructure	International shipments	Logistics competence	Tracking & tracing	Timeliness
Germany	4.11	4.00	4.34	3.66	4.14	4.18	4.48
Singapore	4.09	4.02	4.22	3.86	4.12	4.15	4.23
Sweden	4.08	3.88	4.03	3.83	4.22	4.22	4.32
Netherlands	4.07	3.98	4.25	3.61	4.16	4.12	4.41
Luxembourg	3.98	4.04	4.06	3.67	3.67	3.92	4.58
Switzerland	3.97	3.73	4.17	3.32	4.32	4.27	4.20
Japan	3.97	3.79	4.19	3.55	4.00	4.13	4.26
United Kingdom	3.95	3.74	3.95	3.66	3.92	4.13	4.37
Belgium	3.94	3.83	4.01	3.31	4.13	4.22	4.29
Norway	3.93	3.86	4.22	3.35	3.85	4.10	4.35

Also there are financial supports and incentive systems to improve the railway transport and ports. Tax release for railway transport, reduction of infrastructure access fees, and publishing special legislations for the creation of areas where freight transportation from railways to roads can be some examples [18]. The Hamburg port is one of the Germany's and Europe's most advanced ports. It serves a large area that covers; Austria, Denmark

Czech Republic, Germany, Italy, Hungary, Slovakia, Poland, and Switzerland [16]. Also, second country is Singapore having lots of advantages that we mentioned before.

Table 5. Top ten countries LPI rank and score in 2012 [19].

Country	LPI score	Customs	Infrastructure	International shipments	Logistics competence	Tracking & tracing	Timeliness
Singapore	4.13	4.10	4.15	3.99	4.07	4.07	4.39
Hong Kong, China	4.12	3.97	4.12	4.18	4.08	4.09	4.28
Finland	4.05	3.98	4.12	3.85	4.14	4.14	4.10
Germany	4.03	3.87	4.26	3.67	4.09	4.05	4.32
Netherlands	4.02	3.85	4.15	3.86	4.05	4.12	4.15
Denmark	4.02	3.93	4.07	3.70	4.14	4.10	4.21
Belgium	3.98	3.85	4.12	3.73	3.98	4.05	4.20
Japan	3.93	3.72	4.11	3.61	3.97	4.03	4.21
United States	3.93	3.67	4.14	3.56	3.96	4.11	4.21
United Kingdom	3.90	3.73	3.95	3.63	3.93	4.00	4.19

In 2012, we can see in Table 5 that Singapore has the highest score again. Hong Kong, China is the second country in LPI rank. Especially, rapid economic growth in China, make this region 'factory of the world'. Also, Hong Kong has the busiest airway and sea way in the world. Cause of these, Hong Kong has large business volume and became attractive center for logistics [20].

One of the countries that have high score is Denmark with the advantage of enhanced seaway, highway and airway. Copenhagen airport is selected as the best airport in Europe seven times in nine years. Also, Copenhagen airport is the North Europe center of the DHL [21]. In 2014, we can see in Table 6 that Germany has the highest LPI score in customs, infrastructure, tracking and tracing and timeliness. Netherlands is the second country and Belgium has a very close score to Netherlands'. Belgium has a very substantial port named Antwerp Port which is the second biggest port in Europe. Not only for Belgium, but also for Europe this port has huge importance [23].

In 2016, we can see in Table 7 that Germany is the first country with the highest LPI score again. Also, Luxembourg has a very close score to Germany. Luxembourg which is in the center of Europe has enhanced transportation and communication network, international airway. Thus of timeliness score is so high. Also they invest in logistics education too. The Luxembourg Centre for Logistics forms part of the Global Supply Chain and Logistics Excellence Network, in cooperation with the Massachusetts Institute of Technology in Boston, USA [25]. After then, Sweden, Netherlands, Singapore were ranked.

Table 6. Top ten countries LPI rank and score in 2014 [22].

Country	LPI score	Customs	Infrastructure	International shipments	Logistics competence	Tracking & tracing	Timeliness
Germany	4.12	4.10	4.32	3.74	4.12	4.17	4.36
Netherlands	4.05	3.96	4.23	3.64	4.13	4.07	4.34
Belgium	4.04	3.80	4.10	3.80	4.11	4.11	4.39
United Kingdom	4.01	3.94	4.16	3.63	4.03	4.08	4.33
Singapore	4.00	4.01	4.28	3.70	3.97	3.90	4.25
Sweden	3.96	3.75	4.09	3.76	3.98	3.97	4.26
Norway	3.96	4.21	4.19	3.42	4.19	3.50	4.36
Luxembourg	3.95	3.82	3.91	3.82	3.78	3.68	4.71
United States	3.92	3.73	4.18	3.45	3.97	4.14	4.14
Japan	3.91	3.78	4.16	3.52	3.93	3.95	4.24

Table 7. Top ten countries LPI rank and score in 2016 [24].

Country	LPI score	Customs	Infrastructure	International shipments	Logistics competence	Tracking & tracing	Timeliness
Germany	4.23	4.12	4.44	3.86	4.28	4.27	4.45
Luxembourg	4.22	3.90	4.24	4.24	4.01	4.12	4.80
Sweden	4.20	3.92	4.27	4.00	4.25	4.38	4.45
Netherlands	4.19	4.12	4.29	3.94	4.22	4.17	4.41
Singapore	4.14	4.18	4.20	3.96	4.09	4.05	4.40
Belgium	4.11	3.83	4.05	4.05	4.07	4.22	4.43
Austria	4.10	3.79	4.08	3.85	4.18	4.36	4.37
United Kingdom	4.07	3.98	4.21	3.77	4.05	4.13	4.33
Hong Kong, China	4.07	3.94	4.10	4.05	4.00	4.03	4.29
United States	3.99	3.75	4.15	3.65	4.01	4.20	4.25

The last LPI analysis was performed in 2018 by World Bank. In Table 8, we can see that Germany is the first country with highest score, after than Sweden, Belgium, Austria is ranked. Also Japan, Singapore and United Kingdom are in the top 10 list with high LPI scores.

Table 8. Top ten countries LPI rank and score in 2018 [26].

Country	LPI score	Customs	Infrastructure	International shipments	Logistics competence	Tracking & tracing	Timeliness
Germany	4.20	4.09	4.37	3.86	4.31	4.24	4.39
Sweden	4.05	4.05	4.24	3.92	3.98	3.88	4.28
Belgium	4.04	3.66	3.98	3.99	4.13	4.05	4.41
Austria	4.03	3.71	4.18	3.88	4.08	4.09	4.25
Japan	4.03	3.99	4.25	3.59	4.09	4.05	4.25
Netherlands	4.02	3.92	4.21	3.68	4.09	4.02	4.25
Singapore	4.00	3.89	4.06	3.58	4.10	4.08	4.32
Denmark	3.99	3.92	3.96	3.53	4.01	4.18	4.41
United Kingdom	3.99	3.77	4.03	3.67	4.05	4.11	4.33
Finland	3.97	3.82	4.00	3.56	3.89	4.32	4.28

3.2 Overall LPI for Best Ten Countries

In this section, the aggregated LPI scores are examined to determine the best performing countries in logistics sector. Below in Table 9, the top place countries are listed, and at the right hand side, overall LPI list is given.

Table 9. Overall LPI for top ten countries.

Rank	LPI 2007	LPI 2010	LPI 2012	LPI 2014	LPI 2016	LPI 2018	Overall LPI
1	Singapore	Germany	Singapore	Germany	Germany	Germany	Germany
2	Netherlands	Singapore	Hong Kong	Netherlands	Luxemburg	Netherlands	Netherlands
3	Germany	Sweden	Finland	Belgium	Sweden	Sweden	Sweden
4	Sweden	Netherlands	Germany	United Kingdom	Netherlands	Belgium	Belgium
5	Austria	Luxemburg	Netherlands	Singapore	Singapore	Singapore	Singapore
6	Japan	Sweden	Denmark	Sweden	Belgium	United Kingdom	United Kingdom
7	Switzerland	Japan	Belgium	Norway	Austria	Japan	Japan
8	Hong Kong	United Kingdom	Japan	Luxemburg	United Kingdom	Austria	Austria
9	United Kingdom	Belgium	United States	United States	Hong Kong	Hong Kong	Hong Kong
10	Canada	Norway	United Kingdom	Japan	United States	United States	United States

According to table we can see that Germany is the top performing country. When the logistics sector Germany is examined, it is obvious that it is much more developed than the other countries. The port of Hamburg -which is the Europe's most advanced port-, is so important to Germany. This port serves a large area that covers Czech Republic, Austria, Germany, Denmark, Hungary, Poland, Italy, Switzerland and Slovakia. It is among the world's most flexible and high performing ports. It handles ship calls around 9,000 yearly, almost 300 berths and 43 km of quay in total for seagoing vessels, over 2,300 freight trains weekly, having four container and three cruise terminals and around 50 facilities focus to deal with RoRo and break-bulk and all various type of bulk cargoes. In addition, there are 7,300 logistics companies within the city district. Nearly 136.5 million tons of cargo carried through the quay walls of largest seaport in Germany in 2017 including 8.8 million standard containers. Hamburg has the third place considering largest container port in Europe and it occupies the 18th place on the world with regard to largest container ports list [27]. In Germany, where harbor policies are highly developed, the back areas of ports are strengthened by road and rail. The customs work effectively which is very important for international logistics. The investments for infrastructure are done in cooperation with the private sector.

Netherlands is the second best performing country in the list. The port policies of the country are very tight. It is observed that there is a strong business partnership between port operators and state policies. The investments are carried out in parallel with government policies. Also Rotterdam and Amsterdam port are so important to Netherlands. The Rotterdam is the largest port in Europe with 450 million tons of annual throughput capacity. Sweden and Belgium also has high LPI scores. Belgium has well developed railway operations. The Antwerp port is a good example for Belgium logistics activities. It is regarded as the transition door to Europe. The port has strong transport connections and EU barcoded goods arriving in the port [9].

Over the past few years, LPI top 10 countries endured nearly consistent and generally they are the European high-income countries. Among the top 30 performing countries, 24 are OECD members. These countries are dominant in their supply chains [26].

3.3 Worst Ten Countries' LPI from 2007 to 2018

In 2007 we can see in Table 10 that, the lowest LPI score belongs to Afghanistan. Especially tracking and tracing score is too low among all countries. Because of the war environment which lasted for many years, most of the infrastructures were ruined. The lack of a coast forces the transportation to be done by only highway and airway. And the tough territory of the region also effects the shipments. These are the reasons for the lowest LPI score. Timor-Leste –having an internal crisis in 2006- is another worst performer in 2007 LPI list [28]. But in the following LPI lists, Timor-Leste does not take place in the worst 10 because of improving crude oil trade. Rwanda and Myanmar also takes place in the worst 10 list because of focusing on their internal disorders.

In 2010 we can see in Table 11 that again, generally low income countries have the lowest LPI score. That year Somalia, Eritrea, Rwanda were ranked in latest of the list. Iraq had a too low score because of the war which lasted for many years.

Table 10. Bottom ten countries LPI rank and score in 2007 [15].

Country	LPI score	Customs	Infrastructure	International shipments	Logistics competence	Tracking & tracing	Timeliness
Algeria	2.06	1.60	1.83	2.00	1.92	2.27	2.82
Guyana	2.05	1.95	1.78	1.80	1.95	2.35	2.50
Chad	1.98	2.00	1.80	1.83	1.82	1.91	2.56
Niger	1.97	1.67	1.40	1.80	2.00	2.00	3.00
Sierra Leone	1.95	1.58	1.83	1.82	1.91	2.00	2.64
Djibouti	1.94	1.64	1.92	2.00	2.00	1.82	2.30
Tajikistan	1.93	1.91	2.00	2.00	1.90	1.67	2.11
Myanmar	1.86	2.07	1.69	1.73	2.00	1.57	2.08
Rwanda	1.77	1.80	1.53	1.67	1.67	1.60	2.38
Timor-Leste	1.71	1.62	1.67	1.50	1.60	1.67	2.25
Afghanistan	1.21	1.30	1.10	1.22	1.25	1.00	1.38

Table 11. Bottom ten countries LPI rank and score in 2010 [17].

Country	LPI score	Customs	Infrastructure	International shipments	Logistics competence	Tracking & tracing	Timeliness
Burkina Faso	2.23	2.22	1.89	1.73	2.02	2.77	2.77
Sudan	2.21	2.02	1.78	2.11	2.15	2.02	3.09
Nepal	2.20	2.07	1.80	2.21	2.07	2.26	2.74
Iraq	2.11	2.07	1.73	2.20	2.10	1.96	2.49
Guinea-Bissau	2.10	1.89	1.56	2.75	1.56	1.71	2.91
Cuba	2.07	1.79	1.90	2.32	1.88	2.03	2.41
Rwanda	2.04	1.62	1.62	2.88	1.85	1.99	2.05
Namibia	2.02	1.68	1.71	2.20	2.04	2.04	2.38
Sierra Leone	1.97	2.17	1.61	2.33	1.52	1.73	2.33
Eritrea	1.70	1.50	1.35	1.63	1.88	1.55	2.21
Somalia	1.34	1.33	1.50	1.33	1.33	1.17	1.38

It can be seen in 2012 that, Burundi had the lowest LPI score. Also Djibouti, Haiti, Nepal, Sudan had too low LPI score. Again, Iraq was in the bottom ten countries' list. The battle had ended in 2011, but the effects of the war still continued. Also in Haiti and Nepal, put all their efforts to recover from the effects of earthquakes [29].

Table 12. Bottom ten countries LPI rank and score in 2012 [19].

Country	LPI score	Customs	Infrastructure	International shipments	Logistics competence	Tracking & tracing	Timeliness
Iraq	2.16	1.75	1.92	2.38	2.19	1.86	2.77
Comoros	2.14	2.00	1.94	1.81	2.20	2.20	2.70
Eritrea	2.11	1.78	1.83	2.63	2.03	1.83	2.43
Sudan	2.10	2.14	2.01	1.93	2.33	1.89	2.31
Congo, Rep	2.08	1.80	1.27	1.94	2.15	2.35	2.90
Sierra Leone	2.08	1.73	2.50	1.85	1.98	2.14	2.35
Nepal	2.04	2.20	1.87	1.86	2.12	1.95	2.21
Chad	2.03	1.86	2.00	2.00	2.00	1.57	2.71
Haiti	2.03	1.78	1.78	1.94	1.74	2.15	2.74
Djibouti	1.80	1.72	1.51	1.77	1.84	1.73	2.19
Burundi	1.61	1.67	1.68	1.57	1.43	1.67	1.67

Table 13. Bottom ten countries LPI rank and score in 2014 [22].

Country	LPI score	Customs	Infrastructure	International shipments	Logistics competence	Tracking & tracing	Timeliness
Gabon	2.20	2.00	2.08	2.58	2.25	1.92	2.31
Yemen, Rep	2.18	1.62	1.87	2.35	2.21	2.21	2.78
Cuba	2.18	2.17	1.84	2.47	2.08	1.99	2.45
Sudan	2.16	1.87	1.90	2.23	2.18	2.42	2.33
Djibouti	2.15	2.20	2.00	1.80	2.21	2.00	2.74
Syrian Arab Republic	2.09	2.07	2.08	2.15	1.82	1.90	2.53
Eritrea	2.08	1.90	1.68	1.90	2.23	2.01	2.79
Congo, Rep	2.08	1.50	1.83	2.17	2.17	2.17	2.58
Afghanistan	2.07	2.16	1.82	1.99	2.12	1.85	2.48
Congo, Dem. Rep	1.88	1.78	1.83	1.70	1.84	2.10	2.04
Somalia	1.77	2.00	1.50	1.75	1.75	1.75	1.88

In 2014, we can see in Table 13 that Somalia has the lowest score. Syrian Arab Republic was added to bottom ten countries because of the civil war. If Syrian Arab Republic LPI rank is examined it can be seen that, in 2010 it ranked as the 80th country, in 2012 it ranked as the 92nd country and in 2014 it ranked as the 155th country. Especially in quality of logistics services Syrian Arab Republic has lowest score among six dimensions.

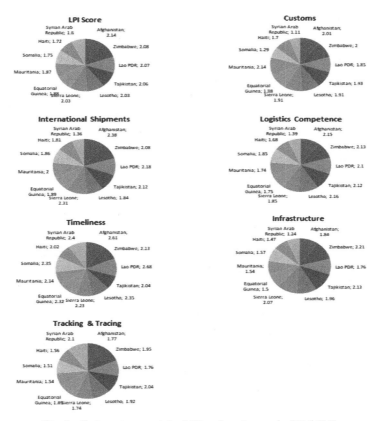

Fig. 1. Bottom ten countries LPI rank and score in 2016 [24].

In 2016, we can see that the lowest LPI score belong to Syrian Arab Republic due to the civil war again. In 2014, its rank was 155th, but in 2016 the logistics performance was worse than a year ago. Other countries are generally middle-income and low-income countries. Their performance in six dimensions not good. Also they have one in common that all of them deal with internal crisis [30].

In 2018, again Afghanistan has the lowest LPI score and rank. Also, Angola, Burundi, Eritrea, Haiti which are low income countries in the bottom ten countries list.

Table 14. Bottom ten countries LPI rank and score in 2018 [26].

Country	LPI score	Customs	Infrastructure	International shipments	Logistics competence	Tracking & tracing	Timeliness
Gabon	2.16	1.96	2.09	2.10	2.07	2.07	2.67
C.A.R	2.15	2.24	1.93	2.30	1.93	2.10	2.33
Zimbabwe	2.12	2.00	1.83	2.06	2.16	2.26	2.39
Haiti	2.11	2.03	1.94	2.01	2.19	2.05	2.44
Libya	2.11	1.95	2.25	1.99	2.05	1.64	2.77
Eritrea	2.09	2.13	1.86	2.09	2.17	2.17	2.08
Sierra Leone	2.08	1.82	1.82	2.18	2.00	2.27	2.34
Niger	2.07	1.77	2.00	2.00	2.10	2.22	2.33
Burundi	2.06	1.69	1.95	2.21	2.33	2.01	2.17
Angola	2.05	1.57	1.86	2.20	2.00	2.00	2.59
Afghanistan	1.95	1.73	1.81	2.10	1.92	1.70	2.38

3.4 Overall LPI for Worst Ten Countries

In above Table 15, the aggregated worst LPI performances of the countries can be seen. The worst performing ten countries are generally lower-middle –income and low-income countries. These countries have fragile economies and generally they are afflicted by natural disasters, internal disturbances, political unrest or they face problems regarding their geographical locations that restrain the connections to global supply chains [26]. For example, Iraq, Afghanistan and Syrian Arab Republic have effected badly due to the battles lasted for years. Their infrastructure, communication systems and customs were affected very much.

Table 15. Overall LPI for bottom ten countries.

LPI 2007	LPI 2010	LPI 2012	LPI 2014	LPI 2016	LPI 2018	Overall LPI
Guyana	Sudan	Comoros	Yemen, Repl	Zimbabwe	Central Afr. Repl	Gabon
Chad	Nepal	Eritrea	Cuba	Laos Republic	Zimbabwe	Iraq
Niger	Iraq	Sudan	Sudan	Tajikistan	Haiti	Angola
Sierra Leone	Guinea-Bissau	Congo, Repl	Djibouti	Lesotho	Libya	Zimbabwe

(*continued*)

Table 15. (*continued*)

LPI 2007	LPI 2010	LPI 2012	LPI 2014	LPI 2016	LPI 2018	Overall LPI
Djibouti	Cuba	Sierra Leone	Syrian Arab Repl	Sierra Leone	Eritrea	Eritrea
Tajikistan	Rwanda	Nepal	Eritrea	Equatorial Guinea	Sierra Leone	Syr. Arab Repl
Myanmar	Namibia	Chad	Congo, Repl	Mauritania	Nigar	Sierra Leone
Rwanda	Sierra Leone	Haiti	Afghanistan	Samolia	Burundi	Afghanistan
Timor-Leste	Eritrea	Djibouti	Congo, Repl	Haiti	Angola	Haiti
Afghanistan	Somalia	Burundi	Somalia	Syrian Arab Repl	Afghanistan	Somalia

4   Analysis of Turkey's LPI

In this section, Turkey's Logistics performance between 2007 and 2018 briefly is analyzed. Tables which show the rank and score of LPI helped us to comment on Turkey's
logistics performance comparison across years. Below in Table 16, all the LPI scores
that belong to Turkey are given.

Table 16. Turkey's LPI scores [26].

Year	LPI rank	LPI score	Customs	Infrastructure	International shipments	Logistics competence	Tracking & tracing	Timeliness
2012	27	3.51	3.16	3.62	3.38	3.52	3.54	3.87
2014	30	3.50	3.23	3.53	3.18	3.64	3.77	3.68
2016	34	3.42	3.18	3.49	3.41	3.31	3.39	3.75
2010	39	3.22	2.82	3.08	3.15	3.23	3.09	3.94
2007	34	3.15	3.00	2.94	3.07	3.29	3.27	3.38
2018	47	3.15	2.71	3.21	3.06	3.05	3.23	3.63

In 2007, Turkey ranked as the 34th out of 150 countries with a score of 3.15. The
Lowest score was in infrastructure and the highest score was in the timeliness. In 2010,
Turkey ranked as the 39th with a score of 3.22. If we compare LPI scores of 2007 and
2010, we can say that score of "Customs" and "Tracking and Tracing" performed better.
In 2012, Turkey has achieved its best performance and ranked as 27th. After the 2009
crisis which happened around the world, many European countries economies were
affected badly. Also, these negative effects reflected to their logistics sectors. But this
crisis didn't affect Turkey' economy as much as the European countries. Even in year

2010 Turkey's economy grew 9, 2% and in year 2011 economy grew 8, 8%. Thus in 2012, Turkey's logistics performance score so high and ranked as 27th [1]. In 2014, Turkey ranked as the 30th with a close score to 2012 LPI. In 2014, logistics competence, tracking and tracing, timeliness and customs scores were better than 2012. In 2016, Turkey ranked as the 34th among 160 countries. If we compare LPI rank across years, there is no change in the rank of 2007 and 2016. In 2007 LPI score was 3.15 and rank was 34, but in year 2016 the LPI score was 3.42 and the rank was 34. Despite of increased score, rank didn't get better. The reason can be that most of the countries understood the importance of logistics and made improvements on their logistics performance dimensions. In 2018, Turkey has displayed its worst performance. 2016–2018 rankings showed that, none of the criteria rank increased. According to 2018 report, our biggest problem was in the "Customs". Waiting time in border gates, coordination problems, custom bureaucracy and lack of information technologies in customs may have affected this score.

Although the LPI is a very comprehensive and reflective approach for the logistics sector, it has some drawbacks that need to be improved. For instance, since the LPI is formed by a survey of logistics professionals, it may cause a skewed rating. In addition, or instead of survey approach, quantitative approaches and data should be used. For different proposals to improve the LPI, the reader is referred to Özceylan et al. [31] and Beysenbaev and Dus [10].

5 Conclusion

The LPI is a highly important parameter for countries to analyze their rank among other countries. If the score can be understood in detail, a country can enhance the logistics service and result in a better position/wealth. In addition, LPI plays a vital role in revealing the logistics-related problems of the countries, determining the reform priorities in the public-private sector, and implementing the reforms as soon as possible. Thus, firstly the LPI is defined and the calculation methodology is tried to explain. Later, the best and worst 10 LPI scores in all World Bank reports between 2007 and 2018 are examined. Under favor of the LPI tables, some comments are added about the best and worst performances for each year. In the last section we analyzed Turkey's logistics performance score. It is determined that generally high income countries are in the top performer list. Germany ranked as the first, due to its quality infrastructure and advanced ports. Also Netherlands and Belgium are in the top performer list and they both have important ports in Europe. Hong Kong and Japan are close to important points of production and they also benefit from shipping opportunities. In addition, generally low-income countries are in the worst performer list. In summary, Syria decreased in the list because of the civil war which started in 2011. Also Iraq was in the worst performer list because of the war environment which lasted for years. Also Afghanistan faced the war and most of its infrastructures were ruined. Also other countries in the list are low income countries, their infrastructure quality, tracking, timeliness performance were not good. When it comes to Turkey, which ranked 30th in 2007 (the first LPI), dropped to 39th in 2010, but managed to rise to 27th in 2012, especially with the improvements made in the field of customs. But since 2014, again it is observed that Turkey has drawn a steady decline profile, and this time it was ranked 34th in 2016 and ranked 47th in

2018. It has achieved its lowest logistic performance in 2018. LPI scores showed that, none of the criteria increased from 2016 to 2018. According to the results in 2018, the lowest score was in the customs and highest score was in the timeliness. Turkey should use its advantages of geopolitical position, ports and should invest on information technologies. Improving customs processes will not only ensure that Turkey will reach a good point in the Index, but also it will increase the power of Turkish logistics companies in competition with foreign logistics companies. In the grand scheme of things, it is determined that the income rates of the countries affected the LPI scores as expected. Many worst performing countries are located near to oceans, which is the cheapest transportation mode. But most of them had internal crisis in their countries. So it can be understood that the security directly affects the logistics activities. There would still be a "worst performer 10 country" in any case but it the security levels were better the performance gap would be narrower. This paper can be extended with the publication of new LPI 2020 report, and also the impact of COVID-19 on countries should be statistically investigated.

References

1. Yaprakli, T.Ş., Ünalan, M.: The global logistics performance index and analysis of the last ten years logistics performance of Turkey. J. Econ. Admin. Sci. Univ. Ataturk **31**(3), 589–606 (2017)
2. Ekici, Ş.Ö., Kabak, Ö., Ülengin, F.: Improving logistics performance by reforming the pillars of Global Competitiveness Index. Transp. Policy **81**, 197–207 (2019)
3. Sergi, B.S., D'Aleo, V., Konecka, S., Szopik-Depczyńska, K., Dembińska, I., Ioppolo, G.: Competitiveness and the logistics performance index: the ANOVA method application for Africa, Asia, and the EU regions. Sustain. Cities Soc. **69**, 102845 (2021)
4. Martí, L., Puertas, R., García, L.: The importance of the logistics performance index in international trade. J. Appl. Econ. **46**(24), 2982–2992 (2014)
5. Civelek, M.E., Uca, N., Çemberci, M.: The mediator effect of logistics performance index on the relation between global competitiveness index and gross domestic product. Eur. Sci. J. **11**(13), 368–375 (2015)
6. Çakır, S.: Measuring logistics performance of OECD countries via fuzzy linear regression: measuring logistics performance. J. Multi-Criteria Decis. Anal. **24**(3), 177–186 (2016)
7. Gani, A.: The logistics performance effect in international trade. Asian J. Shipping Logistics **33**(4), 279–288 (2017)
8. Rezaei, J., van Roekel, W.S., Tavasszy, L.: Measuring the relative importance of the logistics performance index indicators using Best Worst Method. Transp. Policy **68**, 158–169 (2018)
9. Tabak, C., Yıldız, K.: Turkey's logistics impact compared to the Netherlands, Germany and Belgium. Int. J. Logistics Syst. Manage. **31**(1), 1–19 (2018)
10. Beysenbaev, R., Dus, Y.: Proposals for improving the Logistics Performance Index. Asian J. Shipping Logistics **36**(1), 34–42 (2020)
11. Göçer, A., Özpeynirci, Ö., Semiz, M.: Logistics performance index-driven policy development: an application to Turkey. Transp. Policy (2020). https://doi.org/10.1016/j.tranpol.2021.03.007
12. Çemberci, M., Civelek, M.E., Canbolat, N.: The moderator effect of global competitiveness index on dimensions of logistics performance index. Procedia. Soc. Behav. Sci. **195**, 1514–1524 (2015)
13. World Bank. https://lpi.worldbank.org/about. Accessed 27 Mar 2019

14. Arvis, J.F., Saslavsky, D., Ojala, L., Shepherd, B., Busch, C., Raj, A.: Connecting to compete 2014: trade logistics in the global economy, The Logistics Performance Index and Its Indicators, World Bank, Washington, DC (2014)
15. World Bank, https://lpi.worldbank.org/international/global/2007. Accessed 27 Mar 2019
16. OEC, https://atlas.media.mit.edu/en/profile/country/sgp/. Accessed 27 Mar 2019
17. World Bank, https://lpi.worldbank.org/international/global/2010. Accessed 27 Mar 2019
18. Deutsche Bank Research, Logistics in Germany: Only modest growth in the near term, https://www.dbresearch.com/PROD/RPS_EN-PROD/PROD0000000000452421/Logistics_in_Germany%3A_Only_modest_growth_in_the_ne.pdf. Accessed 27 Mar 2019
19. World Bank, https://lpi.worldbank.org/international/global/2012. Accessed 27 Mar 2019
20. Honk Kong Economy Research, Logistics Industry in Hong Kong, http://hong-kong-economy-research.hktdc.com/business-news/article/Hong-Kong-Industry-Profiles/Logistics-Industry-in-Hong-Kong/hkip/en/1/1X000000/1X0018WG.htm. Accessed 27 Mar 2019
21. Copenhagen Airport Website, https://www.cph.dk/en/about-cph/press/news/2016/10/condenast-traveler-cph-recognized-as-the-worlds-7th-best-international-airport. Accessed 27 Mar 2019
22. World Bank, https://lpi.worldbank.org/international/global/2014. Accessed 27 Mar 2019
23. Port of Antwerp, https://www.portofantwerp.com/en. Accessed 27 Mar 2019
24. World Bank, https://lpi.worldbank.org/international/global/2016. Accessed 27 Mar 2019
25. University of Luxembourg, https://wwwen.uni.lu/fdef/luxembourg_centre_for_logistics_and_supply_chain_management. Accessed 27 Mar 2019
26. World Bank, https://www.worldbank.org/en/news/feature/2018/07/24/from-parts-to-products-why-trade-logistics-matter. Accessed 27 Mar 2019
27. Port of Hamburg, https://www.hafen-hamburg.de/. Accessed 27 Mar 2019
28. Van der Auweraert, P.: The 2006 crisis in Timor-Leste: a brief backgrounder. In: Ending the 2006 Internal Displacement Crisis in Timor-Leste: Between Humanitarian Aid and Transitional Justice, IOM, Geneva (2012). https://doi.org/10.18356/0f94cb96-en
29. The World Factbook, https://www.cia.gov/library/publications/the-world-factbook/geos/sl.html. Accessed 27 Mar 2019
30. Encyclopedia Britannica, https://www.britannica.com/topic/Nepal-earthquake-of-2015. Accessed 27 Mar 2019
31. Özceylan, E., Çetinkaya, C., Erbaş, M., Kabak, M.: Logistic performance evaluation of cities in turkey: a GIS-based multi-criteria decision analysis. Transp. Res. Part A **94**, 323–337 (2016)

High Level Petri Nets Application for Reliability Visualization on Multi Echelon Supply Chain

Hamid Esmaeeli🆔 and Matin Aleahmad$^{(\boxtimes)}$ 🆔

Industrial Engineering Group, Islamic Azad University of North Tehran Branch, Tehran, Iran
m.aleahmad@iau-tnb.ac.ir

Abstract. The supply chain management is one of the important processes in every organization. The agile and economical supply chains become the competitive advantage for successful organizations. For developing an effective supply chain system, it is necessary for organizations to model and simulate their supply chain before they start to run it. There are many methods for simulating the supply chain and logistics systems, in this research the High-Level Petri Nets (HLPN) presented as an effective method for simulating a supply chain system; in this research the researchers presented a four-echelon supply chain system for two types of products. The results show the high performance and easiness of using of High-level petri net (HLPN) method. In this research the researchers were able to follow every scenarios that can happened to the materials from suppliers to consumers. The components of high-level petri nets (HLPN) can help the researchers for modeling and simulating the supply chain and logistic complex systems in the way of developing the agile and low-cost supply chains.

Keywords: High-level petri nets · Reliability · Visualizing · Multi echelon · Supply chain management

1 Introduction

Now a days the organizations competition is on delivery the products and services efficient and effective to their customers and this is the reason of making logistics and supply chains very important to the organizations. Depending on the organization's size, variety of products and quantity of products, verity of customers and quantity of customers and the organization's strategies, there are few or many suppliers that deliver the goods and materials to the organization for producing and delivering the products to their targeted consumers. This chain of goods and services that transfer from a supplier to the next supplier until the consumer called supply chain and the management of this supply chain called supply chain management (SCM) [1].

1.1 Supply Chain Management

Supply chain management is the progress of managing the flow of goods and services at the right time, right place and the right cost with the best quality for the consumers [2]. The

© Springer Nature Switzerland AG 2021
Z. Molamohamadi et al. (Eds.): LSCM 2020, CCIS 1458, pp. 43–52, 2021.
https://doi.org/10.1007/978-3-030-89743-7_3

verity of consumers and customer's demands, the globalization of the businesses make the organizations smaller and more agile, this is the reason organizations cooperation for delivering the value for the consumers and the customers [3].

"The maximum strength of a chain is equal to the resistant of its weakest ring."

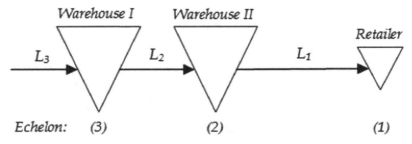

Fig. 1. Three echelons supply chain management

The above quote shows the importance of supply chain management. The stages transformation of raw materials to the final product and the delivery of this product in every stage called echelon. In the (Fig. 1) the three echelons supply chain presented. Supply chain management is the combination of qualitative and quantitative scientific methods. There are several types of qualitative and quantitative methods for managing the supply chain [4–9]. In this research the researchers use the high-level petri nets (HLPN) to model the four echelons supply chain. The HLPN is a graphical-mathematical method for modelling the complex systems with concurrent behavior [10].

1.2 High-Level Petri Nets

High-level petri nets (HLPN) originated from the formal petri nets (PN) method that presented by Carl Adam Petri [11]. Petri net (PN) is the combination usage of graphical and mathematical method for system modelling. The (PN) combined with four components; Places, Transitions, Arcs and tokens that presented in (Fig. 2).

The petri net originated from bipartite graph theory. In bipartite graph theory the nodes separated to two types. In bipartite graph theory every homogenous node does not related with arcs. In petri nets places and transitions are known as nodes; this rule presented as (Fig. 3).

The formal petri net (PN) presented with 4 tuples:

PN = (P,T,F,M0).

P = number of places.

T = number of transitions.

F = number of arcs; $F \subseteq (P \times T) \cup (T \times P)$.

M0 is a matrix that shown tokens quantity that held in each place at beginning of modelling that called initial marking. After every fire the number of M0 will change to M1, [11].

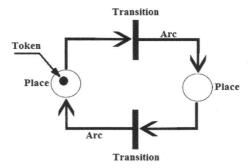

Fig. 2. Components of petri net.

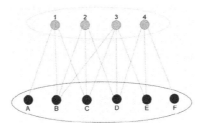

Fig. 3. Bipartite Graph.

Formal petri nets are not compatible to model the complex timed base system, the (HLPN) presented for modeling complex systems with time [12–16]. High-level petri nets (HLPN) combined with many types of petri nets that categorized as stochastic Petri nets (SPN), timed petri nets (TPN), colored petri nets (CPN) and etc. [17–26].

In this article the researchers used the stochastic petri nets as High-Level petri nets to simulate a four-echelon supply chain system that stated 2 types of products.

2 Literature Review

There are many research presented in logistics and supply chain management that used different methods and tools for evaluating and modelling. In this section some of these research were presented in evaluating supply chain management and logistic systems by (HLPN). Some outputs presented a formalized Petri net method for modeling the supply problem of agricultural products. This Petri net model divided supply chain into separate links and presented a quantitative analysis of the operational tasks cycle and the operational efficiency of each link. The stream of information in the whole of supply chain has been realized. thus, to ensuring the benefits of each farmer production, the previous risks in this business that a single farmer needs to bear were reduced and the operational efficiency of the entire supply chain were improved; the presented model shown as (Fig. 4) [27].

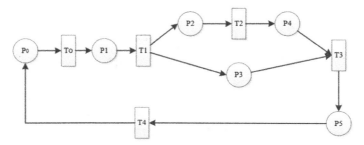

Fig. 4. The Petri net model divided supply chain into different links

In a research A colored Petri net (CPN) model developed for configure supply chains. The model used to synthesize product and process concerns into process configuration of supply chain. The industrial application of this model reported the high potential of the CPN modelling formalism and configurating models for supply chain system development; the colored petri net model shown as (Fig. 5) [28].

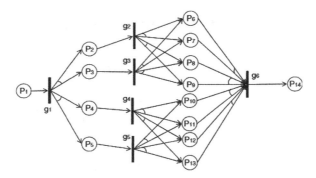

Fig. 5. The industrial supply chain modeled by colored petri net

In a research modular property First-Order Hybrid Petri Net (FOHPN) has been used to model a Closed-Loop Supply Chain Design (CLSCND) process. The outputs depict that FOHPN are high performance modeling CLSCND and applicable as an efficient tool at both tactical and strategic levels of decision making. As the research shown how FOHPN can be extended to diagnose financial, operational, environmental network's performance measures simultaneously at different managerial decision-making levels. The outputs are exclusively persuading for industrial researchers and practitioners who can apply same methodology in their network's performance evaluation and educated management decisions-making based on the results and the impact of their selected supply chain and manufacturing strategies; the model presented as (Fig. 6) [29].

In a research a Colored Petri Nets model purposed for evaluating a bus network performance by using (MAX+) Algebra. The main novelty in this method was the integration of CPN with (MAX+) equations in order of buses timetables and boarding evaluation, passenger's descendant and waiting times; the model presented as (Fig. 7) [30].

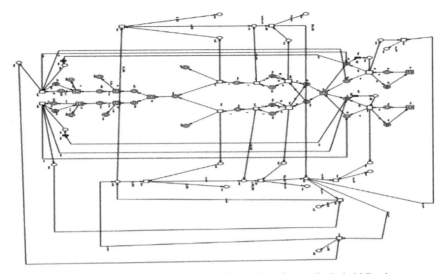

Fig. 6. Closed-loop supply chain model based on first-order hybrid Petri nets

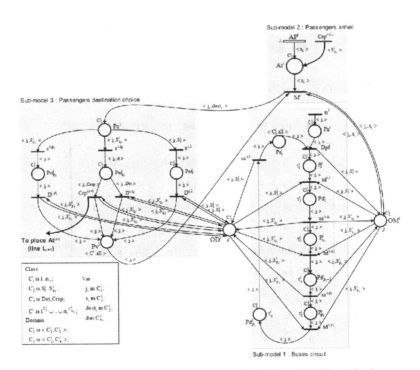

Fig. 7. The bus network model presented by CPN and MAX + Algebra

3   Proposed High-Level Petri Net Model for Four-Echelon Supply Chain

In this research the discussion based on four-echelon supply chain model based on two products. The echelons divided to suppliers, manufacturers, retailers and customers. The four-echelon supply chain presented as (Fig. 8).

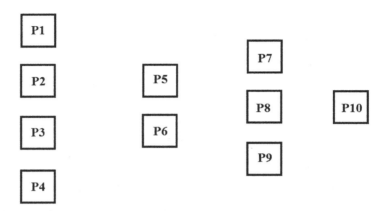

Suppliers   Manufacturers   Retailers   Customers

Fig. 8. The four-echelon supply chain

The probability of material flow between every node presented as (Fig. 9) matrix.

	P1	P2	P3	P4	P5	P6	P7	P8	P9	P10
P1	0	0	0	0	0.5	0.5	0	0	0	0
P2	0	0	0	0	0.4	0.55	0	0	0	0
P3	0	0	0	0	0.5	0.5	0	0	0	0
P4	0	0	0	0	0.6	0.4	0	0	0	0
P5	0.02	0.03	0.04	0.03	0.04	0	0.21	0.42	0.2	0
P6	0.05	0.04	0.04	0.05	0	0.03	0.27	0.33	0.18	0
P7	0	0	0	0	0.13	0.11	0	0	0	0.73
P8	0	0	0	0	0.09	0.12	0	0	0	0.77
P9	0	0	0	0	0.11	0.17	0	0	0	0.71
P10	0	0	0	0	0	0	0.1	0.13	0.15	0

Fig. 9. Material flow matrix

The consumption of consumers calculated as the equation number 1.

$$1 - (0.1 + 0.13 + 0.15) = 0.62 \tag{1}$$

The disposal of each node calculated as Eqs. (2–6)

$$P5(Disposal) = 1 - (0.02 + 0.03 + 0.04 + 0.03 + 0.04 + 0.21 + 0.42 + 0.2) = 0.01 \tag{2}$$

$$P6(Disposal) = 1 - (0.05 + 0.04 + 0.04 + 0.05 + 0.03 + 0.27 + 0.33 + 0.18) = 0.01 \tag{3}$$

$$P7(Disposal) = 1 - (0.13 + 011 + 0.73) = 0.03 \tag{4}$$

$$P8(Disposal) = 1 - (0.09 + 0.12 + 0.77) = 0.02 \tag{5}$$

$$P9(Disposal) = 1 - (0.11 + 0.17 + 0.71) = 0.01 \tag{6}$$

The Disposal matrix presented as (Fig. 10):

$$\begin{matrix} \textbf{P1} & \textbf{P2} & \textbf{P3} & \textbf{P4} & \textbf{P5} & \textbf{P6} & \textbf{P7} & \textbf{P8} & \textbf{P9} & \textbf{P10} \\ \begin{bmatrix} 0 & 0 & 0 & 0 & 0.01 & 0.01 & 0.03 & 0.02 & 0.01 & 0 \end{bmatrix} \end{matrix}$$

Fig. 10. Disposal matrix of the 4 echelon supply chain

The (HLPN) model presented as (Fig. 11):

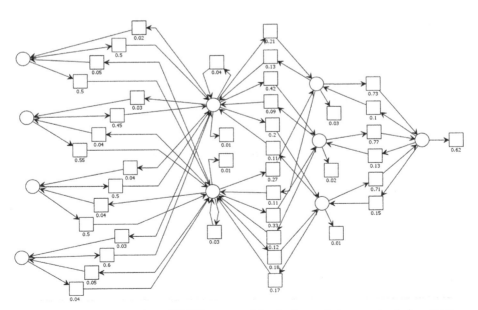

Fig. 11. High-level petri net (HLPN) presented for the four-echelon supply chain system

As it presented the 62% of the raw materials can consummated by the consumers and the 8% of raw materials are disposed by the system stations and more than 30% of the system performance used for the reworking.

4   Conclusion

The importance of supply chain management is necessary every organization. In 21st century the agile supply chains become competitive advantage in the market competition. There are many methods for modeling and simulating the supply chain; in this research the (HLPN) method presented as a mathematical-graphical method for modeling and simulating the four-echelon supply chain system. The results show the high performance of this modeling method and the simplicity of this modeling method. There are many benefits in using this method that some of them are:

1- The user-friendly graphical presentation
2- The relations divided to output and input arcs that makes the rules easy for researchers
3- The different objective of different nodes; (circles) as places and (triangles) as transitions can assist the researchers to model the system more accurate and effective.

And many other benefits that prove the performance of this method for modeling many kind of systems like logistics and any type of supply chain systems.

References

1. Lambert, D.M., Enz, M.G.: Issues in supply chain management: progress and potential. Ind. Market. Manage. 1–16 (2017).https://doi.org/10.1016/j.indmarman.2016.12.002. ISSN 0019 8501
2. Banchuen, P., Sadler, I., Shee, H.: Supply chain collaboration aligns order-winning strategy with business outcomes. IIMB Manage. Rev. **29**(2), 109–121 (2017).https://doi.org/10.1016/j.iimb.2017.05.001. ISSN 0970 3896
3. Hacklin, F., Björkdahl, J., Wallin, M.W.: Strategies for business model innovation: how firms reel in migrating value. Long Range Plann. **51**(1), 82–110 (2018). https://doi.org/10.1016/j.lrp.2017.06.009. ISSN 0024 6301
4. Sundarakani, B., Ajaykumar, A., Gunasekaran, A.: Big data driven supply chain design and applications for blockchain: an action research using case study approach. Omega **102**, 102452 (2021).https://doi.org/10.1016/j.omega.2021.102452. ISSN 0305 0483
5. Younis, A., Sundarakani, B., O'Mahony, B.: Green supply chain management and corporate performance: developing a roadmap for future research using a mixed method approach. IIMB Manage. Rev. (2019). https://doi.org/10.1016/j.iimb.2019.10.011. ISSN 0970 3896
6. Taifouris, M., Martín, M., Martínez, A., Esquejo, N.: Simultaneous optimization of the design of the product, process, and supply chain for formulated product. Comput. Chem. Eng. 107384 (2021). https://doi.org/10.1016/j.compchemeng.2021.107384. ISSN 0098 1354
7. Pazhani, S., Mendoza, A., Ramkumar Nambirajan, T.T., Narendran, K.G., Olivares-Benitez, E.: Multi-period multi-product closed loop supply chain network design: a relaxation approach. Comput. Ind. Eng. **155**, 107191 (2021). https://doi.org/10.1016/j.cie.2021.107191. ISSN 0360 8352

8. Reza Alikhani, S., Torabi, A., Altay, N.: Retail supply chain network design with concurrent resilience capabilities. Int. J. Product. Econ. **234**, 108042 (2021). https://doi.org/10.1016/j. ijpe.2021.108042. ISSN 0925 5273

9. Kain, R., Verma, A.: Logistics management in supply chain – an overview. In: Materials Today: Proceedings, vol. 5, no. 2, pp. 3811–3816 Part 1 (2018).https://doi.org/10.1016/j. matpr.2017.11.634. ISSN 2214 7853

10. Cabac, L., Haustermann, M., Mosteller, D.: Software development with Petri nets and agents: approach, frameworks and tool set. Sci. Comput. Programm. 157, 56–70 (2018). https://doi. org/10.1016/j.scico.2017.12.003. ISSN 0167 6423

11. Van der Aalst W.M.P.: Petri Nets. In: Liu, L., Özsu, M. (eds.) Encyclopedia of Database Systems. Springer, New York (2016). https://doi.org/10.1007/978-1-4899-7993-3

12. Ding, J., Chen, X., Sun, H., Yan, W., Fang, H.: Hierarchical structure of a green supply chain. Comput. Ind. Eng. **157**, 107303 (2021). https://doi.org/10.1016/j.cie.2021.107303

13. Elidrissi, H.L., Nait-Sidi-Moh, A., Tajer, A.: Modular design for an urban signalized inter-sections network using synchronized timed Petri nets and responsive control. Proc. Comput. Sci. **170**, 458–465 (2020). https://doi.org/10.1016/j.procs.2020.03.089. ISSN 1877 2509

14. Komenda, J., Lai, A., Soto, J.G., Lahaye, S., Boimond, J.L.: Modeling of safe time Petri nets by interval weighted automata∗∗Partially supported by RVO 67985840 and by GACR grant GC19–06175J. IFAC-PapersOnLine **53**(4), 187–192 (2020). https://doi.org/10.1016/j.ifacol. 2021.04.018. ISSN 2405–8963

15. Di Marino, E., Su, R., Basile, F.: Makespan optimization using timed Petri nets and mixed integer linear programming problem. IFAC-PapersOnLine **53**(4), 129–135 (2020). https:// doi.org/10.1016/j.ifacol.2021.04.073. ISSN 2405 8963

16. Mecheraoui, K., Lomazova, I.A., Belala, N.: A Petri net extension for systems of concur-rent communicating agents with durable actions. J. Parallel Distribut. Comput. **155**, 14–23 (2021)https://doi.org/10.1016/j.jpdc.2021.04.011. ISSN 0743 7315

17. Boubeta-Puig, J., Díaz, G., Macià, H., Valero, V., Ortiz, G.: MEdit4CEP-CPN: an approach for complex event processing modeling by prioritized colored petri nets. Inf. Syst. 81, 267–289 (2019). https://doi.org/10.1016/j.is.2017.11.005. ISSN 0306 4379

18. Basile, F., Chiacchio, P., Coppola , J.: IdentifyTPN: a tool for the identification of time Petri nets. IFAC-PapersOnLine 50(1), 5843–5848 (2017). ISSN 2405 8963

19. Bevilacqua, M., Ciarapica, F., Mazzuto, G.: Timed coloured petri nets for modelling and managing processes and projects. Proc. CIRP. **67**, 58–62 (2018). https://doi.org/10.1016/j.pro cir.2017.12.176

20. Esmaeeli, H., Aleahmad, M.: Bottleneck detection in job shop production by high-level Petri nets. In: 2019 15th Iran International Industrial Engineering Conference (IIIEC). 2019 15th Iran International Industrial Engineering Conference (IIIEC), January 2019. https://doi.org/ 10.1109/iiiec.2019.8720639

21. Chahrour, N., Nasr, M., Tacnet, J.-M., Bérenguer, C.: Deterioration modeling and maintenance assessment using physics-informed stochastic Petri nets: application to torrent protection structures. Reliab. Eng. Syst. Saf. **210**, 107524 (2021). https://doi.org/10.1016/j.ress.2021. 107524. ISSN 0951 8320

22. Li, X.-Y., Liu, Y., Lin, Y.-H., Xiao, L.-H., Zio, E., Kang, R.: A generalized petri net-based modeling framework for service reliability evaluation and management of cloud data cen-ters, Reliability Engineering & System Safety, Volume 207, 2021. ISSN **107381**, 0951–8320 (2021). https://doi.org/10.1016/j.ress.2020.107381

23. Assaf, G., Heiner, M., Liu, F.: Colouring fuzziness for systems biology. Theoret. Comput. Sci. **875**, 52–64 (2021). https://doi.org/10.1016/j.tcs.2021.04.011.ISSN 0304 3975

24. Zeinalnezhad, M., Chofreh, A.G., Goni, F.A., Hashemi, L.S., Klemeš, J.J.: A hybrid risk anal-ysis model for wind farms using Coloured Petri Nets and interpretive structural modelling., Energy **229**, 120696 (2021). https://doi.org/10.1016/j.energy.2021.120696.ISSN 0360 5442

25. Ben Mesmia, W., Escheikh, M., Barkaoui, K.: DevOps workflow verification and duration prediction using non-Markovian stochastic Petri nets. J. Softw. Evolut. Process **33**(3) (2020). https://doi.org/10.1002/smr.2329
26. Marsan, M.A.: Stochastic Petri nets: an elementary introduction. In: Rozenberg, G. (ed.) APN 1988. LNCS, vol. 424, pp. 1–29. Springer, Heidelberg (1990). https://doi.org/10.1007/3-540-52494-0_23
27. Liu, J.-P., Wu, R.-G.: A petri net-based supply chain system. Int. J. Online Eng. (iJOE). **14**, 28 (2018). https://doi.org/10.3991/ijoe.v14i11.9502
28. Zhang, L., You, X., Jiao, J., Helo, P.: Supply chain configuration with coordinated product, process and logistics decisions: an approach based on petri nets. Int. J. Prod. Res. **47**, 6681–6706 (2009). https://doi.org/10.1080/00207540802213427
29. Outmal, I., Kamrani, A., Abouel Nasr, E.S., Alkahtani, M.: Modeling and performance analysis of a closed-loop supply chain using first-order hybrid Petri nets. Adv. Mech. Eng. **8**(5), 168781401664958 (2016). https://doi.org/10.1177/1687814016649584
30. Mahjoub, Y.I., El-Alaoui, E., Nait-Sidi-Moh, A.: Modeling a bus network for passengers transportation management using colored Petri nets and (max, +) algebra. Proc. Comput. Sci. **109**, 576–583 (2017). https://doi.org/10.1016/j.procs.2017.05.344

How Does Social Media Marketing Affect Customer Response Through Brand Equity as a Mediator in Commercial Heavy Vehicles Industry?

Seyed Babak Ebrahimi[✉], Ali Bazyar, and Nazila Shadpourtaleghani

K. N. Toosi University of Technology, Tehran, Iran
B_Ebrahimi@kntu.ac.ir

Abstract. Adopting the theoretical perspective of a consumer based model of brand equity, objective of this study is to analyze the impact of social media marketing activities (SMMAs) on customer response through brand equity in heavy commercial vehicles industry. Data was collected by deploying a questionnaire in which the truck owners were the respondents. An empirical research was conducted with a total number of 384 drivers who used social media and structural equation modeling (SEM) was used to perform data analysis. The outcome shows the importance of focusing on SMMAs, resulting tremendous increase in brand image and brand awareness which directly enhances brand commitment and customer willingness to pay a price premium. The main contribution of the paper is delineating the impacts of each SMMAs components on brand commitment and price premium which may be used as essential data on expansion of heavy commercial vehicles industry pricing strategies and marketing strategies.

Keywords: Brand equity · Social media marketing activities · Price premium · Brand commitment

1 Introduction

1.1 Social Media Marketing Activities

Millions of individuals are connected around the globe, and this cannot endure unless using Social media, which is making conventional methods of uncovering data unfashionable. It is becoming both well-matched and extra influential, ushering many companies to consider it as a key asset in their plans [1]. In business, social media marketing (SMM) uses the benefit of social platforms to upsurge business brand awareness along with expanding of its clients. Main objective is to establish material enthralling enough that users hand them out to others in their social networks. The usage of social media platforms as marketing means are quickly expanding, it reaches to over two third of the Internet users, enabling diverse opportunities for brand and reputation building [2]. It's crucial for managers and researchers to understand the significance of such networks in

© Springer Nature Switzerland AG 2021
Z. Molamohamadi et al. (Eds.): LSCM 2020, CCIS 1458, pp. 53–69, 2021.
https://doi.org/10.1007/978-3-030-89743-7_4

marketing [3]. In spite conducting immense researches about significance of SMM in different grounds, commonly focusing on influences of SMMAs on behavioral intents or client's contentment, few studies are devoted to brand equity while its impacts are considered on customer response [4–6].

1.2 Brand Equity

Methods of estimating different types of brands and their worth for various aims are available, yet this estimation is still a challenge to marketing professionals. Being a sociocultural singularity, brand equity reaches further than a simple item label; but it could act as figurative connotation that a brand longs for [7]. In case of lake of information on a firm, brand equity can assess the consumers. It is engraved in shoppers' recollection like a sole advantage distinguishing brand from others by merging numerous of its aspects. Brand equity is a mixture of brand awareness with its image [7]. Brand awareness is a feature, informing individuals and making them familiar with a the name, leading to remembering or identifying the brand [8, 9]. Keller in 1993 specified brand image as an overall insight of a brand, located in a client's memory along with a mixture of various recollections, making this aspect of a brand an important marketing element.

1.3 Customer Response

In this study customer response is a combination of commitment and price premium. Although there are various views on measurement and importance of these two, most scholars agree that both have a direct impact on how a customer responses towards a particular brand [7, 10, 11].

Purpose
In a nutshell, purpose of this study is to cover some missed grounds on previous studies, prioritize the importance of SMMAs' elements and develop a complete framework that projects how SMMAs influence customer response along with brand equity in commercial heavy vehicles business. This research starts with a theoretical background for SMMAs, brand equity and its elements and two different aspects of customer response. Continued by giving a justification for the empirical methods with explanations of results. After describing management practices of the research, finally, it settles the discussion with intimations of the outcomes for further investigations.

2 Theoretical Background

2.1 Social Media Marketing Activities (SMMAs)

Content sharing and joint work, could not be possible unless with existence of an online platform that facilitates these interactions [1, 12]. Offering chances to access customers is a role that social media plays when an individual relationship is stablished with customers [13]. Based on a study on segmentation of social media users, ranging from active participants to lucky users, merely one percent generates a new content and nine percent have interactions by writing comments, the rest of the users just use it by luck and watch whatever is posted [1]. Although scholars agree on importance of SMMAs [14], there are different classifications on its components. According to Kim and Ko's (2012) study of some luxurious brands there are five elements namely trendiness, interaction, entertainment, word-of-mouth (WOM) and customization contributing to SMMAs. But in some other studies the main components are social response and activity, providing information, promotion and selling, communication and support for daily life or in some others the main elements are information display, information usage, response to customer, content suitability, unique differentiation and customer participation [1]. According to literature, this study divides SMMAs into WOM, entertainment, interaction, trendiness and customization.

Entertainment. An outcome of enjoyment and recreational activities, entertainment is acquired through a social media encounter [4]. If hedonic view of social media is considered, people using it are gratification seekers and in need of acquiring enjoyment, thus the virtual platform is designed to fulfill their interests [15, 16]. Based on various studies entertainment in social media is introduce as a significant component and a solid motive for users [1, 4, 17, 18]. To give some illustrations, [19] Courtois et al. (2009) suggest that content loading on social media is driven by entertainment seeking which is the result of relaxation and escapism. Entertainment is a motive to participate in social networks [18] and it is an asset for consuming user-generated content [20]. According to Muntinga et al. in 2011; pleasure, leisure, and amusement activities are some pivotal reasons for users in consumption of brand related content. Based on foregoing, entertainment in social media could be defined as a driving force and a dense motive for users.

Interaction. With respect to the fact that social media create a platform which enables users to discuss and interact with one a another [21]; interaction offer insights into users with concurring opinion to discuss specific products and/or brands [17]. User-generated content in such an atmosphere is derived by these interactions which are deeply altering the dynamics of communications between brand and customers [22]. Manthiou et al. (2013) pointed out the importance of posing unique content, reflecting the consumers profile in the best way possible, being open and completely active in discussions and leading thoughts in order to encourage interaction, such a situation can upsurge credibility and attraction. Based on nature of interaction [23] proposed profile based and content-based as two groups building social media. Being interested in user, a profile based social media such as Instagram and LinkedIn, specifically emphasizes on individual members encouraging them to connect with the explicit topics. Contrarily, other groups of platforms aiming to connect users with the content of a certain profile, only fixate on contexts, arguments or statements on the posted content. Some good illustration of it could be Flickr, Pinterest, and YouTube. In this study interaction is considered as sharing of information as well as exchanging of users' opinion in social media.

Trendiness. The most up to dated news and the hottest discussions are published throughout social media, making it a powerful search channel for desired products [4].

This freshness of the information, can lead to consumer's trust to those social medias and also it reliefs users who have a tendency to trust data received from social media more than that of gained through corporate-sponsored promotions or marketing activities [24]. New information can trigger four different aspects of motivation. Knowledge gained from other users' expertise, information which is obtained by reading product reviews before purchasing, being updated with the latest news on one's social environment and getting inspired by the hottest trending issues [17]. Choosing a new outfit by the appearance of others wearing it, specifically by celebrities, is a good illustration of the fact that newest trending subjects in social media can be inspiring, so in essence trendiness is the newest data obtained about a product or a service [4]. According to afore discussion, in this study trendiness is defined as the act of spreading latest information on brand.

Customization. The amount to which a service can be adapted by the demand of customers to overcome their desire is represented by customization [1]. The fact that customization is built on interaction with individual users, makes it completely different from ordinary advertisement aids. Customer satisfaction and feeling of being in power is created through customization which is used as a strategy providing every single user the authority to adjust the information from various sources [25]. Individuality can be conveyed by customization and customer's ability of personalization of social media, resulting to solider brand commitment and brand loyalty [26]. Client's brand engagement as well as their loyalty to the brand is affected by customization [27] and ultimately consumer's engagement could mold their purchasing intentions [28]. Contingent on extend of customization, posts are divided into two groups, a customized message which only aims an individual user, and a broadcast which is for anyone who is interested [23]. Having access to huge amount of data on different customers shopping preferences and buying habits is the cause that customization is progressively important [27]. In this research, customization is regarded as the degree to which a social media offers personalized search of data.

Word of Mouth (WOM). Pre-purchase product information that majorly influence customers' purchase intentions [29] can easily be obtained by searching through the Internet. This can provide a chance for customers to distribute their knowledge, and it is called electronic word of mouth [30]. To customers, user generated WOM has higher reliability, importance and affinity than information provided by company owned social media [31]. Through these channels customers' ideas, comments and sentiments are spread to their friends and associates limitlessly [5]. Opinion seeking, opinion giving and opinion passing are three viewpoints that are subjected when studying electronic WOM applications in social media. Customer who seeks for user's advice and opinion before purchasing has high level of opinion seeking. On the other hand an opinion giver is the one who tries to be influential on others purchasing behavior. Finally in social media, information flow expedites through online forwarding as explicit aspect of electronic WOM [32]. Therefore, WOM in this study is defined as the information passing and expressing opinion on social media.

2.2 Brand Equity

As a set of brand possessions, brand equity is accountabilities associated with company's title which increases the value delivered by the product [33], and it plays an essential marketing role since the 90s [34]. To have a clearer view on its formation, it is required to conduct researches on brand equity mainly focusing on consumers, perceiving added value as the result of brand [33]. Although there has been lots of studies on brand equity and most of them found it a complex construct [11] having several extents [35], yet petite unity exists on its dimensionality [4] resulting different approaches and definitions of the brand equity by numerous researchers [33, 34, 36, 37]. Based on Keller and Lehmann (2006) studies, there are three views on brand equity, financial based, business based and customer based. In financial based viewpoint, brand equity generates extra cash flow in comparison to non-branded ones [10]. On the other hand based on Keller definition (1993), perception of a brand as well as the responses to that, forms customer based brand equity. Which is a valuable asset aiding the marketing plans and advertisement strategies [37]. In this study the examination of brand equity is founded on Keller's (1993) consumer-based perspective.

Brand Awareness. Considered as an important concept in marketing and consumer behavior, brand awareness happens as a result of customers becoming informed and accustomed to a brand, brand awareness is defined whether decision-makers can recall, recognize or distinguish it [8]. It is an essential asset for communications processes [9]. When consumer knows a brand, chance of choosing it will be much higher.[38] Even though customers may have merely a little knowledge about the product itself, still they can purchase the product just by remembering the title of that brand [39]. Brand awareness can be used as a purchase decision heuristic and it has an important effect on consumer choice [9]. When choosing a product or service among a bunch of brands that interest customers it can play an important part [39]. Based on the literature, brand awareness in this study is client's capability to identify a specific branded product in diverse situations.

Brand Image. In brand management, it is presumed that forming a satisfactory and a robust brand image leads to maintainable competitive advantages with considerable financial earnings [40]. It's a tool for customers to get familiar to the product, minimize the risk and gain satisfaction [41]. Customer loyalty can be achieved through positive brand image [40]. It is a tool to show off the benefits that firms provide in order to fulfill customer desires [42]. Strong associations to brand in customers' point of view can be obtained through positive brand image [11]. Consequently it can be advanced by clients experiencing products as well as services [43]. In this study, brand image is defined as connotations the brands have in clients' minds [44].

2.3 Customer Response

Social media is becoming an effective channel, linking consumers and marketers, through which companies impose influence on customers' response [45], which in this research, is separated into behavioral reactions and emotional reactions, considering brand equity as a mediator, the impacts of SMMAs on price premium as a behavioral reaction and brand commitment as an emotional reaction is investigated. Not only commitment is an important outcome, due to it relation to behavioral reactions, eventually it can have monetary implications for the business [46].

Commitment. Permanent desire to last the relationship with a brand is stated as commitment [47]. In today's competitive market, reaching brand commitment is undoubtedly one of the definitive objectives for brands [46, 48]. Consumers are eager to endure an emotional connection with the brand that creates a sincere and pleasant feeling [47]. Based on previous studies, providing inspiring and exceptional brand experiences along with building a decent brand image in minds of consumers, could end in long-term commitment [49]. Customers with great brand commitment would have solider emotional connection with the brand [50]. On the other hand satisfied customers will be committed to buy the same brand and they will become loyal to that brand as well [46].

Price Premium. In the academic economics works, price premiums are defined as additional charges paid over reasonable prices that are vindicated by the value of a product or service, the main reason driving price premiums is consumers' craving for a specific superiority of a product or service [51]. Typically high quality products seller puts a price that is higher than average price of same quality. Price premium is conceptualized as prices that yield beyond normal profits, and brands can charge a premium amount for risk reduction of customer purchase and shortening the purchase procedure [52]. Price premium is customers' inclination to pay an exceptional price for a specific brand. Also it was initiated in the pricing and customer behavior field; precisely calculating customers' purchasing behavior and eventually supporting businesses to advance their pricing policies [53]. Price premium in this study is defined as how important brand equity impacts on clients' tendency to spend a price premium.

3   Research Design

3.1   Research Model

After prioritizing all components of SMMAs based on their importance, primary purpose of this research is to explore the impact of SMMAs on empowerment of brand awareness and its image as elements of brand equity and their effects on customer response towards a brand which consist brand commitment and customer tendency to spend a price premium. According to previous literature Fig. 1 depicts research conceptual model.

3.2   Hypotheses

Main objective of any marketing plan is to upsurge sales and profitability, organizations are connected with customers through marketing channels to inform them of their products and services and generate attention to their offering [5, 23]. As a pivotal component of marketing, brand image was introduced in the 1950s [40], a positive one is molded

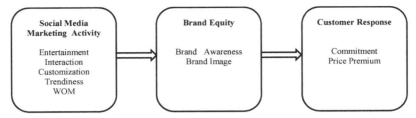

Fig. 1. Conceptual model

when experiences connected to that specific name are satisfactory, robust, and exceptional in consumers' minds [7]. Based on literature and previous studies on SMMAs influences on equity of brand [1, 4, 5, 54], the following hypotheses are presented in this study.

H1. Brand awareness is positively influenced by SMMAs.

H2. Brand image is positively influenced by SMMAs.

Brand image is a means for consumers to identify a product or a service, comprehend purchase risks, appraise its superiority [41]. Consequently its effects on price premium and commitment is inevitable. In marketing literature, after brand image the concept of brand equity was introduced as a pivotal portion in the 1980s [40]. The following hypotheses on effects of brand equity elements on customer response main components are drawn from the studies of [1, 4, 5, 7, 10, 11].

H3. Price premium is positively influenced by brand awareness.

H4. Price premium is positively influenced by brand image.

H5. Brand commitment is positively influenced by brand awareness.

H6. Brand commitment is positively influenced by brand image.

3.3 Sample Design

By designing a quantitative survey aiming customers of heavy truck industries with considerable investments in SMM, and by utilizing SEM, the authors empirically study the influence of SMMAs on clients' commitment to the brand and tendency to pay price premium with brand equity as a mediator. As it was one of authors' purposes to calculate SMM's impact on dissimilar customer, the survey was conducted with convenience sampling method, five fairly diverse customer populations in different parts of Iran were chosen, allowing authors to enhance generalization of the study's outcomes. Based on Cochran's formula a sample of 384 people for this study was defined. Considering earnings and education, the research samples are harmonious with population distribution. Also it consist of 3.13% women and 96.88% men, with 43 years as an average, the majority of after sales services experience with the current brand was less than 3 years, which shows most of the population merely visit in their warranty or extended warranty period. Table 1 depicts demographic features of the research samples.

In January, 2018, 400 questionnaires were distributed, from 392 returned questionnaires, eight were incomplete, leaving 384 questionnaires for analysis. Respondents stated their views on a classic five-point Likert scale divided the responses from one ("strongly disagree") to five ("strongly agree"). Table 2 reveals questionnaire items.

Table 1. Demographic features of sample

Classification		Frequency	Composition ratio
Gender	Male	372	96.88%
	Female	12	3.13%
After Sales services experience with the current brand	Less than 1 year	107	27.86%
	1 to 2 years	91	23.7%
	2 to 3 years	89	23.18%
	3 to 5 years	51	13.28%
	More than 5 years	46	11.98%
Age	Less than 20	1	0.26%
	21 to 30	57	14.84%
	31 to 40	127	33.07%
	41 to 50	110	28.65%
	51 to 60	75	19.53%
	More than 60	14	3.65%

For validating the research instrument, different types of validity were tested. For content validity, the questionnaire was distributed among marketing experts and academics of related field, and it got their approval for relevance and bearing of questions associated with each variable. Prior testing the model, in order to check the existence of an association between the observed variables and related latent constructs, a CFA was administered by employing SPSS 24.0 and AMOS 24.0. It was approved based on KMO measure of sampling adequacy, being over 0.7 [55], along with the Bartlett Test of Sphericity showing the correlations are commonly significant at $p < 0.001$. Moreover, internal consistency was measured by Cronbach's Alfa for checking the reliability of the instrument, the result was 96.9 confirming that of the questionnaire [55]. Values presented in Table 3 indicate KMO test results as well as that of Cronbach's Alpha.

As shown in Table 4 below, the composite construct reliabilities (CCR) are more than 0.7 [56]. Average variance extracted (AVE) numbers being over 0.5 approves convergent validity [56]. Furthermore, the corresponding inner constructs' squared correlation coefficients is less than the AVE for each construct. Also with assessment of goodness of fit indexes to evaluate the overall model fit, the following results were produced: $x^2/df = 1.622$, goodness-of-fit $(GFI) = 0.905$, adjusted goodness-of-fit index $(AGFI) = 0.881$, root mean residual $(RMR) = 0.036$, root mean square of approximation $(RMSEA) = 0.04$. In conclusion according to the test outcomes the questionnaires were settled to be hand out.

Table 2. Questionnaire items

Concept	Variable	Item
Social media marketing activities	Entertainment	I have fun when using this brand's social media
		I find contents presented in this brand's social media enjoyable
	Interaction	It is possible to share information in this brand's social media
		It is possible to have discussion with others in this brand's social media
		Delivering my viewpoint through this brand's social media is easy
	Word of mouth	Positive comment regarding this brand's social media will be posted by me
		This brand will be recommended to my social media friends by me
	Customization	Information search is customizable in social media of this brand
		Customized services is provided by social media of this brand
	Trendiness	I find the information presented in this brand's social media new
		I find using this brand's social media popular
Brand equity	Brand awareness	This brand is constantly at the centre of my attention
		I can always differentiate this brand from others
		This brand's logo is easily remembered by me
		This brand's features is quickly remembered by me
	Brand image	My memory concerning this brand is excellent
		I find this brand as a wide ranging adventure
		I find this brand as a frontrunner in the business
		I find this brand client-centric
Customer response	Commitment	I find myself honoured to be this brand's customer
		I look forward for this brand to be prosperous for years to come
		This brand is my most favoured one
		Although other brands are available but I will wait to buy this brand

(continued)

Table 2. (*continued*)

Concept	Variable	Item
	Price premium	Unless this brand's prices upsurge tremendously, I won't shit to another one
		Paying a higher price for this brand is not a concern for me
		Spending a lot more for this brand will not be an issue for me
		Among all brands of heavy commercial vehicles I will pay a lot more for this one

Table 3. KMO and Cronbach's Alpha test results

Item	KMO	Cronbach's Alpha	Sig Bartlett
Social media marketing activities	0.956	0.936	0.000
Brand awareness	0.76	0.812	0.000
Brand image	0.824	0.859	0.000
Brand commitment	0.785	0.814	0.000
Price premium	0.808	0.822	0.000
All variables	0.977	0.969	0.000

Table 4. Result of confirmation factor analysis

Construct	Items	Factor loading	C.R	SMC	AVE	C.C.R
SMMA	EN_1	0.8		0.640	0.926	0.993
	EN_2	0.766	15.312	0.586		
	IN_3	0.771	15.756	0.594		
	IN_2	0.807	18.095	0.652		
	IN_1	0.782	16.893	0.666		
	wom_2	0.747	15.026	0.558		
	wom_1	0.842	19.431	0.709		
	CU_2	0.757	15.286	0.573		
	CU_1	0.764	15.325	0.584		
	Trnd_2	0.853	16.513	0.728		
	Trnd_1	0.764	15.632	0.583		

(*continued*)

Table 4. (*continued*)

Construct	Items	Factor loading	C.R	SMC	AVE	C.C.R
Brand image	BI_1	0.743	15.481	0.551	0.944	0.985
	BI_2	0.815	17.32	0.664		
	BI_3	0.843		0.710		
	BI_4	0.87	18.408	0.757		
Brand awareness	BA_1	0.731		0.535	0.912	0.976
	BA_2	0.736	13.187	0.541		
	BA_3	0.757	13.395	0.574		
	BA_4	0.779	13.506	0.608		
Commitment	CO_1	0.727	13.596	0.529	0.921	0.979
	CO_2	0.734	13.717	0.539		
	CO_3	0.785	15.693	0.616		
	CO_4	0.798		0.636		
Price premium	PrmPrc_1	0.786	14.697	0.617	0.923	0.979
	PrmPrc_2	0.772		0.596		
	PrmPrc_3	0.746	14.797	0.556		
	PrmPrc_4	0.715	14.598	0.511		

4   Empirical Results

In this study, SEM was employed to determine how well the data fit the proposed model. It is a multivariate examination of multivariate regression families that besides providing precise extension to general linear model, it helps the researcher to examine a set of regression equations at the same time. Furthermore in this study path analysis is utilized, it is a method that demonstrates the connection between the different types of variables concurrently. Ultimately, in this study to approve or disprove hypotheses, SEM and path analysis were used. By deploying generalized least squares method the results of model fit indices were within acceptable range ($x^2 = 529.112$, $DF = 313$, $CMIN/DF = 1.69$, $p < 0.001$, $GFI = 0.898$, $AGFI = 0.876$, $RMR = 0.038$, and $RMSEA = 0.042$), approving the goodness of fit of research model used. Table 5 depicts the proposed model fit obtained values and their acceptable range.

As approved by previous studies SMMAs significantly affects brand equity, consumers are expected to recall or recognize the companies which are actively carrying out SMMAs more, compared to others. SMMAs' influence on brand awareness ($\beta = 0.811$, $C.R. = 13.671$, $p < 0.001$) and brand image ($\beta = 0.905$, $C.R. = 16.390$, $p < 0.001$) is found to be significant, which confirms H1 and H2. Also brand image is affected more than brand awareness by brand equity elements. SMMAs' elements are statistically important; the relative significance of SMMAs' elements were topmost in interaction, succeeded by entertainment, WOM, trendiness, and finally customization. Proposing

Table 5. Model fit results

Statistics	Suitable values	Values obtained
x^2/DF	Between 1 and 2	1.69
GFI	$GFI > 0.9$	0.898
AGFI	$\cdot$ $AGFI > 0.85$	0.876
RMSEA	$RMSEA < 0.05$	0.042
RMR	$RMR < 0.05$	0.038

that heavy commercial vehicles customers favored social media that delivered grounds for interaction of ideas since social media is the most convenient place for such collaborations. Furthermore customers desired social media that was entertaining and publish reliable content with the most up to date news. Besides, social media that fits its clients' taste is favored, yet it is essential for a successful SMMAs to have stability amongst all components. The important relation among brand equity, customer commitment and tendency to pay a price premium is approved by numerous previous researches. Analyzing the path for impact of brand image on commitment ($\beta = 0.528$, $CR = 3.591$, $p < 0.001$) and price premium ($\beta = 0.372$, $CR = 2.176$, $p < 0.05$) shows significant and positive effect, supporting H4 and H6. In parallel as it is approved by previous studies, especially in industries who provide intangible services, even a slight improvement in brand image will have a significant result in customer response. The effect of brand awareness was significant on both customer willingness to pay price premium ($\beta = 0.588$, $CR = 3.051$, $p < 0.01$) and commitment ($\beta = 0.410$, $CR = 2.517$, $p < 0.05$) supporting both H3 and H5. This suggest that customers who have high awareness of a brand are more loyal to that brand and liable to pay price premium. Table 6 present standardized estimates (path

Table 6. Hypothesis results

Hypothesis	Item	Critical ratio	Standardized estimates	Result
Brand awareness is positively influenced by SMMAs	H1	13.671	0.966	Accepted
Brand image is positively influenced by SMMAs	H2	16.390	0.976	Accepted
Price premium is positively influenced by brand awareness	H3	3.051	0.597	Accepted
Price premium is positively influenced by brand image	H4	2.176	0.417	Accepted
Brand commitment is positively influenced by brand awareness	H5	2.517	0.419	Accepted
Brand commitment is positively influenced by brand image	H6	3.591	0.595	Accepted

coefficient), critical ratios and accordingly hypothesis results. Figure 2 depicts research model.

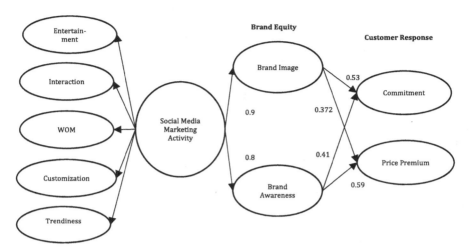

Fig. 2. Hypothesis testing results

5   Conclusions and Limitations

5.1   Conclusion and Discussion

The need of investigating main social media factors affecting customer response is clear based on the several studies [12, 57]. Nevertheless none of the researches have studied SMMAs on customer response through brand equity as a mediator in commercial heavy vehicles industry. Although prior studies acknowledged the vitality of social media components [1, 4, 5] by offering an in-depth investigation and presenting a holistic framework, besides prioritizing SMMAs components based on their relative importance, this research presents exclusive insights about SMMAs impact on brand equity and customer behavior towards a brand in heavy vehicle industry. This study is the first to experimentally investigate connection of heavy vehicle industry SMMAs and client's behavior with respect to brand equity as a mediator. Also it proposes a more precise model for describing the effects of SMMAs on two main aspects of customer reactions, which are being committed to the brand and willing to pay more for it. More over this study points out the huge impact of SMMAs on brand equity elements, which is consistent with the results from [1, 4, 5], furthermore it empirically test the effect of focusing on brand equity that can escalates customer positive responses to heavy vehicle truck brands in terms of commitment and price premium.

5.2   Managerial Implications

Helping with evaluating the effectiveness and capability of businesses' marketing actions is the primary managerial implication of current research. According to the findings, businesses should allocate exceptional devotion to all SMM elements. In a nutshell brands should promote contents with highest interaction and exquisite entertainment which are completely up-to-date thorough adjustable platform that leads to positive WOM. Technology is growing rapidly and if firm managers don't try to implement the newest marketing strategies in their line of work, others easily take the advantage. The results show that SMMAs have huge impacts on brand equity which lead to improving of brand image and brand awareness, also it will directly soars customer's response toward the brand. This article studies customer response from two viewpoints, price premium as behavioral response and commitment as emotional response. Firms should have different channels in social media, as each can serve specific aspect. Besides, they are significant and cost effective means not only should firms utilize the SMMAs as an image building instrument, but they also should use these powerful tools for building emotional connections with brand [5]. Clients look forward to see their favored brand's presence in social media segment, nevertheless based on the recent studies if the social media is managed by one other than the firm, the acceptance rate of the users will increase rapidly [58, 59].

5.3 Limitations and Recommendations for Further Studies

This study is accompanied by a number of restrictions, hence it might need further investigations. The outcome of the research could have been affected by convenience sampling method. Yet, any self-administered questionnaire experiences such inclinations. For further studies other socio-demographic variables could be investigated. Also using longitudinal survey method is recommended, since the researcher only measured consumers' perceptions and objectives at a single point of time, additional investigations are suggested because individual's insights are likely to alter as they gain more experience over time. One of the core limitations of the study is generalizability over heavy commercial vehicles industry, not being precisely applicable in other businesses, the overall concept of outcomes consequently, are recommended to be confirmed in other contexts. Finally this research merely focuses on brand awareness and its image as two aspects of brand equity and for further investigations and re-evaluations of the brand equity creation process, other aspects, namely, brand associations and perceived quality can be employed.

References

1. Seo, E.J., Park, J.W.: A study on the effects of social media marketing activities on brand equity and customer response in the airline industry. J. Air Transp. Manag. **66**, 36–41 (2018). https://doi.org/10.1016/j.jairtraman.2017.09.014
2. Correa, T., Hinsley, A.W., de Zúñiga, H.G.: Who interacts on the Web?: The intersection of users' personality and social media use. Comput. Human Behav. **26**, 247–253 (2010). https://doi.org/10.1016/j.chb.2009.09.003
3. Kumar, A., Bezawada, R., Rishika, R., Janakiraman, R., Kannan, P.K.: From social to sale: the effects of firm-generated content in social media on customer behavior. J. Mark. **80**, 7–25 (2016). https://doi.org/10.1509/jm.14.0249

4. Godey, B., et al.: Social media marketing efforts of luxury brands: Influence on brand equity and consumer behavior. J. Bus. Res. **69**, 5833–5841 (2016). https://doi.org/10.1016/j.jbusres. 2016.04.181
5. Kim, A.J., Ko, E.: Do social media marketing activities enhance customer equity? An empirical study of luxury fashion brand. J. Bus. Res. **65**, 1480–1486 (2012). https://doi.org/10.1016/ j.jbusres.2011.10.014
6. Michopoulou, E., Moisa, D.G.: Hotel social media metrics: The ROI dilemma (2019)
7. Keller, K.L.: Conceptualizing, measuring, and managing customer-based brand equity. J. Mark. **57**, 22 (1993). https://doi.org/10.2307/1252054
8. Chi, C., Gursoy, D., Chen, J.: Theoretical examination of destination loyalty formation (2014)
9. Barreda, A.A., Bilgihan, A., Nusair, K., Okumus, F.: Generating brand awareness in Online Social Networks (2015)
10. de Oliveira, M.O.R., Silveira, C.S., Luce, F.B.: Brand equity estimation model. J. Bus. Res. **68**, 2560–2568 (2015). https://doi.org/10.1016/j.jbusres.2015.06.025
11. Keller, K.L., Lehmann, D.R.: Brands and branding: research findings and future priorities (2006)
12. Zollo, L., Filieri, R., Rialti, R., Yoon, S.: Unpacking the relationship between social media marketing and brand equity: The mediating role of consumers' benefits and experience. J. Bus. Res. **117**, 256–267 (2020). https://doi.org/10.1016/j.jbusres.2020.05.001
13. Kelly, L., Kerr, G., Drennan, J.: Avoidance of advertising in social networking sites. J. Interact. Advert. **10**, 16–27 (2010). https://doi.org/10.1080/15252019.2010.10722167
14. Cheung, M.L., Pires, G.D., Rosenberger, P.J., Leung, W.K.S., Ting, H.: Investigating the role of social media marketing on value co-creation and engagement: an empirical study in China and Hong Kong. Australas. Mark. J. **29**, 118–131 (2021). https://doi.org/10.1016/j.ausmj. 2020.03.006
15. Manthiou, A., Chiang, L., Tang, L.R.: Identifying and responding to customer needs on Facebook fan pages (2013)
16. Felix, R., Rauschnabel, P.A., Hinsch, C.: Elements of strategic social media marketing: a holistic framework (2017)
17. Muntinga, D.G., Moorman, M., Smit, E.G.: Introducing COBRAs. Int. J. Advert. **30**, 13–46 (2011). https://doi.org/10.2501/IJA-30-1-013-046
18. Park, N., Kee, K.F., Valenzuela, S.: Being immersed in social networking environment (2009)
19. Courtois, C., Mechant, P., De Marez, L., Verleye, G.: Gratifications and seeding behavior of online adolescents. J. Comput. Commun. **15**, 109–137 (2009). https://doi.org/10.1111/j. 1083-6101.2009.01496.x
20. Shao, G.: Understanding the appeal of user-generated media: a uses and gratification perspective (2009)
21. Dabbous, A., Barakat, K.A.: Bridging the online offline gap: assessing the impact of brands' social network content quality on brand awareness and purchase intention. J. Retail. Consum. Serv. **53**, 101966 (2020). https://doi.org/10.1016/j.jretconser.2019.101966
22. Kaplan, A.M., Haenlein, M.: Users of the world, unite! the challenges and opportunities of Social Media. Bus. Horiz. **53**, 59–68 (2010). https://doi.org/10.1016/j.bushor.2009.09.003
23. Zhu, Y.-Q., Chen, H.-G.: Social media and human need satisfaction: implications for social media marketing. Bus. Horiz. **58**, 335–345 (2015). https://doi.org/10.1016/j.bushor.2015. 01.006
24. Mangold, W.G., Faulds, D.J.: Social media: the new hybrid element of the promotion mix. Bus. Horiz. **52**, 357–365 (2009). https://doi.org/10.1016/j.bushor.2009.03.002
25. Ding, Y., Keh, H.T.: A re-examination of service standardization versus customization from the consumer's perspective (2016)

26. Martin, K., Todorov, I.: How will digital platforms be harnessed in 2010, and how will they change the way people interact with brands? J. Interact. Advert. **10**, 61–66 (2013). https://doi.org/10.1080/15252019.2010.10722170

27. Shanahan, T., Tran, T.P., Taylor, E.C.: Getting to know you: social media personalization as a means of enhancing brand loyalty and perceived quality (2019)

28. Hollebeek, L.D., Glynn, M.S., Brodie, R.J.: Consumer brand engagement in social media: conceptualization, scale development and validation. J. Interact. Mark. **28**, 149–165 (2014). https://doi.org/10.1016/j.intmar.2013.12.002

29. Zhang, R., Tran, T.: Helping e-commerce consumers make good purchase decisions: a user reviews-based approach (2009)

30. Murtiasih, S., Siringoringo, H.: How word of mouth influence brand equity for automotive products in Indonesia (2013)

31. Gruen, T.W., Osmonbekov, T., Czaplewski, A.J.: eWOM: The impact of customer-to-customer online know-how exchange on customer value and loyalty. J. Bus. Res. **59**, 449–456 (2006). https://doi.org/10.1016/j.jbusres.2005.10.004

32. Chu, S., Kim, Y.: The Review of Marketing Communications Determinants of consumer engagement in electronic word-of-mouth (eWOM) in social networking sites. Int. J. Advertis. (2011)

33. Dedeoğlu, B.B., Van Niekerk, M., Weinland, J., Celuch, K.: Re-conceptualizing customer-based destination brand equity. J. Destin. Mark. Manag. **11**, 211–230 (2019). https://doi.org/10.1016/j.jdmm.2018.04.003

34. Lim, Y., Weaver, P.A.: Customer-based brand equity for a destination: the effect of destination image on preference for products associated with a destination brand. Int. J. Tour. Res. **16**, 223–231 (2012). https://doi.org/10.1002/jtr.1920

35. Datta, H., Ailawadi, K.L., van Heerde, H.J.: How well does consumer-based brand equity align with sales-based brand equity and marketing-mix response? (2017)

36. Chatzipanagiotou, K., Christodoulides, G., Veloutsou, C.: Managing the consumer-based brand equity process: a cross-cultural perspective (2019)

37. Thakshak: Analysing customer based airline brand equity: Perspective from Taiwan (2018)

38. Abou-Shouk, M., Soliman, M.: The impact of gamification adoption intention on brand awareness and loyalty in tourism: the mediating effect of customer engagement. J. Destin. Mark. Manag. **20**, 100559 (2021). https://doi.org/10.1016/j.jdmm.2021.100559

39. Foroudi, P.: Influence of brand signature, brand awareness, brand attitude, brand reputation on hotel industry's brand performance. Int. J. Hosp. Manag. **76**, 271–285 (2019). https://doi.org/10.1016/j.ijhm.2018.05.016

40. Dirsehan, T., Kurtuluş, S.: Measuring brand image using a cognitive approach: Representing brands as a network in the Turkish airline industry. J. Air Transp. Manag. **67**, 85–93 (2018). https://doi.org/10.1016/j.jairtraman.2017.11.010

41. Nagar, K.: Modeling the effects of green advertising on brand image: investigating the moderating effects of product involvement using structural equation. J. Glob. Mark. **28**, 152–171 (2015). https://doi.org/10.1080/08911762.2015.1114692

42. Merz, M.A., He, Y., Vargo, S.L.: The evolving brand logic: a service-dominant logic perspective. J. Acad. Mark. Sci. **37**, 328–344 (2009). https://doi.org/10.1007/s11747-009-0143-3

43. Lahap, J., Ramli, N.S., Said, N.M., Radzi, S.M., Zain, R.A.: A study of brand image towards customer's satisfaction in the malaysian hotel industry. Proc. Soc. Behav. Sci. **224**, 149–157 (2016). https://doi.org/10.1016/j.sbspro.2016.05.430

44. Anggraeni, A.: Rachmanita: effects of brand love, personality and image on word of mouth; the case of local fashion brands among young consumers. Proc. Soc. Behav. Sci. **211**, 442–447 (2015). https://doi.org/10.1016/j.sbspro.2015.11.058

45. Sheng, J.: Being active in online communications: firm responsiveness and customer engagement behaviour (2019)
46. Gligor, D., Bozkurt, S., Russo, I.: Achieving customer engagement with social media: a qualitative comparative analysis approach. J. Bus. Res. **101**, 59–69 (2019). https://doi.org/10.1016/j.jbusres.2019.04.006
47. Erciş, A., Ünal, S., Candan, F.B., Yıldırım, H.: The effect of brand satisfaction, trust and brand commitment on loyalty and repurchase intentions. Procedia - Soc. Behav. Sci. **58**, 1395–1404 (2012). https://doi.org/10.1016/j.sbspro.2012.09.1124
48. Trigeorgis, L., Baldi, F., Katsikeas, C.S.: Valuation of brand equity and retailer growth strategies using real options. J. Retail. (2021). https://doi.org/10.1016/j.jretai.2021.01.002
49. Girardin, F., Guèvremont, A., Grohmann, B., Malär, L., Morhart, F.: Brand authenticity: An integrative framework and measurement scale (2015). https://www.sciencedirect.com/science/article/abs/pii/S1057740814001089
50. Kang, A.: Brand love – moving beyond loyalty an empirical investigation of perceived brand love of indian consumer. Arab Econ. Bus. J. **10**, 90–101 (2015). https://doi.org/10.1016/j.aebj.2015.04.001
51. Jeong, E.H., Jang, S.C. (Shawn): Price premiums for organic menus at restaurants: what is an acceptable level? (2019)
52. Dwivedi, A., Nayeem, T., Murshed, F.: Brand experience and consumers' willingness-to-pay (WTP) a price premium: mediating role of brand credibility and perceived uniqueness. J. Retail. Consum. Serv. **44**, 100–107 (2018). https://doi.org/10.1016/j.jretconser.2018.06.009
53. Singh, G., Pandey, N.: The determinants of green packaging that influence buyers' willingness to pay a price premium. Australas. Mark. J. **26**, 221–230 (2018). https://doi.org/10.1016/j.ausmj.2018.06.001
54. Chae, H., Shin, J., Ko, E.: The effects of usage motivation of hashtag of fashion brands' image based sns on customer social participation and brand equity: focusing on moderating effect of SNS involvement (2016)
55. Hair, J.F., Black, W.C., Babin, B.J., Anderson, R.E.: Pearson New International Edition: Multivariat Data Analysis (2014)
56. Fornell, C., Larcker, D.F.: Structural equation models with unobservable variables and measurement error: algebra and statistics. J. Mark. Res. **18**, 382 (1981). https://doi.org/10.2307/3150980
57. Kozinets, R.V., de Valck, K., Wojnicki, A.C., Wilner, S.J.: Networked narratives: understanding word-of-mouth marketing in online communities. J. Mark. **74**, 71–89 (2010). https://doi.org/10.1509/jmkg.74.2.71
58. Dolega, L., Rowe, F., Branagan, E.: Going digital? The impact of social media marketing on retail website traffic, orders and sales. J. Retail. Consum. Serv. **60**, 102501 (2021). https://doi.org/10.1016/j.jretconser.2021.102501
59. Huerta-Álvarez, R., Cambra-Fierro, J.J., Fuentes-Blasco, M.: The interplay between social media communication, brand equity and brand engagement in tourist destinations: an analysis in an emerging economy. J. Destin. Mark. Manag. **16**, 100413 (2020). https://doi.org/10.1016/j.jdmm.2020.100413

A Study on IOT Applications and Technologies in Logistics

Arman Behnam⬛, Ali Sarkeshikian⬛, and Mohammad Ali Shafia(✉)⬛

Iran University of Science and Technology, Tehran, Iran
`arman_behnam@ind.iust.ac.ir, omidshafia@iust.ac.ir`

Abstract. One of the most trending state-of-the-art technologies due to broadening usages of logistics is Internet of Things (IOT) applications which has been enabled by new expandable platforms which are occupied in many applications like transportation, smart city, smart home and etc. To broaden this field of study there is need to classify and group all documents to find useful patterns and speed up and accurate future studies in order to achieve to greatest results. In this paper, some analysis to fulfil the pattern of studies and determine latest IOT applications in logistics is done. Results demonstrate that new inclines in IOT applications with logistics are embedded in airports and railways and some technologies like WSN, RFID and GIS are top of useful devices in this direction.

Keywords: IoT · Logistic · Bibliometric · Internet of Things

1 Introduction

Bibliometrics evaluates the impact of research outputs using quantitative measures. Bibliometrics complements qualitative indexes of research impact such as peer review, funding received. Together they assess the quality and impact of research.

Bibliometrics is used to process needed research outputs impact pieces of evidence when applying for jobs, research funding, find new and extruding areas of research, identify potential research colleagues and specify journals in which is appropriate to publish.

One of the total state-of-the-art technology due to broadening usages of logistics is IoT applications which have been enabled by new expandable platforms which are engaged in many fields such as transportation and smart city. Basically, IoT is an embedded devices platform are connected to the internet, hence they collect and exchange data with each other through a network. It makes interaction, collaboration and, learning among devices possible just like humans do.

The IoT facilitates the improvement of numerous industry-oriented and user-specific IoT applications. IoT applications enable two paired device and human-device interactions in a reliable and robust manner, while devices and networks provide connectivity. IoT applications on devices need to ensure that data have been received and acted upon properly with appropriate timing. For instance, transportation and logistics applications monitor the transported goods' status. During transportation, the conservation status

© Springer Nature Switzerland AG 2021
Z. Molamohamadi et al. (Eds.): LSCM 2020, CCIS 1458, pp. 70–83, 2021.
https://doi.org/10.1007/978-3-030-89743-7_5

(e.g., temperature, humidity) is always under monitoring to avoid spoilage when the connection is out of range [9].

IoT applications are expected to provide some objects with connectivity and intelligence. It is widely being deployed, in various fields, namely: smart home applications, health care, smart cities, agriculture, automation. One of the most related them in below will is revealed.

Through Industrial Automation, speeding developments are happening, as well as the quality of products, which are the critical components for a greater Return on Investment. Today with IoT technologies, one could re-design products and other processes to get better performance in both cost and customer satisfaction. So, IoT all in this field includes the following domains, such as the smart grid, factory digitalization, self-checkouts and inventory management, operation management and predictive maintenance, packaging optimization, logistics, and supply chain optimization.

IoT develops, makes difference, and keeps going. The IoT has combined hardware and software to make the world smarter. It has been growing at an exponential pace and offers some opportunities for infrastructure and business.

With all subsequent studies, the question of this paper that has not been answered yet is an accurate bibliometric study to find fields of progress in this area and justify answers by some analytics, then knowing which IoT technologies have been embedded yet and which of them are more useful and at last, a roadmap of these findings to lead researchers to better academic studies and developing new IoT approaches and methods.

The rest of this paper starts with reviewing some of the works that have been done in this field of study in Sect. 2, then the method of search in order to conduct the study will be determined in Sect. 3, after those some analytics on the results will be done in Sect. 4, results of these analytics will be survived in Sect. 5 and at last conclusion and final scientific findings will be discussed in Sect. 6.

2 Literature Review

Recently, with becoming IoT and its applications a hot topic, some studies in the format of bibliometric and scientometric have been conducted. In [14], Python modules called ScientoPy were developed to execute quantitative analysis that brings insight into research trends by a lead author's country affiliation inquiry, most published authors, top research applications, communication protocols, software processing, and operating systems. Another research in this area by [10], provided an overview of the key concepts related to IoT services development. Several research challenges have been identified, which are expected to become major research trends in the next years. In another study with a more bibliometric approach, [11] introduced a research theme five-cluster overview through IoT that shows the IoT increasing importance, but on the other hand, the studies that acknowledge the IoT applications for organizations and supply chains, and the wider socio-organizational context that needs to be considered. It also highlights the need for alternative theories to be used in order to study IoT-related phenomena.

In terms of IoT applications, due to supply chain management, [3] found out that most studies have focused on conceptualizing the impact of IoT with limited analytical models and empirical studies, and also most studies have focused on the delivery supply chain process and the food and manufacturing supply chains. Another application of IoT s economics that is reviewed by [13], biggest producers of greenhouse gas emissions, including Brazil and Russia, still lack studies in the area. In addition, a disconnection between important industry initiatives and academic research seems to exist that shows it can be useful for all institutions and researchers to figure out potential research gaps and to focus on future investments.

3 Research Methodology

In order to comprehend great researches and results in IoT applications in logistics, the Web of Science is used to search engines for articles to find the most authentic ones. So, the proposed method searched for titles, abstracts, and keywords that the term "Internet of things" AND "Logistics" appearing. As a result, there have been 249 research outputs from 2004 which is the start of trending of IoT technology up to now. that 4 of them are highly cited in this field which are leading other articles to the future of IoT researches.

After analyzing these results by VOS viewer, we accessed new and more comprehensive keywords. According to Table 1, these are: "internet of things", "IoT", "internet", "logistics", "RFID", "industry 4.0", "big data", and "cloud computing". So, then we combined all searches with the format of "Internet of things" + all keywords above to gather a true list about the intersection of IoT and logistics.

Table 1. Initial keywords list for search.

Selected	Keyword	Occurrences ⌄	Total link strength
☑	internet of things	174	1203
☑	iot	109	729
☑	internet	75	651
☑	logistics	61	514
☑	things	51	433
☑	internet of things (iot)	48	351
☑	rfid	37	259
☑	management	34	329
☑	model	29	301
☑	industry 4.0	29	229
☑	technology	23	235
☑	security	23	194
☑	big data	23	186
☑	supply chain	23	179
☑	cloud computing	23	158
☑	systems	18	212
☑	design	18	167

As a result, our study attained to 3434 results from 2004 which is the start of trending of IOT technology combined with logistics up to know that 154 of them are highly cited in this field which are leading other articles to the future of IOT researches. There are about 3119 articles and 84 proceeding papers (conference papers) and also 239 reviews which have the most shares among all types of papers in this field according to Fig. 1.

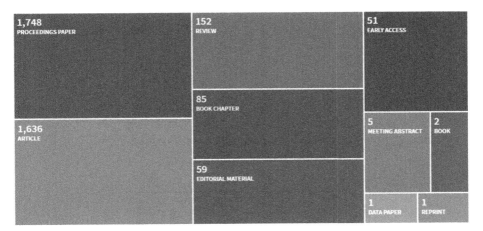

Fig. 1. Types of papers

Another information about our final dataset of search results is explained below:

- Total citation of papers in our dataset is 34823.
- Without self-citations (papers in our dataset cite each other) the total citation will be 32609 that shows papers mentioned in our dataset have a lot of impact on other fields of IoT or logistics and are influential papers.

4   Data Analytics

4.1   Discuss on Years of Research

It's essential to find the trend of academic works in every field grouped by years. According to Fig. 3 First pioneer publications in this field were [5, 9] that brought up the idea of the usage of IoT in many applications like logistics. Some analytics about years of publication are mentioned below:

- Total Publications by Year: Every year has some publications that important thing about this is the total number of publications per year which is shown in Fig. 2 that most active years are 2019 with 1332 and 2018 with 936 articles that make sense and it can be predicted that in 2020 this record will break.
- Sum of Times Cited by Year: Another statistic about years is sum of cites per year that every year differs a lot in this metric with years before (Displayed in Fig. 3).

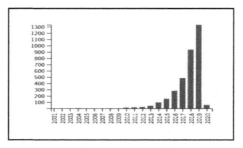

Fig. 2. Total publications by year

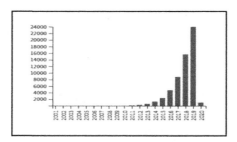

Fig. 3. Sum of times cited by year

4.2   Discuss on Keywords of This Field

The most important part in bibliometric studies is keywords survey including occurrences of key words and their relation with each other. Results of Figure. Shows that most frequent keywords in our research are "Internet", "Internet of things", "IOT", "Things", "Security", "Fog Computing" and "Cloud computing". (Displayed in Table 2).

Table 2. Final keywords list.

Selected	Keyword	Occurrences ∨	Total link strength
☑	internet of things	921	6811
☑	internet	635	5506
☑	big data	533	4373
☑	cloud computing	430	3369
☑	iot	366	2726
☑	internet of things (iot)	362	2614
☑	things	359	3219
☑	security	243	2216
☑	cloud	229	1836
☑	fog computing	228	1598
☑	system	199	1712
☑	management	195	1904
☑	challenges	183	1922
☑	framework	175	1771
☑	architecture	168	1693
☑	model	164	1454
☑	design	160	1503

Our results indicating that both occurrences and relations have the same patterns that can be seen in Figure. Due to Fig. 4, some keywords like "System", "Security" and "Networks", "Cyber-physical systems" and "Big data" have been under study for a long time.

In addition, some keywords like "Edge", "Blockchain", "SVM" and "Energy management" are lately trending keywords of edge technologies in this area.

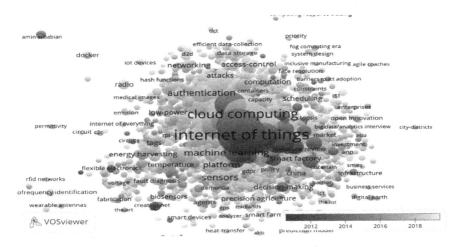

Fig. 4. Keyword's occurrence graph

4.3 Discuss on Authors in This Field

Great authors in field of IOT are divided by two aspects: Number of documents and number of citations.

1. Number of documents: According to Fig. 5 Rajkumar Buyya with 24 publications, Choo Kim-Kwang with 22 and Laurence Yang with 21 are top listed in numbers of documents in this area.

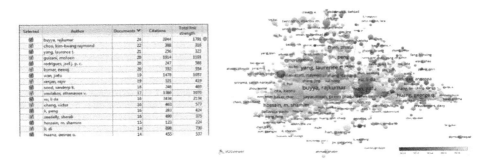

Fig. 5. Number of documents list and graph

2. Number of citations: As shown in Fig. 6, from this point of view, Luigi Atzori with 5073 citations and Antonio Iera with 5049 and Giacomo Morabito with 5041 citations are the best in citations. Also, there are new successful authors like Mohsen Guizani and Amir taherkordi in the plot.

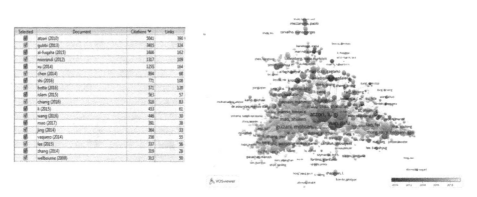

Fig. 6. Number of citations list and graph

4.4 Discuss on Countries

One of the subjects which have been always on the edge is countries with most documents and most citations that are somehow the same as each other in this field. Due to Fig. 7, it can understand that there are 14 groups or clusters that have been working with each other in the meantime. Foremost, due to Table 3, China with 1000 documents and 16760 citations, USA with 580 documents and 19396 citations, and Italy with 183 documents and 9717 citations are academic leaders of this field having the most documents and highest citations by their publication.

Table 3. List of countries.

Selected	Country	Documents	Citations ⌄	Total link strength
✓	usa	580	19396	5474
✓	peoples r china	1000	16760	5500
✓	italy	183	9717	2008
✓	australia	195	6706	2105
✓	england	256	6503	2076
✓	south korea	246	3338	1603
✓	qatar	25	2409	549
✓	canada	146	2281	1196
✓	india	205	2185	1731
✓	spain	184	2182	1156
✓	germany	112	1976	819
✓	sweden	73	1950	818
✓	saudi arabia	111	1857	973
✓	france	83	1418	582
✓	taiwan	129	1214	672
✓	greece	53	1048	398
✓	malaysia	46	1037	632

According to Fig. 7, there are some expert countries like USA, China, England and India which have been working on this area more than others and also newly added countries such as Iraq, Pakistan and Vietnam.

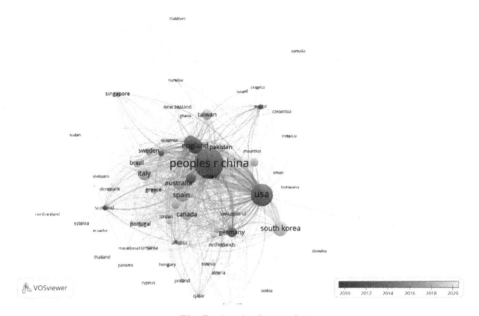

Fig. 7. Graph of countries

4.5   Discuss on Most Important Documents

Finally, have to classify all documents by their citation which is the best metric for evaluating the importance of an article. In order to group them, Fig. 8 will help us to find out which articles have most citation. Note that most of them are not lately released. So, with attention to Table 4, and also Web of Science introduces us three articles: [2, 5] and [1] as hot papers meaning they are on top of the list of noticeable articles to read and cite them in IoT applications field.

Hence, a substantial question will be asked "What are hot papers or some papers that are being cited which are Recently published?". To answer that from Fig. 8 it can find out that [7, 12, 15] and [4] are the most newly noteworthy articles.

Table 4. Important documents list.

Selected	Document	Citations ∨	Links
☑	atzori (2010)	5041	390
☑	gubbi (2013)	3405	324
☑	al-fuqaha (2015)	1666	162
☑	miorandi (2012)	1317	109
☑	xu (2014)	1255	164
☑	chen (2014)	894	68
☑	shi (2016)	771	108
☑	botta (2016)	571	120
☑	islam (2015)	563	57
☑	chiang (2016)	518	83
☑	li (2015)	453	61
☑	wang (2016)	446	30
☑	mao (2017)	391	38
☑	jing (2014)	364	33
☑	vaquero (2014)	356	55
☑	lee (2015)	337	56
☑	zhang (2014)	319	28
☑	welbourne (2009)	313	50

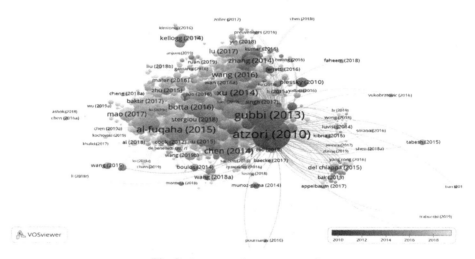

Fig. 8. Important documents graph

5   Results

In this section, we will use bibliometric analysis and some keywords related to logistics that is useful to find all IOT technologies have been utilized in logistics. According to Fig. 4 all useful keywords in this direction in sort of importance are 12 keywords explaining all subjects and academic studies in this field including: "Internet of things", "IoT", "airport", "technology", "inventory", "loading unloading", "Logistic", "port", "TECH", "rail", "road" and "transportation".

In the following, foregoing keywords will combine to determine IOT technologies in logistics. Hence, double words searching method is used to fulfil the main purpose of study. The output of searches are many articles and other kind of papers. These results will be used to find keywords that describes the technologies that have intersection with different types of logistics with purpose of answering the paper question. All results are shown in appendix A. In the next step, some patterns and connections between these technologies should be extracted.

There is not a lot of connection between our results per search, because of length of words. Totally:

1. As can be seen, "airport" + "internet of things" has the most technologies which means most of utilization of IOT has been in airport area and its facilities and after that, "Logistic" + "Internet of things" gives us the most technologies as might have guessed. So, some analytics on this will be conducted and as shown in Table 5 and Fig. 9, there are some significant keywords known as airport IOT technologies introduced here. Most important ones are "radio frequency machines", "Cloud computing", "Distributed computer systems" and some new technologies here are "Citizen science", "Distributed data stream". Continuously, for all searches, these analytics will be survived and all results are in Appendix A (Table A).

2. The most frequent technology that appears in 6 searches is the Wireless sensor network (WSN) defining a group of spatially dispersed and specified sensors for monitoring and recording the physical conditions of the environment and organizing the collected data at a central location. WSNs measure environmental conditions like temperature, sound, pollution levels, humidity, wind, etc. The next technology that includes 5 searches is Radio-frequency identification (RFID) which is a form of wireless connection that utilizes the use of electromagnetic or electrostatic coupling in the radio frequency portion of the electromagnetic spectrum to uniquely identify an object. After that is Geographic Information System (GIS) with 3 appearance which is an arrangement designed to receive, store, manipulate, analyze, control, and present structured and unstructured data. All mentioned, are devices that make IoT cycle possible.

3. From all keywords that are generated from the search results, it might have understood that most applications of IoT in logistics are in fields of:

a. "Traffic control": using devices to intelligently managing traffic by optimizing times of traffic lights or warning officers to reduce the risk of accidents and mass of cars. Managing parking lots and monitoring systems in format of internet network is also be done to optimize this process.

b. "Automation": in this area using smart sensors and remote-controlled devices, speed of work and reliability will rise, and also the quality of products will get better under certain infrastructures like smart inventory control, decision support system, barcodes, smart maintenance, building smart expert systems, using NFCs, smart online fault detection systems, etc.

c. "Transportation": Managing Vehicles and containers with intelligent scheduling and timetabling is the true form of the familiar form of logistics that doesn't need any modeling or optimization using old methods. In fact, instead of them, sensors will gather a large amount of data every moment and will help this process more accurate and with less estimation or risks.

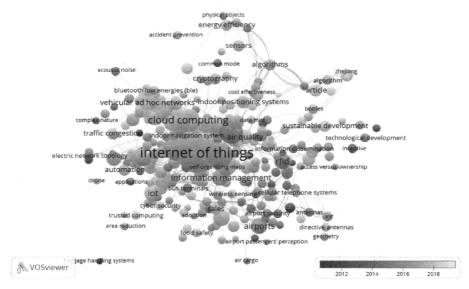

Fig. 9. "Airport" + "internet of things" keywords graph

Table 5. "Airport" + "internet of things" keywords list.

Selected	Keyword	Occurrences ∨	Total link strength
☑	internet of things	35	516
☑	internet	16	268
☑	internet of things (iot)	16	226
☑	radio frequency identification (rfid)	12	222
☑	cloud computing	12	179
☑	big data	10	151
☑	surveys	9	143
☑	internet of thing (iot)	8	122
☑	artificial intelligence	7	126
☑	iot	7	82
☑	mobile telecommunication systems	6	123
☑	distributed computer systems	6	117
☑	vehicles	6	110
☑	digital storage	6	102
☑	information management	6	101
☑	airports	6	96
☑	supply chains	6	89

In the next step, all analytics discussed earlier to get used to drawing a roadmap that is a great assistant to follow this field and help to concentrate more effectively on important points (Fig. 10). In this figure, it's obvious although three important major fields and four significant time intervals have been in this area, IoT applications have been more utilized in some specific areas and three unequal time phases.

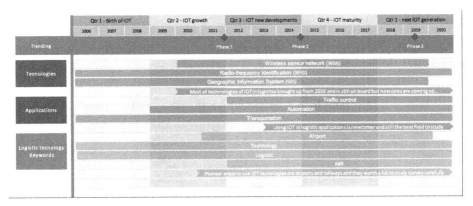

Fig. 10. IOT technologies and applications roadmap

6   Conclusion

In conclusion, from all analytics in intersection of internet of things and logistics area grouping by years, authors, documents, countries and keywords, to wrap up these points in summary, there are some points:

1. Some countries like USA, China, India and England have always been pioneers in this field.
2. Citation in this area doesn't have anything to do with history of that article because of state-of-the-art instinct of this technology. So, most cited articles are for not many years away.
3. Most used keywords of this field are "System", "Security" and "Networks", "Cyber-physical systems" and "Big data".
4. New hot keywords to use for this field are "Edge", "Blockchain", "SVM" and "Energy management".
5. At last, it's a great environment to fulfil or interests in internet of things because of its available infrastructure.
6. Airports and railways are always best places to facilitate with IOT devices and testing experiments in this field and three sections are the antecessors of developing IOT application with logistics approaches which many of technologies have been coming out from these processes including traffic control, automation and transportation.
7. IOT research in logistics contains four stages of progress and with passing every period, more technologies will come out and in future this trend will continue.

Appendix A

In the following, search results of final keywords extracted from paper analytics which explains IOT technologies will be noted.

Table 6. A.1. technology search results part 1.

airport + internet of things	airport + internet of things2	airport + technology	inventory .Internet of things	"loading unloading" + "Internet of things"	Logistic + "Internet of things"
RFID	intelligent system	object detection	RFID	Vehicle scheduling	RFID
intelligent buildings	efficiency	air navigation	scheduling	simultaneous control	mobile devices
energy saving	vehicular cloud	cooling	contex awareness	decision support system	service oriented architecture
energy conservation	security	traffic control	retail store	expert system	wsn
mobile device	ads-b	safety	automatic identification	emission	information exchange
mobile phone	optimization	gis	track and trace	RFID	energy conservation
retail store	vehicle to vehicle communication	simulation	robotic	real time information	temperature and humidties
integration	access control	temperature	inbound logistic	information sharing	information sharing
customer experience	fog	drainage	mobile phone		bar code
DATA MANAGEMENT	face recognition	engines	NFC		food saftey
fingerprint recognition	face identification	noise pollution			cellular telephones
enabling technologies	enviromental monitoring system	fire			near field communication
antennas	baggage handeling	water (waste-surface)			wireless sensor node
wsn	mobile application	groundwater pollution			mobile phone
cyber-physical system	decision support system	accident			mobile application
contex awareness	operational monitoring				expert system
data fusion	sales				waste management
	bsc				GIS
	health information				Traffic control
	lightening quality				fault detection
	air condition				virtual reality
	thermal comfort				data mining
					process control
					maintenance
					image processing
					traceability systems
					radio waves
					container

Table 7. A.2. technology search results part 2.

Logistic + technology	port + IOT	rail + iot	road + iot	road + tech	transportation + iot
image processing	smart phone	electric fault current	traffic light	temperature	internet of vehicle
signal processing	mobile internet	failur analysis	vehicle to vehicle		wsn
intelligent system	GIS	fault detection	ITS		embeded sys
maintenance	robotic	high speed	Routing		decision support sys
assessment method	WSN	contorel	WSN		inventory control
		delay	accident preventin		smart parking
		automation	mobile devices		
		intelligent system	navigation		
		monitoring	traffic monitorog		
		multimodal transportation	traffic control		
		data acquisition	parking		
		embeded sys	time delay		
		cyber physical sys			
		reliability			
		rfid			
		smart sensor			
		vehicular sensor network			
		wireless			
		WSN			
		signal processing			
		condition monitoring			

References

1. Al-Fuqaha, A., Guizani, M., Mohammadi, M., Aledhari, M., Ayyash, M.: Internet of Things: a survey on enabling technologies, protocols, and applications. IEEE Commun. Surv. Tutorials **17**, 2347–2376 (2015)
2. Atzori, L., Iera, A., Morabito, G.: The Internet of Things: a survey. Comput. Netw. **54**, 2787–2805 (2010)
3. Ben-Daya, M., Hassini, E., Bahroun, Z.: Internet of things and supply chain management: a literature review. Int. J. Prod. Res. **57**, 4719–4742 (2017)
4. Cui, Z., Cao, Y., Cai, X., Cai, J., Chen, J.: Optimal LEACH protocol with modified bat algorithm for big data sensing systems in Internet of Things. J. Parallel Distribut. Comput. **132**, 217–229 (2019)
5. Gubbi, J., Buyya, R., Marusic, S., Palaniswami, M.: Internet of Things (IoT): a vision, architectural elements, and future directions. Futur. Gener. Comput. Syst. **29**, 1645–1660 (2013)
6. Jankowski-Mihulowicz, P., Kalita, W., Pawlowicz, B.: Problem of dynamic change of tags location in anticollision RFID systems. Microelectron. Reliab. **48**, 911–918 (2008)
7. Kaplan, A., Haenlein, M.: Siri, Siri, in my hand: Who's the fairest in the land? On the interpretations, illustrations, and implications of artificial intelligence. Bus. Horiz. **62**, 15–25 (2019)
8. Kumar, P.M., Gandhi, U., Varatharajan, R., Manogaran, G., Vadivel, T.: Intelligent face recognition and navigation system using neural learning for smart security in Internet of Things. Clust. Comput. **22**, 7733–7744 (2017)
9. Lee, I., Lee, K.: The Internet of Things (IoT): applications, investments, and challenges for enterprises. Bus. Horiz. **58**, 431–440 (2015)
10. Miorandi, D., Sicari, S., de Pellegrini, F., Chlamtac, I.: Internet of things: vision, applications and research challenges. Ad Hoc Netw. **10**, 1497–1516 (2012)
11. Mishra, D., et al.: Vision, applications and future challenges of Internet of Things. Ind. Manag. Data Syst. **116**, 1331–1355 (2016)
12. Muhuri, P.K., Shukla, A.K., Abraham, A.: Industry 4.0: a bibliometric analysis and detailed overview. Eng. Appl. Artif. Intell. **78**, 218–235 (2019)
13. Nobre, G.C., Tavares, E.: Scientific literature analysis on big data and internet of things applications on circular economy: a bibliometric study. Scientometrics **111**, 463–492 (2017)
14. Ruiz-Rosero, J., Ramirez-Gonzalez, G., Williams, J., Liu, H., Khanna, R., Pisharody, G.: Internet of Things: a scientometric review. Symmetry **9**, 301 (2017)
15. Wan, S., Zhao, Y., Wang, T., Gu, Z., Abbasi, Q.H., Choo, K.-K.R.: Multi-dimensional data indexing and range query processing via Voronoi diagram for internet of things. Futur. Gener. Comput. Syst. **91**, 382–391 (2019)

Suppliers' Evaluation and Ranking in Telecommunication Infrastructure Company Using the TOPSIS Method in an Uncertain Environment

Mahmoud Tajik$^{(\boxtimes)}$ ⓘD, Ahmad Makui ⓘD, and Neda Mansouri ⓘD

Iran University of Science and Technology, Tehran, Iran
Mahmoud_tajik@ind.iust.ac.ir

Abstract. Nowadays, communication networks have a significant role in developing a worldwide dynamic economy. Iran has one of the biggest infrastructure networks in the Mideast and due to the rapid advancement of the communications industry, the customers expect receiving high secure services. On the other hand, the growth of the telecommunication industry has significantly increased the number of projects; hence, outsourcing the activities in this area is highly considered. Therefore, given the sensitivity, supplier selection in this industry is in high importance. The target of this research is to provide an empirical model for supplier selection in a communication company, in which the criteria are determined by experts and conducting studies. In the proposed approach, due to the lack of sufficient information and some linguistic information, decision making is done in fuzzy environment. In order to achieve the weight of the specified indicator, first, paired comparisons between the criteria are performed and the weight of each indicator is computed using the Buckley method and finally fuzzy TOPSIS method is applied to rate the suppliers. Finally, to validate the model presented, real data of the infrastructure telecommunication company is tested.

Keywords: Suppliers selection · Fuzzy theory · Uncertainty · TOPSIS

1 Introduction and Literature Review

In the Mideast, Iran has one of the greatest telecommunication networks. The highest expansion rate of telecommunication network in the region belongs to Iran due to its increase in the credit of phones and cell phones [1]. The swift growth of this industry in Iran has boosted its influence. It plays a significant role in revolutionizing the dynamic economy [2]. In the new age, with the widespread development of the telecommunication industry and its use in business and daily life, the customers expect receive safe, reliable and high-quality services has increased. Industrial systems are continuously looking for new solutions and strategies to expand and enhance their competitive benefits. Outsourcing is one of these approaches that can contribute to more competition [3]. Outsourcing helps organizations focus on the key factors that influence their core operations. However, some issues such as selection method and supplier criteria are important

© Springer Nature Switzerland AG 2021
Z. Molamohamadi et al. (Eds.): LSCM 2020, CCIS 1458, pp. 84–99, 2021.
https://doi.org/10.1007/978-3-030-89743-7_6

in outsourcing [4]. To ensure that companies benefit from all potential resource offers, the relation between supplier and buyer need to be effectively managed. Regarding the decline in value-added activities offered within companies, suppliers are increasingly becoming key factors in gaining competitive advantage in the companies. As a result, in order to reach and maintain competitive benefits, companies strive to build long-term relations with their most crucial suppliers and to vigorously enhance their productivity and performance [5].

The process of supplier selection (SP) is one of the most essential phases of manufacturing and procurement management for many companies. Suppliers are vital for every company because of their key position in company triumph; and their selection is salient in supply chain management ([6] and [7]). Occasionally, different quantitative and qualitative indicator, such as cost, technical capability, lead time, quality, and resilience, impress the SP problem [1]. Choosing the wrong supplier can cause problems in the fiscal and practical position of the company. Choosing the right suppliers remarkably decreases procurement costs, promotes market competition and increases ultimate consumer satisfaction [8]. The company has to make a right choice from a number of suppliers with different characteristics. Given the competitive characteristics and different strengths and weaknesses of suppliers, choosing the right one is a difficult task for decision makers and requires careful analysis and evaluation.

SP depends on several indicator while there is no ideal for this operation. Each SP method is various, so noticeably companies have several options to choose the selection method based on their production, forecast, indicator and industry [7]. Various methods have been developed for this purpose in the past. Analytical models of supplier assessment ranging from simple weighted scoring to complicated mathematical programming approaches have been employed to such problems [3]. The following is a scrutinizing of some articles on SP from the beginning up to now.

Analyzing the selection indicator and measuring supplier efficiency has been the main subject of many academics and buying experts since the 1960s. In 1966, G. W. Dickson [9] examined 50 factors that various authors consider important in the process of SP. Dickson later summarized the set to 23 factors by studying the purchasing agencies considering their overall importance in deciding on SP [10]. In 1994 Mandal and Deshmukh [11] developed Interpretive Structural Modeling (ISM) as a method of group judgment to recognize and synopsize relationship between SP indicator through a graphical model. They suggest separating the dependent indicator from the independent indicator to help the buyer. Data Envelopment Analysis (DEA) has also been used in SP. Weber is the first person to discuss the application of DEA in SP in a lot papers [12].

Over the past years, many decision-making models have been developed for SP. The decision-making operations becomes difficult when the values are ambiguous. So, the use of fuzzy sets to describe uncertainties in various factors clarifies the complicated structure of the decision-making operation. Alternatively stated, using verbal preferences can be very appropriate for ambiguous condition. Önüt et al. (2009) exerted the fuzzy TOPSIS Method (TM) for SP. They obtained criteria weights using paired comparisons based on fuzzy ANP. The results of this research help company to select the most suitable supplier among the options [8]. In 2014, a new fuzzy decision-making framework integrating Quality Function Deployment (QFD) and Data Envelopment Analysis (DEA) was introduced for SP. The proposed method analyzes inaccurate data using linguistic variables [13].

Ahmadi et al. (2016) employed a two-stage methodology based on Analytical Hierarchy Process (AHP) and developed the GRA to assess and chose the best tolerable supplier in the telecommunication industry, bearing in mind financial, social and environmental indicator. First, they defined sustainability criteria based on the experts' opinions and literature. AHP was then used to evaluate the selected criteria. In addition, they applied an improved Gray Relational Analysis (GRA) technique to sort suppliers based on their sustainability performance [1]. Azimi Fard et al. (2018) first identified four indicators for sustainable SP in the steel industry through a literature review. Then the weight of each criteria was gained using AHP. Lastly, the best supplier countries in the second and third level of the Iran steel industry have been established by using TOPSIS [14].

In 2018 Avasthi et al. proffered a framework based on the integrated fuzzy AHP-VIKOR approach for selecting a universal sustainable supplier. This issue is examined in two steps. The first, the fuzzy AHP is implemented to generate the weight of the indicator for tolerable SP and in the second step, the fuzzy VIKOR is applied to evaluate the supplier's performance according the indicator [15]. Babbar et al. (2018) offered a new mathematical model for selecting a set of suppliers and ordering quantities. Considering the significance of green concerns, both qualitative and quantitative indicator have been considered in this study. The presented model consists of two stages called a two-stage QFD and a multi-objective stochastic mathematical model. Stochastic (scenario) approach guides managing uncertainty in the order assignment process. In addition, the model uses trapezoidal fuzzy numbers [16]. Manerba et al. proposed a two-phase stochastic programming model with the influence of different sources of uncertainty. They focused on cases where only the price of the products or the demand for the products were stochastic [17]. A comprehensive overview of the most recent SP methods can also be established in papers [18, 19] and [7].

According to above mentioned states, in this study, appropriate criteria have been identified based on the studies conducted and the opinion of the telecommunication industry experts. Owing to the lack of complete connection to information and some linguistic information, decision making has been made in a fuzzy environment. To determine the weight of the described indicator, paired comparisons between criteria are first performed and weighted using the Buckley method. Finally, the fuzzy TM is devised to rank the suppliers and decision making in the fuzzy environment. The actual data obtained from the Telecommunication Infrastructure Company is also used to validate the model.

The structure of this article is as follows: Sect. 2 considers the research methodology. Section 3 proposes a case study and solution results. And eventually, Sect. 4 contains the results and suggestions.

2   A Subsection Sample

As mentioned earlier, the aim of this article is to offer a solution and a framework for selecting and ranking the best suppliers in a telecommunication company. For this purpose, a general classification of SP methods is presented in Fig. 1.

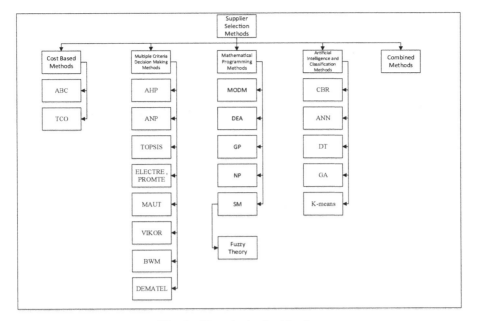

Fig. 1. SP methods classification

A decision-making approach involves defining and weighing criteria and then evaluating suppliers according to the criteria selected. In order to select the criteria in this paper, we have studied the literature in this field. Then, using experts' ideas, the main and most applicable criteria are screened from the research literature. Due to the specificity of the criteria to determine their importance and weight relative to each other, the AHP method of Buckley has been used. The alternatives are then ranked using the TM in the fuzzy environment. Figure 2 presents the conceptual framework used in this paper. The following section describes the concepts required in the proposed research and approach.

2.1 Weighting the Criteria

A fuzzy multiple indicator decision-making problem, can be shown briefly in a matrix as below:

$$
\tilde{D}_{ij} = \begin{matrix} & A_1 \ A_2 \ A_n \\ \begin{matrix} I_1 \\ I_2 \\ \cdot \\ \cdot \\ \cdot \\ I_m \end{matrix} & \begin{bmatrix} \tilde{y}_{11} & \tilde{y}_{12} & \dots & \tilde{y}_{1n} \\ \tilde{y}_{21} & \tilde{y}_{22} & \dots & \tilde{y}_{2n} \\ \cdot & \cdot & \cdot & \cdot \\ \cdot & \cdot & \cdot & \cdot \\ \cdot & \cdot & \cdot & \cdot \\ \tilde{y}_{m1} & \tilde{y}_{m2} & \dots & \tilde{y}_{mn} \end{bmatrix} \end{matrix} \quad \tilde{W}_i = \begin{bmatrix} \tilde{w}_1 \\ \tilde{w}_2 \\ \cdot \\ \cdot \\ \cdot \\ \tilde{w}_m \end{bmatrix}
$$

The significance of indicator weights can be obtained by paired comparisons [20]. In this research, we apply Buckley AHP fuzzy method [21] to gain weight of each criteria.

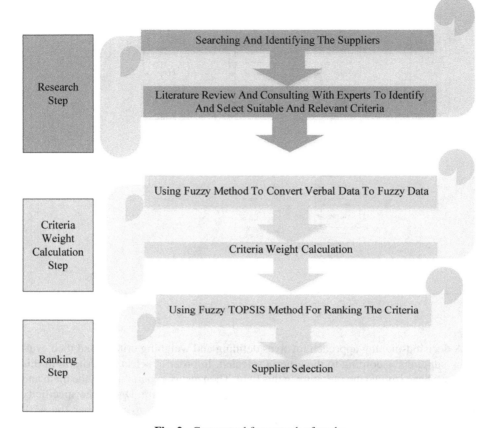

Fig. 2. Conceptual framework of study

In order to employ this method for weighting the indicator, linguistic variables are used in accordance with Table 1 and then the matrix of paired comparisons of the criteria is formed.

Table 1. Verbal scale related to the importance of criteria weights comparing each other

Linguistic value	The corresponding triangular fuzzy number
Equal importance	(1,1,1)
Weak importance	(2,3,4)
Strong importance	(4,5,6)
Highly strong importance	(6,7,8)
Extreme importance	(8,9,10)

According to the Buckley method [21], the matrix of fuzzy paired comparisons of the indicator is first formed and the weight of each indicator is calculated using a geometric mean approach according to the Eq. 1–2.

$$\tilde{e}_i = (\tilde{e}_{i1} \bullet \tilde{e}_{i2} \bullet ... \bullet \tilde{e}_{im})^{1/m}, \tilde{w}_i = \tilde{e}_i \bullet (\tilde{e}_1 \oplus \tilde{e}_2 \oplus ... \oplus \tilde{e}_m)^{-1}$$
$$i = 1, 2, ..., m$$

(1-2)

2.2 Weighting the Criteria

Fuzzy theory was first proposed by Professor Lotfizadeh [22] in 1965. Since its inception, this theory has expanded and deepened and employed in various aspects. According to the research [23] the concepts of fuzzy set theory applied in this article are as below.

Definition 1: Each fuzzy number can be considered as a function with a unique set-in domain, which is usually a set of real numbers, and its scope covers the range of non-negative real numbers between 0 and 1. The membership function used in this study is the triangular membership function, shown in Fig. 3. A triangular fuzzy number is denoted as (l/m, m/u) or (l, m, u). The variables l, m and u demonstrate the smallest possible value, the most probable value, and the biggest possible value of a fuzzy event, respectively.

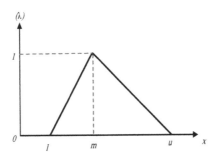

Fig. 3. Triangular fuzzy number

Each triangular fuzzy number (l, m, u) can be shown in function 2–2 [24]:

$$U_{\tilde{a}}(x) = \begin{cases} 0, x \le l \\ \dfrac{x - l}{m - l}, l \le x \le m \\ \dfrac{u - x}{u - m}, m \le x \le u \\ 0, x \ge u \end{cases}$$

(2-2)

Definition 2: The distance between two fuzzy triangular numbers $\tilde{a}_1 = (l_1, m_1, u_1)$ and $\tilde{a}_2 = (l_2, m_2, u_2)$ is calculated via below equation:

$$DT(\tilde{a}_1, \tilde{a}_2) = \sqrt{\frac{1}{3}\left[(l_1 - l_2)^2 + (m_1 - m_2)^2 + (u_1 - u_2)^2\right]}$$

(2-3)

Definition 3: Below equations exist between two fuzzy triangular numbers and: (Suppose k as a real number).

$$\tilde{a}_1 \oplus \tilde{a}_2 = (l_1 + l_2, m_1 + m_2, u_1 + u_2) \tag{2-4}$$

$$\tilde{a}_1 \ominus \tilde{a}_2 = (l_1 - l_2, m_1 - m_2, u_1 - u_2) \tag{2-5}$$

$$\tilde{a}_1 \otimes \tilde{a}_2 = (l_1 \times l_2, m_1 \times m_2, u_1 \times u_2) \tag{2-6}$$

$$\frac{\tilde{a}_1}{\tilde{a}_2} = \left(\frac{l_1}{l_2}, \frac{m_1}{m_2}, \frac{u_1}{u_2}\right) \tag{2-7}$$

$$\tilde{a}_1 \otimes k = (l_1 \times k, m_1 \times k, u_1 \times k) \tag{2-8}$$

2.3 Fuzzy TOPSIS

TOPSIS is one of the multiple indicator decisions making methods presented by Huang and Yun [25] in 1981 [26]. This method is based on the fact that the chosen alternative must have the largest distance from the negative ideal solution (ie, the solution that maximizes the cost indicator and minimizes the benefit indicator) and the shortest distance from the positive ideal solution (That is, a solution that maximizes the benefit indicator) [27]. In classical TOPSIS, the ratings and weights of the indicator are accurately determined. In fuzzy TOPSIS, all ratings and weights are described using linguistic variables [28]. In this paper, triangular fuzzy numbers are exerted for ease of calculation.

The fuzzy TOPSIS steps are [8]:

Step 1: The linguistic values are selected by the decision maker for the indicator in accordance with Table 2.

Table 2. Verbal scale related to suppliers' performance against the criteria and their fuzzy equivalents [29].

Linguistic scale	Corresponding triangular fuzzy number
Very Low (VL)	(0,0,2.5)
Low (L)	(0,2.5,5)
Good (G)	(2.5,5,7.5)
High (H)	(5,7.5,10)
Excellent (EX)	(7.5,10,10)

Normalized matrix in obtained by Eq. 2-9 in which $\tilde{x}_i^- = \left(l_i^-, l_i^-, l_i^-\right)$, $\tilde{x}_{ij} = \left(l_{ij}, m_{ij}, u_{ij}\right)$, $\tilde{x}_i^* = \left(u_i^*, u_i^*, u_i^*\right)$:

$$\tilde{P}_{ij} = \left[\tilde{p}_{ij}\right]_{m \times n}$$

$$\tilde{p}_{ij} = \begin{cases} \dfrac{\tilde{x}_{ij}}{\tilde{x}_i^*} = \left(\dfrac{l_{ij}}{u_i^*}, \dfrac{m_{ij}}{u_i^*}, \dfrac{u_{ij}}{u_i^*} \right), u_i^* = \max_j u_{ij} (benefit \text{ criteria}) \\[3mm] \dfrac{\tilde{x}_i^-}{\tilde{x}_{ij}} = \left(\dfrac{l_i^-}{l_{ij}}, \dfrac{l_i^-}{m_{ij}}, \dfrac{l_i^-}{u_{ij}} \right), l_i^- = \max_j l_{ij} (\cos t \text{ criteria}) \end{cases} \tag{2-9}$$

shows the fuzzy values assigned to the alternatives and introduces assigned normalized numbers.

Step 2: The weighted normalized decision-making matrix is computed using the Eq. 2–9:

$$\tilde{v} = \left[\tilde{v}_{ij} \right]_{s \times t} i = 1, 2, 3, ..., s \text{ and } j = 1, 2, ..., t$$

$$\text{where } \tilde{v}_{ij} = \tilde{p}_{ij} \times w_i \tag{2-10}$$

According to Eq. 2–10, w_i introduces the weight assigned to each indicator and shows weighted normalized numbers.

Step 3: The ideal positive and negative responses are determined using Eqs. 2–11 and 2–12.

$$B^{star} = \left\{ v_1^+, v_2^+, ..., v_i^+ \right\} = \left\{ (\max_j v_{ij} \ni i \in \psi_b), (\min_j v_{ij} \ni i \in \psi_c) \right\} \tag{2-11}$$

$$B^{nadir} = \left\{ v_1^-, v_2^-, ..., v_i^- \right\} = \left\{ (\min_j v_{ij} \ni i \in \psi_b), (\max_j v_{ij} \ni i \in \psi_c) \right\} \tag{2-12}$$

In a way that is the set of positive criteria and is the set of negative criteria.

Step 4: The distance from each alternative to positive and negative ideals are computed using Eqs. 2–13 and 2–14.

DT_j^{star} is the distance between alternatives and positive ideal and DT_j^{nadir} is the distance between alternatives and negative ideal.

$$DT_j^{star} = \sum_{i=1}^{m} DT \left(\tilde{v}_{ij}, \tilde{v}_i^+ \right), j = 1, 2, ..., t \tag{2-13}$$

$$DT_j^{nadir} = \sum_{i=1}^{m} DT \left(\tilde{v}_{ij}, \tilde{v}_i^- \right), \quad j = 1, 2, ..., t \tag{2-14}$$

Step 5: Finally, the index of relative proximity to the ideal response is calculated using Eq. 2–15.

$$cc_j = \frac{DT_j^{nadir}}{DT_j^{star} + DT_j^{nadir}} \quad j = 1, 2, ..., t \tag{2-15}$$

Step 6: In this stage, considering the index of relative proximity, alternatives are ranked from most proximity to least proximity.

3 Case Study

3.1 Introducing Telecommunication Infrastructure Company

Telecommunication Infrastructure Company (www.tic.ir) is a state corporation under the Ministry of Communication and Information Technology. The company operates in the field of telecommunications infrastructure and Internet bandwidth, and practically monopolizes the import and distribution of Internet bandwidth throughout the country. The Telecommunications Infrastructure Company, as the custodian of the telecommunication network in the country and the broker of the Ministry of Communication and Information Technology, is also responsible for switching and providing inter-provincial and international telecommunications operators.

Many construction projects, research projects and information technology projects are carried out by Telecommunications Infrastructure Company, which in most cases are outsourced and carried out by volunteer contractors. However, given the large number of projects underway and the greater focus on projects with higher valued amounts, unfortunately, careful monitoring is not implemented on smaller projects (values less than 25 billion Rials). More precisely, in this organization the selection of suppliers for small projects is done solely on the basis of price indicator and other influential indicators in SP is not considered. For this reason, it has been observed that most small projects encounter problems such as project delays, contractor withdrawals due to unreasonable pricing, and contractor underperformance due to increased inflation in procurement of goods. For this reason, it has been attempted in this research to consider SP process in terms of influential criteria in SP according to experts in this field and to select suppliers in more precise and reasonable conditions. According to surveys conducted, five suppliers are able to implement the standards mandatory to encounter the needs of the company, which this research intends to rank these five suppliers. Due to the respect of the rights of these providers and the confidentiality of the use of the information, their names are not mentioned and only the mathematical notations used to express them. The notations are shown in Table 3.

Table 3. Suppliers' introduction

Supplier	1	2	3	4	5
Notation	A_1	A_2	A_3	A_4	A_5

3.2 Paired Comparisons Matrix Development and Criteria Weighting

In order to select the criteria, previous literature in this field has been studied. Then, using the experts' opinions, the main and most applicable criteria were screened from the literature. Table 4 introduces the criteria.

Table 4. Criteria notations

Notation	Criteria
I_1	Operational skills
I_2	Considering safety during work
I_3	The quality of goods/services [9]
I_4	Offered price
I_5	Using new technologies in the activity context [7]
I_6	Considering environmental criteria [7]
I_7	Reliability/credibility [14]
I_8	Assets and fund [9]
I_9	Goods/services guarantee [9]
I_{10}	On-time delivery [9]
I_{11}	Warranties
I_{12}	Previous cooperation experience in the organization [9]
I_{13}	Employee's training
I_{14}	Reputation and well positioned in the industry [9]
I_{15}	Management Capabilities [9]
I_{16}	The potential for future cooperation

As mentioned before, we employ the Buckley method to specify the weight of above indicator. In this step, the results of implementing the Buckley method and the weights obtained for the indicator are reported in Table 5.

3.3 Supplier Ranking Using Fuzzy TM

In order to employ the fuzzy TM, six stages outlined in Sect. 2–2 shall be implemented and the decision regarding suppliers' selection must be made. In this section, the results of the six phases are presented.

Step 1: After calculating the weight of the criteria, a decision matrix for each supplier should also be formed. In this paper, the status of each supplier against the criteria is evaluated using linguistic variables, the results of which are presented in Table 6.

Then the linguistic variables used in Table 6 are changed to fuzzy numbers using the information in Table 2 and finally the fuzzy decision matrix is formed. The corresponding consequences are reported in Table 7.

Table 5. Normalized fuzzy weighting criteria

Criteria	Weight
I_1	(0.30098,0.37081,0.32822)
I_2	(0.28059,0.35164,0.36777)
I_3	(0.25485,0.33314,0.41201)
I_4	(0.27741,0.32743,0.39516)
I_5	(0.30921,0.36789,0.3229)
I_6	(0.2747,0.33121,0.39409)
I_7	(0.27102,0.3338,0.39518)
I_8	(0.33165,0.3923,0.27605)
I_9	(0.27513,0.33222,0.39265)
I_{10}	(0.24333,0.30782,0.44885)
I_{11}	(0.27453,0.32753,0.39793)
I_{12}	(0.27149,0.32481,0.4037)
I_{13}	(0.27025,0.32351,0.40625)
I_{14}	(0.29313,0.34431,0.36256)
I_{15}	(0.26602,0.32095,0.41302)
I_{16}	(0.26949,0.32234,0.40816)

Step 2: After developing the fuzzy decision matrix, the next phase is to reckon the weighted normalized decision matrix using Eq. (2-10), the results of which are demonstrated in Table 8.

Step 3: At this phase the ideal positive (B^{star}) and negative (B^{nadir})solution is calculated and the results are shown in Table 9.

Step 4: At this step, according to the weighted normalized fuzzy decision matrix, the distance of each supplier from positive to negative ideal is calculated through the Eq. (2-13) and (2-14). The results are presented in Table 10. It is to note that DT_j^{star} is the distance between each alternative and positive ideal and DT_j^{nadir} is the distance between each alternative and negative ideal.

Step 5: In this phase, the relative adjacency index to the ideal answer is promoted by the Eq. (2-15).

Step 6: Finally, in this section, according to the proximity coefficient obtained for each alternative, the eventual rating of the suppliers is made and the decision on the selection of suppliers is made at this step.

Table 6. Verbal variables related to the performance of the alternatives against the criteria.

	A$_1$	A$_2$	A$_3$	A$_4$	A$_5$
I$_1$	G	G	G	H	G
I$_2$	L	G	L	L	H
I$_3$	G	H	EX	H	H
I$_4$	G	G	VL	L	G
I$_5$	L	G	L	L	L
I$_6$	VL	VL	L	H	G
I$_7$	G	G	H	H	G
I$_8$	G	H	G	G	H
I$_9$	VL	VL	G	G	G
I$_{10}$	L	L	L	G	G
I$_{11}$	H	H	G	H	H
I$_{12}$	VL	L	G	G	G
I$_{13}$	VL	L	L	G	G
I$_{14}$	G	L	G	G	G
I$_{15}$	G	G	G	H	G
I$_{16}$	G	VL	VL	G	L

Table 7. Conversion of verbal variables to fuzzy numbers related to the performance of alternatives against indicator.

Criteria	Weight	A$_1$	A$_2$	A$_3$	A$_4$	A$_5$
I$_1$	(0.30098,0.37081,0.32822)	(2.5,5,7.5(	(2.5,5,7.5)	(2.5,5,7.5)	(5,7.5,10)	(2.5,5,7.5)
I$_2$	(0.28059,0.35164,0.36777)	(0,2.5,5)	(2.5,5,7.5)	(0,2.5,5)	(0,2.5,5)	(5,7.5,10)
I$_3$	(0.25485,0.33314,0.41201)	(2.5,5,7.5)	(5,7.5,10)	(7.5,10,10)	(5,7.5,10)	(5,7.5,10)
I$_4$	(0.27741,0.32743,0.39516)	(2.5,5,7.5)	(2.5,5,7.5)	(0,0,2.5)	(0,2.5,5)	(2.5,5,7.5)
I$_5$	(0.30921,0.36789,0.3229)	(0,2.5,5)	(2.5,5,7.5)	(0,2.5,5)	(0,2.5,5)	(0,2.5,5)
I$_6$	(0.2747,0.33121,0.39409)	(0,0,2.5)	(0,0,2.5)	(0,2.5,5)	(5,7.5,10)	(2.5,5,7.5)
I$_7$	(0.27102,0.3338,0.39518)	(2.5,5,7.5)	(2.5,5,7.5)	(5,7.5,10)	(5,7.5,10)	(2.5,5,7.5)
I$_8$	(0.33165,0.3923,0.27605)	(2.5,5,7.5)	(5,7.5,10)	(2.5,5,7.5)	(2.5,5,7.5)	(5,7.5,10)
I$_9$	(0.27513,0.33222,0.39265)	(0,0,2.5)	(0,0,2.5)	(2.5,5,7.5)	(2.5,5,7.5)	(2.5,5,7.5)
I$_{10}$	(0.24333,0.30782,0.44885)	(0,2.5,5)	(0,2.5,5)	(0,2.5,5)	(2.5,5,7.5)	(2.5,5,7.5)
I$_{11}$	(0.27453,0.32753,0.39793)	(5,7.5,10)	(5,7.5,10)	(2.5,5,7.5)	(5,7.5,10)	(5,7.5,10)
I$_{12}$	(0.27149,0.32481,0.4037)	(0,0,2.5)	(0,2.5,5)	(2.5,5,7.5)	(2.5,5,7.5)	(2.5,5,7.5)
I$_{13}$	(0.27025,0.32351,0.40625)	(0,0,2.5)	(0,2.5,5)	(0,2.5,5)	(2.5,5,7.5)	(2.5,5,7.5)
C$_{14}$	(0.29313,0.34431,0.36256)	(2.5,5,7.5)	(0,2.5,5)	(2.5,5,7.5)	(2.5,5,7.5)	(2.5,5,7.5)
I$_{15}$	(0.26602,0.32095,0.41302)	(2.5,5,7.5)	(2.5,5,7.5)	(2.5,5,7.5)	(5,7.5,10)	(2.5,5,7.5)
I$_{16}$	(0.26949,0.32234,0.40816)	(2.5,5,7.5)	(0,0,2.5)	(0,0,2.5)	(2.5,5,7.5)	(0,2.5,5)

Table 8. Weighted normalized decision matrix

Criteria	A_1	A_2	A_3	A_4	A_5
I_1	(0.167,0.454,0.548)	(0.167,0.454,0.548)	(0.167,0.454,0.548)	(0.335,0.619,0.731)	(0.167,0.454,0.548)
I_2	(0,0.195,0.409)	(0.156,0.430,0.614)	(0,0.195,0.409)	(0,0.195,0.409)	(0.312,0.587,0.819)
I_3	(0.141,0.408,0.688)	(0.283,0.556,0.917)	(0.425,0.742,0.917)	0.283,0.556,0.917)	(0.283,0.556,0.917)
I_4	(0.206,0.534,0.880)	(0.206,0.534,0.880)	(0,0,0.293)	(0,0.243,0.586)	(0.206,0.534,0.880)
I_5	(0,0.273,0.479)	(0.229,0.601,0.719)	(0,0.273,0.479)	(0,0.273,0.479)	(0,0.273,0.479)
I_6	(0,0,0.219)	(0,0,0.219)	(0,0.184,0.439)	(0.305,0.553,0.878)	(0.152,0.405,0.658)
I_7	(0.150,0.409,0.660)	(0.150,0.409,0.660)	(0.301,0.557,0.880)	(0.301,0.557,0.880)	(0.150,0.409,0.660)
I_8	(0.184,0.480,0.461)	(0.369,0.655,0.615)	(0.184,0.480,0.461)	(0.184,0.480,0.461)	(0.369,0.655,0.615)
I_9	(0,0,0.291)	(0,0,0.291)	(0.204,0.542,0.874)	(0.204,0.542,0.874)	(0.204,0.542,0.874)
I_{10}	(0,0.228,0.666)	(0,0.228,0.666)	(0,0.228,0.666)	(0.180,0.502,0.999)	(0.180,0.502,0.999)
I_{11}	(0.305,0.547,0.886)	(0.305,0.547,0.886)	(0.152,0.401,0.664)	(0.305,0.547,0.886)	(0.305,0.547,0.886)
I_{12}	(0,0,0.299)	(0,0.241,0.599)	(0.201,0.530,0.899)	(0.201,0.530,0.899)	(0.201,0.530,0.899)
I_{13}	(0,0,0.301)	(0,0.240,0.603)	(0,0.240,0.603)	(0.200,0.528,0.905)	(0.200,0.528,0.905)
I_{14}	(0.217,0.562,0.807)	(0,0.255,0.538)	(0.217,0.562,0.807)	(0.217,0.562,0.807)	(0.217,0.562,0.807)
I_{15}	(0.148,0.393,0.690)	(0.148,0.393,0.69)	(0.148,0.393,0.690)	(0.296,0.536,0.920)	(0.148,0.393,0.690)
I_{16}	(0.200,0.526,0.909)	(0,0,0.303)	(0,0,0.303)	(0.200,0.526,0.909)	(0,0.239,0.606)

Table 9. Ideal positive and negative responses

B^{star}	B^{nadir}
0.73123909	0.1676373
0.81934964	0
0.91793001	0.14194415
0.88038919	0
0.71939875	0
0.87800306	0
0.88041914	0.15095185
0.65550861	0.18472119
0.87478481	0
0.99999561	0
0.88656079	0.15290947
0.89940698	0
0.90508285	0
0.80775364	0
0.92018475	0.14816895
0.90934819	0

Table 10. Distance from ideal

Distance from positive ideal					Distance from negative ideal				
DT_1^{star}	DT_2^{star}	DT_3^{star}	DT_4^{star}	DT_5^{star}	DT_1^{nadir}	DT_2^{nadir}	DT_3^{nadir}	DT_4^{nadir}	DT_5^{nadir}
9.105469	8.714257	8.59128	7.076982	7.15571	5.458372	5.910817	6.04489	7.9245	7.774188

Table 11. Closeness coefficient

Closeness coefficient				
cc5	cc4	cc3	cc2	cc1
0.520713	0.528248	0.41301	0.404156	0.374789

Table 12. Suppliers' Ranking

Supplier	Ranking
A_5	First
A_4	Second
A_3	Third
A_2	Fourth
A_1	Fifth

4 Conclusion

Iran owns one of the largest telecommunications lines in the Middle East, and its development has increased the expectation of customers in the telecommunications industry. Rapid development of this industry, leads to increase the projects needed to facilitate the activities of this industry; hence, outsourcing operations are being considered. In this manner, SP is very essential given the industry's sensitivity. In this research, a pragmatic model for SP in the communications industry is presented. As there is no sufficient information and some data are linguistic, decision making in fuzzy environment is presented. Selected criteria are determined by experts in this field and weighted by the Buckley method after paired comparisons matrix. Finally, these criteria are ranked using fuzzy TM. In order to validate the approach presented, real data obtained from the Telecommunications Infrastructure Company have been used which showed important results in this regard. For example, the results show that suppliers who were previously selected because of their low price, after using proposed approach, deviate from the first rank and have very little chance of being selected. The results also show that despite the lack of sufficient and reliable information, it is possible to rank and evaluate the status of each supplier relative to others.

Different future research directions can be suggested to enrich this field. For example, proposing a Combined or innovative approaches ca to evaluate and rank can be an interesting avenue. Also, future research may be aimed at considering the input parameters under uncertainty.

References

1. Badri Ahmadi, H., Hashemi Petrudi, S.H., Wang, X.: Integrating sustainability into supplier selection with analytical hierarchy process and improved grey relational analysis: a case of telecom industry. Int. J. Adv. Manuf. Technol. **90**(9–12), 2413–2427 (2016). https://doi.org/10.1007/s00170-016-9518-z
2. Chatterjee, K., Kar, S.: Supplier selection in Telecom supply chain management: a Fuzzy-Rasch based COPRAS-G method. Technol. Econ. Dev. Econ. **24**(2), 765–791 (2018)
3. Büyüközkan, G., şakir Ersoy, M.: Applying fuzzy decision making approach to IT outsourcing supplier selection. System **2**, 2 (2009)
4. Razmi, J., Rafiei, H., Hashemi, M.: Designing a decision support system to evaluate and select suppliers using fuzzy analytic network process. Comput. Ind. Eng. **57**(4), 1282–1290 (2009)
5. Glock, C.H., Grosse, E.H., Ries, J.M.: Reprint of 'Decision support models for supplier development: Systematic literature review and research agenda.' Int. J. Prod. Econ. **194**, 246–260 (2017)
6. Wagner, S.M., Johnson, J.L.: Configuring and managing strategic supplier portfolios. Ind. Mark. Manag. **33**(8), 717–730 (2004)
7. Taherdoost, H., Brard, A.: Analyzing the process of supplier selection criteria and methods. Procedia Manuf. **32**, 1024–1034 (2019)
8. Önüt, S., Kara, S.S., Işik, E.: Long term supplier selection using a combined fuzzy MCDM approach: a case study for a telecommunication company. Expert Syst. Appl. **36**(2), Part 2, 3887–3895 (2009)
9. Dickson, G.W.: An analysis of vendor selection systems and decisions. J. Purch. **2**(1), 5–17 (1966)
10. Weber, C.A.: Supplier selection using multi-objective programming: a decision support system approach. Int. J. Phys. Distrib. Logist. Manag. **23**(2), 3–14 (1993)
11. Mandal, A., Deshmukh, S.G.: Vendor selection using interpretive structural modelling (ISM). Int. J. Oper. Prod. Manag. **14**(6), 52–59 (1994)
12. De Boer, L., Labro, E., Morlacchi, P.: A review of methods supporting supplier selection. Eur. J. Purch. supply Manag. **7**(2), 75–89 (2001)
13. Karsak, E.E., Dursun, M.: An integrated supplier selection methodology incorporating QFD and DEA with imprecise data. Expert Syst. Appl. **41**(16), 6995–7004 (2014)
14. Azimifard, A., Moosavirad, S.H., Ariafar, S.: Selecting sustainable supplier countries for Iran's steel industry at three levels by using AHP and TOPSIS methods. Resour. Policy **57**, 30–44 (2018)
15. Awasthi, A., Govindan, K., Gold, S.: Multi-tier sustainable global supplier selection using a fuzzy AHP-VIKOR based approach. Int. J. Prod. Econ. **195**, 106–117 (2018)
16. Babbar, C., Amin, S.H.: A multi-objective mathematical model integrating environmental concerns for supplier selection and order allocation based on fuzzy QFD in beverages industry. Expert Syst. Appl. **92**, 27–38 (2018)
17. Manerba, D., Mansini, R., Perboli, G.: The capacitated supplier selection problem with total quantity discount policy and activation costs under uncertainty. Int. J. Prod. Econ. **198**, 119–132 (2018)

18. Chai, J., Ngai, E.W.T.: Decision-making techniques in supplier selection: recent accomplishments and what lies ahead. Expert Syst. Appl. **140**, 112903 (2020)
19. Rashidi, K., Noorizadeh, A., Kannan, D., Cullinane, K.: Applying the triple bottom line in sustainable supplier selection: a meta-review of the state-of-the-art. J. Clean. Prod. **269**, 122001 (2020)
20. Hsu, H.-M., Chen, C.-T.: Fuzzy credibility relation method for multiple criteria decision-making problems. Inf. Sci. (Ny) **96**(1–2), 79–91 (1997)
21. Buckley, J.J.: Fuzzy hierarchical analysis. Fuzzy sets Syst. **17**(3), 233–247 (1985)
22. Zadeh, L.A.: Fuzzy Sets. Inf. Control **8**(3): 338–353 (1965). https://doi.org/10.1016/S0019-9958(65)
23. Kaufmann, A., Gupta, M.M.: Fuzzy Mathematical Models in Engineering and Management Science. Elsevier Science Inc., Amsterdam (1988)
24. Dhiman, H.S., Deb, D.: Fuzzy TOPSIS and fuzzy COPRAS based multi-criteria decision making for hybrid wind farms. Energy **202**, 117755 (2020)
25. Yoon, K., Hwang, C.L.: TOPSIS (technique for order preference by similarity to ideal solution)–a multiple attribute decision making, w: Multiple attribute decision making–methods and applications, a state-of-the-at survey. Springer Verlag, Berlin (1981)
26. Mina, H., Kannan, D., Gholami-Zanjani, S.M., Biuki, M.: Transition towards circular supplier selection in petrochemical industry: a hybrid approach to achieve sustainable development goals. J. Clean. Prod. **286**, 125273, (2021)
27. Montanari, R., Micale, R., Bottani, E., Volpi, A., La Scalia, G.: Evaluation of routing policies using an interval-valued TOPSIS approach for the allocation rules. Comput. Ind. Eng. **156**, 107256 (2021)
28. Zouggari, A., Benyoucef, L.: Simulation based fuzzy TOPSIS approach for group multi-criteria supplier selection problem. Eng. Appl. Artif. Intell. **25**(3), 507–519 (2012)
29. Lima Jr, F.R., Osiro, L., Carpinetti, L.C.R.: A comparison between Fuzzy AHP and Fuzzy TOPSIS methods to supplier selection. Appl. Soft Comput. **21**, 194–209 (2014)

A Study on the Performance of Grey Wolf Optimizer

Aybike Özyüksel Çiftçioğlu$^{(\boxtimes)}$ (iD)

Faculty of Engineering, Department of Civil Engineering,
Manisa Celal Bayar University, Manisa, Turkey
aybike.ozyuksel@cbu.edu.tr

Abstract. In recent years there is an enormous increase in the emergence of non-deterministic search methods. The effective way of animals in problem-solving (like discovering the shortest path to the food source) has been examined by scientists and swarm intelligence has become a research field that imitates the behaviour of animals in swarm. The moth-flame optimization (MFO) algorithm, salp swarm algorithm (SSA), firefly algorithm (FFA), bat (BAT) algorithm, cuckoo search (CS) algorithm, genetic algorithm (GA), and grey wolf optimizer (GWO) are some of the swarm intelligence based non-deterministic methods. In the present study, the seven methods above are investigated separately. Five mathematical functions are resolved individually by these seven methods. Each algorithm is run 30 times in each benchmark function. The performances of these optimization methods are evaluated and compared within each function individually. Performances of algorithms over convergence are compared by plotting convergence rate graph and boxplots of methods for each function. Considering most of the functions, GWO is observed to be stronger than other algorithms.

Keywords: Comparison · Optimization · Stochastic search methods

1 Introduction

Since the ancient times, people being have tried to keep the discomforts, pain and, outgoings at minimum, economize the energy because of the limited sources in nature and maximize the profit. It is necessary to decide the optimal way to realize this phenomenon. The optimization process is related to reaching the best result of a given problem while satisfying certain constraints [10, 11, 14].

A common engineering optimization problem can be identified as follows; minimize

$$f(x), \ x = \{x_1, x_2, \ldots, x_n\} \tag{1}$$

subjected to

$$gi(x) \leq 0, \ i = 1, 2, \ldots, p \tag{2}$$

© Springer Nature Switzerland AG 2021
Z. Molamohamadi et al. (Eds.): LSCM 2020, CCIS 1458, pp. 100–116, 2021.
https://doi.org/10.1007/978-3-030-89743-7_7

$$hj(x) = 0, j = 1, 2, \ldots, m \tag{3}$$

$$L_{xk} \leq x \leq U_{xk}, k = 1, 2, \ldots, n \tag{4}$$

where, $f(x)$ signifies objective function, x represents the decision solution vector, n is the whole number of decision variables. U_{xk} and L_{xk} and are the upper and lower bound of each decision variable, respectively. p denotes inequality restraints' number and m denotes equality restraints' number [1, 4, 5].

Optimization, often described as finding the best solution for a definite problem [11]. Today, different optimization algorithms are used to solve many problems [9, 12]. Some of those: moth-flame optimization (MFO) algorithm [16], salp swarm algorithm (SSA), firefly algorithm (FFA), bat (BAT) algorithm, cuckoo search (CS) algorithm, genetic algorithm (GA), grey wolf optimizer (GWO), etc.

Meta-heuristic methods are can be considered as two basic classes: evolutionary [20] and swarm intelligence-based [3] methods. Evolutionary process is the adaptation of animals or plants to their environment in order to increase their chances of survival in the environment. Evolutionary Algorithms are created based on evolutionary principles like the collective learning process in populations of individuals. Besides, they are interdisciplinary fields of study associated with biology, artificial intelligence, numerical optimization, and almost all engineering disciplines. Evolutionary computation was first introduced in the 1960s by Rechenberg and was used to optimize the actual parameters of aircraft wings [2]. Genetic Algorithm developed by John Holland [8] is the first and best-known method presented in the evolutionary class. It can be explained as the method of simulating the evolutionary process on the computer and random search technique. The other evolutionary algorithms in the literature is Evolutionary Programming [7] that randomly transformed state transformation diagrams in the finite state machine and selected the best.

Swarm intelligence optimization algorithms have a larger group than optimization methods based on the evolutionary process. Swarm intelligence-based optimization algorithms were created by examining and imitating the movements of creatures such as birds, fish, bees, etc. Some of them are: grey wolf optimizer (GWO) simulates the hunting behaviour of grey wolves is a meta-heuristic optimization method developed by Mirjalili [18]. The basic parts of the grey wolf algorithm are the stages of encircling prey, hunting and attacking prey. Salp swarm algorithm (SSA) offered by Mirjalili [17] is a novel metaheuristic algorithm that models the swarming behaviour of salps in deep oceans. In oceans, salps often create a swarm called salp chain. The basic aim for this attitude is probably achieving better motion by using foraging and rapid coordinated changes. Moth-flame optimization (MFO) method developed by Mirjalili [16] is a recent nature-inspired optimization method. It models the navigation process of moths called transverse orientation in nature. The bat algorithm (BAT) is offered by Yang in 2010 that imitates bats' ability to navigate with sound (echo) [22]. With these abilities, bats can catch their prey, avoid obstacles and stay in the dark. Bats emit a loud sound and listen to the echo of that sound returning from surrounding objects. They use the difference between the sound they make and the time the echo comes. They calculate the distance to the target, the type of prey, and even the speed of movement of the prey.

The firefly algorithm (FFA) [21] is developed based on the blinking characteristics of fireflies. According to the firefly algorithm: all fireflies are hermaphrodites. Therefore, each can affect the other. In addition, the attractiveness is directly proportional to the glow of the firefly, and both decrease as the distance increases. In this study, the GWO algorithm is adapted to five different benchmark functions selected from the literature and a total of seven different optimization algorithms are compared over these functions.

2   Grey Wolf Optimizer

Wolf colony has a rigorous, organized system. The grey wolf optimization algorithm is proposed by imitating the hunting behaviour and social behaviour of grey wolves. Regarding the social hierarchy, grey wolves are classified as alpha, beta, delta, and omega [18]. The wolves separate the task exactly and keep their steps coherent when they are hunting. A few artificial wolves are detected to investigate the prey. When the investigating wolves discover the prey, they inform the position of the prey to the other artificial wolves by howl. The other artificial wolves get close to the prey and surround it. The assignment rule of the wolf colony is to yield the food to the strong wolf first and then to the weak wolf [6, 13, 15, 19].

The pseudo code of the GWO algorithm is presented in Fig. 1.

Initialize the grey wolf population X_i ($i = 1, 2, ..., n$)
Initialize a, A, and C
Calculate the fitness of each search agent
X_α=the best search agent
X_β=the second best search agent
X_δ=the third best search agent
while ($t <$ Max number of iterations)
   for each search agent
            Update the position of the current search agent by equation (3.7)
   end for
   Update a, A, and C
   Calculate the fitness of all search agents
   Update X_α, X_β, and X_δ
   $t=t+1$
end while
return X_α

Fig. 1. Pseudo code of the GWO algorithm [18].

3   Design Examples

3.1   Function 1 (Schwefel 2.22)

The first function is continuous, convex, unimodal, non-differentiable and separable. Dimension of the function is 30. This minimization function has one global minimum f(x*) = 0 at x* = (0,...,0), is implemented to examine the performance of the GWO. Function mathematical definition as follows;

$$\min f(x) = \sum_{i=1}^{n} |x_i| + \prod_{i=1}^{n} |x_i| \tag{5}$$

side constraints;

$$-10 \le x_i \le 10 \tag{6}$$

This function is solved by using seven different algorithm methods (MFO, SSA, FFA, BAT, CS, GA, GWO). The optimum designs of the function achieved by metaheuristic techniques are tabulated in Table 1. The design histories of each algorithm are displayed in Fig. 2. The optimum solution belongs to the GWO which is −3.29E-20. Besides, GWO presents significantly lower worst (3.61E-18) and average (9.56E-19) values for this benchmark function. The algorithm that follows GWO with a value of 0.021075 in finding the best result is MFO. When evaluated for average and worst values, SSA is the second that finds the lowest value after GWO. In addition, the Bat algorithm shows the worst performance by finding the highest results in best, worst and average values in the function. Furthermore, the boxplots of the algorithms are illustrated in Fig. 3. From the figure, it can be understood that all algorithm boxplots perform very close except BAT algorithm.

Table 1. Optimal results for function 1.

	MFO	SSA	FFA	BAT	CS	GA	GWO
1	100.0001	1.516516	26.95364	64.99872	4.47246	24.80908	1.58E-18
2	20.00426	0.352063	35.53781	55.8063	8.667319	20.15373	2.30E-19
3	50.0014	0.858895	31.45082	606.8716	3.846369	26.67873	4.93E-18
4	30.00193	0.753203	30.50091	58.6459	7.478827	21.27676	2.02E-18
5	10.01135	0.259874	2.782699	65.40107	6.589571	21.86882	6.88E-19
6	40.00207	0.356355	0.284074	67.04129	6.212948	31.44663	2.41E-19
7	40.00168	2.270353	27.76694	72.43219	5.34957	25.60062	4.97E-19
8	20.00737	1.75596	7.894253	42.48482	7.796042	19.27472	4.30E-19
9	10.00532	1.545848	1.312558	62.56324	7.368403	19.50736	3.28E-18
10	30.017	1.263358	36.29246	109.1789	5.120388	18.59425	2.10E-19
11	60.00536	0.258922	0.179179	68.41408	4.199672	25.2828	3.54E-19
12	0.020292	0.394039	47.12112	75.94642	6.481624	29.89988	5.16E-19
13	0.02842	0.643379	24.70372	104.9872	6.08156	26.64182	6.17E-19
14	30.00303	0.864842	1.997773	74.57441	8.429324	22.75124	2.23E-18

(continued)

Table 1. (*continued*)

	MFO	SSA	FFA	BAT	CS	GA	GWO
15	20.01656	0.262527	40.9454	24893.18	4.776235	31.97988	2.62E-19
16	10.01225	0.402131	5.482987	57.01707	7.571104	25.02382	4.61E-19
17	10.01046	3.226768	0.139893	73.6874	6.265984	28.70503	8.88E-19
18	20.00226	0.825209	0.52348	314.5031	4.373244	18.29054	3.83E-19
19	50.00296	0.306236	22.69973	75.46483	4.858459	20.08453	1.66E-18
20	50.00219	0.049447	9.446288	62.97256	4.360125	24.37613	4.67E-19
21	50.0024	3.956519	1.069143	80.96539	5.602348	26.44668	5.36E-19
22	0.088665	0.602087	0.435493	51.88861	4.992596	28.86921	1.05E-18
23	40.03906	0.610621	4.634661	221.9743	10.01414	27.12361	1.95E-19
24	30.00485	1.426648	0.886865	99.4694	6.493815	24.94889	1.66E-18
25	20.0212	0.729363	0.391311	33.5566	6.015979	23.40752	1.25E-19
26	70.00182	2.205536	3.567193	146.2781	6.705933	35.39886	2.32E-19
27	50.00328	1.608083	141.9703	56.25357	10.87672	29.2543	3.13E-19
28	40.00556	3.473804	22.10462	87.36544	4.797415	23.77836	1.02E-18
29	60.00181	0.990076	47.69598	62.7259	10.14901	22.83839	1.76E-18
30	40.00266	0.150296	0.120864	858.3201	7.647985	25.1295	5.95E-19
Best	0.021075	0.06607	0.218615	32.64853	3.354007	16.15048	**3.29E-20**
Worst	80.00072	3.720723	56.22065	1.01E + 08	12.10318	31.16218	**3.61E-18**
Average	35.39479	1.038932	17.89048	5428650	5.935988	24.81677	**9.56E-19**

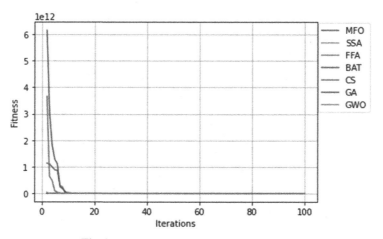

Fig. 2. Convergence graph of the algorithms.

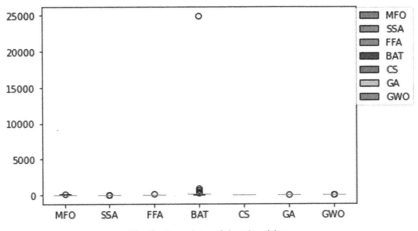

Fig. 3. Boxplots of the algorithms

3.2 Function 2 (Rotated Hyper Ellipsoid)

The second function is continuous, convex, unimodal, differentiable and non-separable. Dimension of the function is 30. This minimization function has one global minimum f(x*) = 0 at x* = (0,...,0), is implemented to examine the performance of the GWO. Function mathematical definition as follows;

$$minf(x) = \sum_{i=1}^{n} \left(\sum_{j=1}^{i} x_j \right)^2 \tag{7}$$

side constraints;

$$-100 \le x_i \le 100 \tag{8}$$

This function is solved by using seven different algorithm methods (MFO, SSA, FFA, BAT, CS, GA, GWO). The optimum designs of the function achieved by metaheuristic techniques are tabulated in Table 2. The optimum solution belongs to the GWO which is 9.09E-10. The best, worst, and average values of GWO are significantly lower than other algorithm results. Therefore, its performance in this benchmark is remarkably better than other algorithms. The design histories of each algorithm are displayed in Fig. 4. It can be seen from this figure that GWO has the best convergence curve. It is observed that the curves of the algorithms other than that of the BAT algorithm are well. Additionally, Fig. 5, where the algorithms and boxplots are presents, shows that the GWO and SSA offer lower and narrower charts.

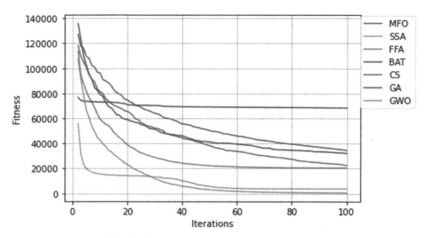

Fig. 4. Convergence graph of the algorithms

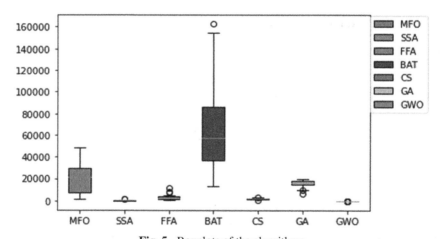

Fig. 5. Boxplots of the algorithms

Table 2. Optimal results for function 2

	MFO	SSA	FFA	BAT	CS	GA	GWO
1	25394.68	631.1904	3359.127	102872.2	1901.971	14742.42	2.13E-06
2	23881.02	674.1033	4561.864	45953.88	2328.936	19596.23	6.22E-05
3	2375.917	928.7059	2687.931	85717.1	3325.648	19674.36	0.000718
4	4936.762	484.1506	4818.69	21258.76	2584.708	15953.83	4.01E-05
5	30355.87	402.9741	2510.04	154780.6	1993.551	18076.75	1.90E-08
6	48366.47	266.8516	2099.925	31250.89	1727.641	17607.62	8.17E-07

(continued)

Table 2. (*continued*)

	MFO	SSA	FFA	BAT	CS	GA	GWO
7	30321.56	184.2937	2604.614	30701.39	1920.712	16245.83	1.46E-05
8	20687.8	306.1065	2856.366	80207.31	2016.055	14697.56	5.94E-05
9	1881.159	717.776	11382.09	66385.19	1609.586	19816.73	3.67E-05
10	12432.99	613.0932	1442.613	34368.73	2001.614	16927.99	0.000368
11	5156.794	79.97077	1237.394	111400.1	2357.887	10131.08	3.93E-06
12	14811.23	172.1102	1508.2	102876.5	2156.294	16154.89	6.98E-05
13	7787.54	361.3541	1875.123	162632.2	2141.879	16999.01	0.000143
14	2500.451	253.2134	1620.331	42918.25	2199.82	15048.64	0.000766
15	21702.32	394.1653	2625.22	48302.78	2122.998	17288.52	0.002246
16	15956.29	201.3314	2214.997	74764.1	2061.324	15899.03	4.67E-06
17	25366.59	138.0095	8157.293	12896.93	1612.848	6426.52	7.97E-05
18	48835.62	229.2225	6976.943	136562.9	2283.248	15343.49	7.56E-05
19	25265.41	320.2438	1092.886	41243.54	2653.831	18456.55	9.44E-06
20	9076.349	119.543	3749.229	41764.49	2222.989	16290.84	6.84E-06
21	38563.85	555.6491	3597.932	68091.06	2363.317	11475.96	1.24E-05
22	41008.96	232.7633	1450.555	31012.52	2095.545	14370.94	7.73E-05
23	4428.072	335.9355	2601.026	86073.37	2831.455	18345.56	1.48E-05
24	41983.71	736.7624	3993.447	35862.19	2946.327	19368.68	0.000103
25	27400.23	280.4144	3062.931	35915.24	2871.018	9695.817	6.77E-06
26	12603.05	362.2473	723.4196	59024.08	2449.414	11980.85	0.001253
27	45310.89	1239.966	1614.546	63151.2	3144.862	17488.37	1.36E-05
28	1654.608	382.1437	2235.61	56267.97	1536.964	15780.41	4.95E-05
29	7500.156	1226.745	4848.142	119390.5	1070.768	19138.54	0.01179
30	25875.83	144.0503	3529.964	49270.59	1501.61	17923.86	1.57E-05
Best	0.021075	0.06607	0.218615	32.64853	3.354007	16.15048	**3.29E-20**
Worst	80.00072	3.720723	56.22065	1.01E + 08	12.10318	31.16218	**3.61E-18**
Average	35.39479	1.038932	17.89048	5428650	5.935988	24.81677	**9.56E-19**

3.3 Function 3 (Griewank)

The third function is continuous, non-convex, unimodal, differentiable and separable. Dimension of the function is 30. This minimization function has one global minimum f(x*) = 0 at x* = (0,...,0), is implemented to examine the performance of the GWO. Function mathematical definition as follows;

$$\max f(x) = \max f(x) = \frac{1}{4000}\sum_{i=1}^{n}x_i^2 - \prod_{i=1}^{n}cos\left(\frac{x_i}{\sqrt{i}}\right) + 1 \qquad (9)$$

side constraints

$$-600 \leq x_i \leq 600 \tag{10}$$

This function is solved by using seven different algorithm methods (MFO, SSA, FFA, BAT, CS, GA, GWO). The optimum designs of the function achieved by metaheuristic techniques are tabulated in Table 3. From the table, SSA, FFA, MFO algorithms follow GWO in terms of best, worst and average results, respectively. The worst performance for the Griewank benchmark belongs to the BAT algorithm. The design histories of each algorithm are displayed in Fig. 6. In terms of convergence curves, the performance of algorithms other than the BAT algorithm is well. With the lowest best, worst, and average values, GWO performs significantly better. Moreover, the boxplots of the algorithms are illustrated in Fig. 7. From the figure it is shown that GWO and FFA have lower and thinner boxplots.

Table 3. Optimal results for function 3

	MFO	SSA	FFA	BAT	CS	GA	GWO
1	0.129027	0.000107	0.041675	147.6326	1.047535	29.37738	0
2	0.188496	0.007484	0.010183	315.0006	1.107411	23.33612	0
3	0.131198	0.007426	0.028569	173.6908	1.146589	37.14877	0
4	0.102617	0.00029	0.024922	231.6857	1.057653	29.45964	0
5	90.50376	0.009875	0.024975	212.8906	1.092712	41.40835	0.01794
6	90.22224	0.007544	0.011222	168.2868	1.114915	24.13591	0.011601
7	90.03703	0.019943	0.013913	304.3709	1.081084	23.39594	0
8	90.05555	0.014815	0.016219	153.4708	1.082188	42.45331	0.012183
9	90.07701	0.000507	0.020709	93.03299	1.122193	34.11912	0
10	0.066239	0.012361	0.027382	291.277	1.058289	43.62264	0
11	0.109719	0.034941	0.017175	132.1806	1.079855	43.51881	0
12	0.080135	0.000244	0.015859	188.6513	1.083322	32.90755	0.01618
13	0.092118	0.012426	0.025699	226.7986	1.108288	25.59694	0.012536
14	0.049272	1.58E-05	0.032651	250.78	1.201121	41.51751	0
15	0.067585	0.007507	0.012399	176.8481	1.091319	45.68797	0
16	0.272151	6.15E-06	0.014775	81.8523	1.092709	16.72173	0
17	0.314074	0.007529	0.023503	140.1036	1.124245	25.54573	0
18	0.14255	0.015656	0.024362	196.6672	1.078451	30.18658	0
19	90.89485	0.017535	0.021758	146.1104	1.088906	39.58998	0
20	90.29152	6.27E-06	0.031478	186.9371	1.104491	30.41208	0
21	0.118346	0.000197	0.022338	139.857	1.095425	15.10102	0
22	180.6521	0.007434	0.010861	146.1057	1.138482	30.78786	0
23	0.050705	5.28E-05	0.00933	198.7393	1.149331	31.46467	0
24	270.7764	0.029537	0.014697	374.9834	1.134832	32.57059	0.010603

(*continued*)

Table 3. (*continued*)

	MFO	SSA	FFA	BAT	CS	GA	GWO
25	0.105262	0.009875	0.039774	233.3331	1.067983	41.27586	0.026937
26	0.040533	7.28E-05	0.01751	193.9358	1.16112	30.81516	0.019203
27	0.174906	0.007767	0.009391	259.5721	1.105891	44.1091	0
28	90.88202	0.029655	0.013279	171.7939	1.123803	36.80873	0.011295
29	0.044997	0.012323	0.028781	128.6408	1.066004	28.06431	0
30	180.3875	4.55E-06	0.011426	178.8575	1.089631	50.67949	0
Best	0.03377	4.15E-06	0.009849	78.66404	1.040285	14.44534	**0**
Worst	90.93665	0.032111	0.041007	286.3749	1.195688	46.47571	**0.024602**
Average	15.18168	0.007614	0.01954	178.1669	1.100452	29.47067	**0.00572**

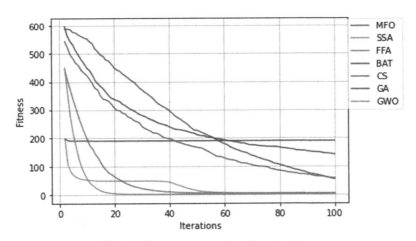

Fig. 6. Convergence graph of the algorithms

3.4 Function 4 (Branin RCOS)

The fourth function is continuous, non-scalable, multimodal, differentiable and non-separable. Dimension of the function is 2. With the global minimum f(x*) = 0.398 at x* = (0,...,0), this minimization function is implemented to examine the performance of the GWO. Function mathematical definition as follows;

$$\min f(x) = \left(x_2 - \frac{5.1}{4\pi^2}x_1^2 + \frac{5}{\pi}x_1 - 6\right)^2 + 10\left(1 - \frac{1}{8\pi}\right)cos x_1 + 10 \qquad (11)$$

and side constraints

$$-5 \le x_1, x_2 \le 5 \qquad (12)$$

This function is solved by using seven different algorithm methods (MFO, SSA, FFA, BAT, CS, GA, GWO). The optimum designs of the function achieved by metaheuristic

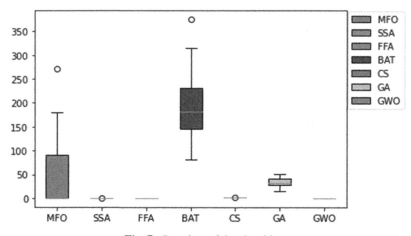

Fig. 7. Boxplots of the algorithms

techniques are tabulated in Table 4. From the table, GWO, MFO, SSA, FFA, BAT and CS find the optimum solution which is 0.397887. MFO, SSA, FFA, BAT, CS algorithms can find the lowest value in terms of worst and average. The worst performance belongs to GA, in the Branin Rcos benchmark. The design histories of each algorithm are displayed in Fig. 8. The convergence curves of all algorithms are generally well. Additionally, the boxplots of the algorithms are illustrated in Fig. 9. The figure shows that the charts of all algorithms except GA are thin and narrow.

Table 4. Optimal results for function 4

	MFO	SSA	FFA	BAT	CS	GA	GWO
1	0.397887	0.397887	0.397887	0.397887	0.397887	0.400016	0.397888
2	0.397887	0.397887	0.397887	0.397887	0.397887	0.398366	0.397889
3	0.397887	0.397887	0.397887	0.397887	0.397887	0.405318	0.397894
4	0.397887	0.397887	0.397887	0.397887	0.397887	0.400932	0.397888
5	0.397887	0.397887	0.397887	0.397887	0.397887	0.398053	0.397889
6	0.397887	0.397887	0.397887	0.397887	0.397887	0.398039	0.397888
7	0.397887	0.397887	0.397887	0.397887	0.397887	0.398425	0.397887
8	0.397887	0.397887	0.397887	0.397887	0.397887	0.403978	0.397887
9	0.397887	0.397887	0.397887	0.397887	0.397887	0.398718	0.397892
10	0.397887	0.397887	0.397887	0.397887	0.397887	0.397928	0.397889
11	0.397887	0.397887	0.397887	0.397887	0.397887	0.404663	0.39789
12	0.397887	0.397887	0.397887	0.397887	0.397887	0.399632	0.397889
13	0.397887	0.397887	0.397887	0.397887	0.397887	0.397888	0.397889

(*continued*)

Table 4. (*continued*)

	MFO	SSA	FFA	BAT	CS	GA	GWO
14	0.397887	0.397887	0.397887	0.397887	0.397887	0.401304	0.397887
15	0.397887	0.397887	0.397887	0.397887	0.397887	0.401259	0.397888
16	0.397887	0.397887	0.397887	0.397887	0.397887	0.406809	0.397888
17	0.397887	0.397887	0.397887	0.397887	0.397887	0.398432	0.397889
18	0.397887	0.397887	0.397887	0.397887	0.397888	0.398136	0.397889
19	0.397887	0.397887	0.397887	0.397887	0.397887	0.400563	0.397889
20	0.397887	0.397887	0.397887	0.397887	0.397887	0.397992	0.397888
21	0.397887	0.397887	0.397887	0.397887	0.397887	0.398088	0.397888
22	0.397887	0.397887	0.397887	0.397887	0.397887	0.398272	0.397889
23	0.397887	0.397887	0.397887	0.397887	0.397887	0.398285	0.397889
24	0.397887	0.397887	0.397887	0.397887	0.397887	0.397963	0.397888
25	0.397887	0.397887	0.397887	0.397887	0.397887	0.398155	0.397888
26	0.397887	0.397887	0.397887	0.397887	0.397887	0.398006	0.397888
27	0.397887	0.397887	0.397887	0.397887	0.397887	0.398453	0.39789
28	0.397887	0.397887	0.397887	0.397887	0.397887	0.398606	0.397891
29	0.397887	0.397887	0.397887	0.397887	0.397887	0.398563	0.397888
30	0.397887	0.397887	0.397887	0.397887	0.397887	0.397937	0.397888
Best	**0.397887**	**0.397887**	**0.397887**	**0.397887**	**0.397887**	0.397974	**0.397887**
Worst	0.397887	0.397887	0.397887	0.397887	0.397887	0.403717	0.4036
Average	0.397887	0.397887	0.397887	0.397887	0.397887	0.399522	0.398169

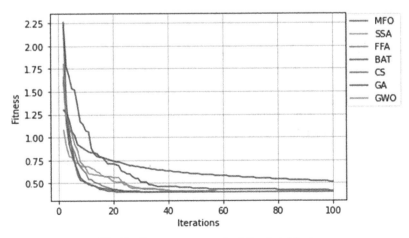

Fig. 8. Convergence graph of the algorithms

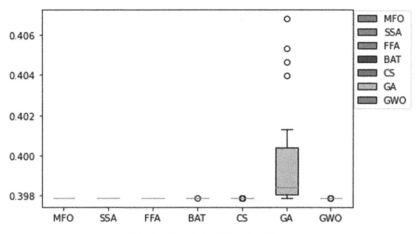

Fig. 9. Boxplots of the algorithms

3.5 Function 5 (Goldstein-Price Function)

The fifth function is non-continuous, non-convex, multimodal, differentiable and non-separable. Dimension of the function is 2. This minimization function has one global minimum f(x*) = 3 at x* = (0, −1), is implemented to examine the performance of the GWO. Function mathematical definition as follows;

$$\min f(x) = \left[1 + (x_1 + x_2 + 1)^2 (19 - 14x_1 + 3x_1^2 - 14x_2 + 6x_1 x_2 + 3x_2^2)\right] x [30 + (2x_1 - 3x_2)^2 x (18 - 32x_1 + 12x_1^2 + 48x_2 - 36x_1 x_2 + 27x_2^2)]$$

(13)

and side constraints;

$$-2 \leq x_1, x_2 \leq 2$$

(14)

This function is solved by using seven different algorithm methods (MFO, SSA, FFA, BAT, CS, GA, GWO). The optimum designs of the function achieved by metaheuristic techniques are tabulated in Table 5. From the table, GWO, MFO, SSA, FFA, BAT and CS can find the optimum solution which is 3. MFO, SSA, FFA, and CS algorithms achieve the lowest result in terms of average and worst values. The BAT algorithm demonstrates the worst performance in the Goldstein-price benchmark. The design histories of each algorithm are displayed in Fig. 10. It can be observed from the figure that except the Bat algorithm, all algorithms have good curves. Furthermore, the boxplots of the algorithms are illustrated in Fig. 11. The figure shows that the charts of all algorithms except BAT are low and narrow.

Table 5. Optimal results for function 5

	MFO	SSA	FFA	BAT	CS	GA	GWO
1	3	3	3	3	3	3.024224	3
2	3	3	3	3	3	3.00504	3.000169
3	3	3	3	3	3	3.007349	3.000046
4	3	3	3	3	3	3.080848	3
5	3	3	3	3	3	3.000156	3.000063
6	3	3	3	3	3	3.001285	3.00003
7	3	3	3	30	3	3.053034	3.000007
8	3	3	3	30	3	3.039473	3.000063
9	3	3	3	3	3	3.045164	3.000016
10	3	3	3	30	3	3.002862	3.000004
11	3	3	3	3	3	3.002542	3.000047
12	3	3	3	3	3	3.002678	3
13	3	3	3	3	3	3.000562	3.000165
14	3	3	3	3	3	3.00069	3.000004
15	3	3	3	3	3	3.030666	3
16	3	3	3	3	3	3.001132	3.000008
17	3	3	3	3	3	3.004862	3.000051
18	3	3	3	3	3	3.00287	3.000016
19	3	3	3	3	3	3.000391	3.000008
20	3	3	3	3	3	3.002428	3.000026
21	3	3	3	3	3	3.021347	3.0001
22	3	3	3	3	3	3.002193	3.000105
23	3	3	3	3	3	3.000413	3.000187
24	3	3	3	3	3	3.001062	3.000027
25	3	3	3	3	3	3.002948	3.000009
26	3	3	3	3	3	3.025873	3.000001
27	3	3	3	3	3	3.001005	3.00011
28	3	3	3	3	3	3.003237	3.000002
29	3	3	3	3	3	3.001177	3.000018
30	3	3	3	3	3	3.036884	3.000029
Best	**3**	**3**	**3**	**3**	**3**	3.000319	**3**
Worst	3	3	3	84	3	3.058533	3.000106
Average	3	3	3	9.3	3	3.015875	3.00003

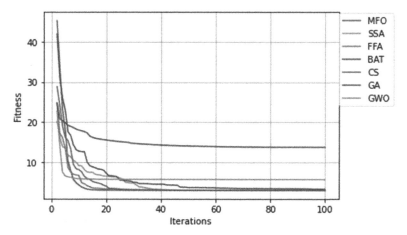

Fig. 10. Convergence graph of the algorithms

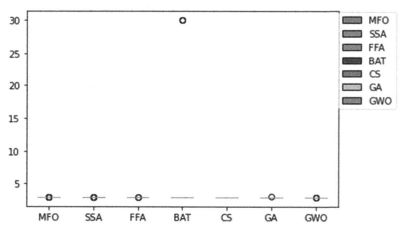

Fig. 11. Boxplots of the algorithms

4   Conclusions

In this study, seven optimization techniques (PSO, ACO, ABC, BO, FFO, GSO, WCO) are examined, and these techniques are used to find the optimum result of five mathematical functions. In the first, second and third benchmark functions, GWO found the lowest values in all the best, worst, and average results. Moreover, in the first and third benchmarks, SSA is the second algorithm in terms of performance ranking, while the BAT algorithm has the worst performance. In the second benchmark, while MFO finds the lowest value in terms of best result, SSA finds the lowest value in terms of worst and average results. As in the first benchmark, the worst performance in this benchmark belongs to the BAT algorithm, too. In the fourth benchmark, all algorithms except GA found the lowest best result. In addition, GA shows the worst performance in this algorithm. In the fifth benchmark, as in the fourth benchmark, all algorithms except GA find the best result. In addition, the algorithm that shows the worst performance in this benchmark in terms of worst and average results is the BAT algorithm. It is observed that the swarm intelligence-based optimization algorithms used are successful in finding the optimum result, and the best algorithm changed according to the problem type. However, considering most of the problems, it can be said that GWO is more powerful than others. In addition, it should be noted that algorithm performance is dependent on the selection of appropriate values for algorithm parameters as a general feature of metaheuristic techniques.

References

1. Panigrahi, B.K., Shi, Y., Lim, M.H.: Handbook of Swarm Intelligence. In: Adaptation, Learning, and Optimization, p. 542 (2011)
2. Beyer, H.-G., Schwefel, H.-P.: Evolution strategies – a comprehensive introduction. Nat. Comput. **1**(1), 3–52 (2002)
3. Blum, C., Li, X.: Swarm Intelligence in Optimization. Springer, Cham (2008). https://doi.org/10.1007/978-3-540-74089-6_2
4. Çarbaş, S., Doğan, E., Erdal, F., Saka, M.P.: Comparison of metaheuristic search techniques in finding solution of optimization problems. In: Second International Symposium on Computing in Science and Engineering, pp. 712–719 (2011)
5. Erdal, F., Doan, E., Saka, M.P.: Optimum design of cellular beams using harmony search and particle swarm optimizers. J. Constr. Steel Res. **67**(2), 237–247 (2011)
6. Faris, H., Aljarah, I., Mirjalili, S., Castillo, P.A., Merelo, J.J.: EvoloPy: an open-source nature-inspired optimization framework in python. In: IJCCI 2016 - Proceedings 8th International Jt. Conference Computer Intelligence, vol. 1, pp. 171–177 (2016)
7. Fogel, L.J., Owens, A.J., Walsh, M.J.: Artificial Intelligence through Simulated Evolution. John Wiley, NY (1966)
8. Holland, J.H.: Genetic algorithms. Sci. Am. **267**(1), 66–73 (1992)
9. Kanwal, S., Hussain, A., Huang, K.: Novel artificial immune networks-based optimization of shallow machine learning (ML) classifiers. Expert Syst. Appl. **165**, 113834 (2021)
10. Khalilpourazari, S., Khalilpourazary, S.: An efficient hybrid algorithm based on Water Cycle and Moth-Flame Optimization algorithms for solving numerical and constrained engineering optimization problems (2019)
11. Khalilpourazari, S., Khalilpourazary, S.: Optimization of production time in the multi-pass milling process via a Robust Grey Wolf Optimizer. Neural Comput. Appl. **29**(12), 1321–1336 (2016). https://doi.org/10.1007/s00521-016-2644-6
12. Khalilpourazari, S., Naderi, B., Khalilpourazary, S.: Multi-Objective Stochastic Fractal Search: a powerful algorithm for solving complex multi-objective optimization problems. Soft. Comput. **24**(4), 3037–3066 (2019). https://doi.org/10.1007/s00500-019-04080-6
13. Khurma, R.A., Aljarah, I., Sharieh, A., Mirjalili, S.: EvoloPy-FS: an Open-source nature-inspired optimization framework in Python for feature selection (2020)
14. Kouka, N., Fdhila, R., Hussain, A., Alimi, A.M.: Dynamic multi objective particle swarm optimization with cooperative agents. In: 2020 IEEE Congress on Evolutionary Computation (CEC), pp. 1–8 (2020)
15. Liu, C., Yan, X., Liu, C., Wu, H.: The wolf colony algorithm and its application. Chinese J. Electron. **20**(2), 212–216 (2011)
16. Mirjalili, S.: Moth-flame optimization algorithm: a novel nature-inspired heuristic paradigm. Knowledge-Based Syst. **89**, 228–249 (2015)

17. Mirjalili, S., Gandomi, A.H., Mirjalili, S.Z., Saremi, S., Faris, H., Mirjalili, S.M.: Salp Swarm Algorithm: a bio-inspired optimizer for engineering design problems. Adv. Eng. Softw. **114**, 163–191 (2017)
18. Mirjalili, S., Mirjalili, S.M., Lewis, A.: Grey wolf optimizer. Adv. Eng. Softw. **69**, 46–61 (2014)
19. Qaddoura, R., Faris, H., Aljarah, I., Castillo, P.A.: EvoCluster: an open-source nature-inspired optimization clustering framework in Python. In: Castillo, P.A., Jiménez Laredo, J.L., Fernández de Vega, F. (eds.) EvoApplications 2020. LNCS, vol. 12104, pp. 20–36. Springer, Cham (2020). https://doi.org/10.1007/978-3-030-43722-0_2
20. Bäck, T.: Evolutionary Algorithms in Theory and Practice: Evolution Strategies, Evolutionary Programming, Genetic Algorithms. Oxford University Press, Oxford (1996)
21. Yang, X.-S.: Firefly algorithms for multimodal optimization. In: Watanabe, O., Zeugmann, T. (eds.) SAGA 2009. LNCS, vol. 5792, pp. 169–178. Springer, Heidelberg (2009). https://doi.org/10.1007/978-3-642-04944-6_14
22. Yang, X.S.: A new metaheuristic bat-inspired algorithm. In: Nature Inspired Cooperative Strategies for Optimization, pp. 65–74. Springer, Cham (2010). https://doi.org/10.1007/978-3-642-12538-6_6

Supplier Selection Through a Hybrid MCDM-QFD Method: A Case Study in Mapna Group

Seyed Hassan Tayyar[1](✉) [iD] and Roya Soltani[2] [iD]

[1] Technology Expert, University of Tehran, Tehran, Iran
tayyar@ut.ac.ir
[2] Department of Industrial Engineering, KHATAM University, Tehran, Iran
r.soltani@khatam.ac.ir

Abstract. Sourcing and procurement are two strategic aspects of any organization whereby they can reduce production and material costs and increase the quality all of which lead to another to increase the organization's profit. Supplier selection is an important task in supply chain management, which helps organizations to achieve this goal. In this paper, a new customer-oriented and qualitative approach is presented to investigate the supplier relationship management in supply chai ns and extract suitable technical criteria for evaluating organization's supply chain requirements. For this purpose, a systematic method including a total quality management (TQM) tool (i.e. Quality Function Deployment (QFD)) and a multi-criteria decision making (MCDM) technique (i.e. Analytic Hierarchy Process (AHP)) is employed to respectively specify technical criteria and the corresponding weights. Then, the suppliers are ranked by employing the simple additive weighting (SAW) method. The proposed method is implemented to a case study taken from Mapna group located in Iran. The results show that, the proposed method decreases the cost of inappropriate selection of suppliers and improve the supplier selection process by choosing suitable evaluation criteria.

Keywords: Supply chain management · Strategic sourcing · Supplier evaluation · Multi-criteria decision making (MCDM) · Quality Function Deployment (QFD)

1 Introduction

Supply chain can be considered as a series of companies which are legally separate but cooperate with each other as a unit and make use of common profits. There are numerous definitions for supply chain in the literature. Herein, some of the most important and common definitions are presented. Some definitions of a supply chain are as follows.

- All activities related to commodity flow and conversion of raw materials to the final product (for consumption), and the corresponding information flows [1].

© Springer Nature Switzerland AG 2021
Z. Molamohamadi et al. (Eds.): LSCM 2020, CCIS 1458, pp. 117–136, 2021.
https://doi.org/10.1007/978-3-030-89743-7_8

- All activities related to commodity flow from the raw material stage to the final customer including systems management, manufacturing, assembly, purchase and sourcing, production time-schedule, order processing, inventory management, storage and customer services [2].
- Supply chain includes any efforts related to supply, production and delivery of a final product from supplier of the supplier to the customer of customer. Four basic processes including programming, sourcing, production and delivery widely describe these efforts which in turn include supply and demand management, raw materials and spare parts sourcing, production and assembly, storage and inventory control, order management, distribution into channels and delivery to the customer [3].
- A supply chain refers to material flow, information flow, payments and services from raw material supplier to the final customers passing through factories and warehouses. The supply chain also includes organizations and processes that produce products, information and services and deliver them to the final customers. The processes include different tasks such as purchase, payment, management of material flow, production planning and control, supplies and inventory control, storage and distribution, delivery and required information systems to control all these tasks [4].

There are also some definitions in the literature for supply chain management, which are:

- Managing all various tasks and procedures which create values for customers [5].
- Coordinating all tasks so that customers can receive high quality, rapid and assured services at the minimum time. Successful coordination, in turn, provides competitive advantages for the company.
- Integrated supply chain management, is an integrated process- oriented approach for supply, production and delivery of goods and services to the customers.
- Increasing economic value through managing coordinated material flow along with information flow from the source to the consumption point.
- Integrating all tasks of the supply chain and the corresponding information flow related to products conversion, flow from raw material stage and extraction to the final status for consumption, and also information related to them through relationship improvement throughout the chain to achieve competitive, reliable and sustainable success [1].

The definition of the supply chain from one view has been taken from how participant organizations are connected to each other. Figure 1 shows a relatively simple supply chain where a company is connected to its suppliers (on the left) and its distributors and customers (on the right). Please note that suppliers might have their own suppliers. In Addition to the material flow, there is also information and financial flow. The flow of money moves against the material flow [4].

A huge amount of resources of organizations is used in supply and procurement processes. Therefore, special attention to the upstream part of the supply chain has high influence on improving the quality of suppliers management and consequently on reduction of the organizational costs. Evaluation and selection of suitable suppliers is the main solution for this challenge.

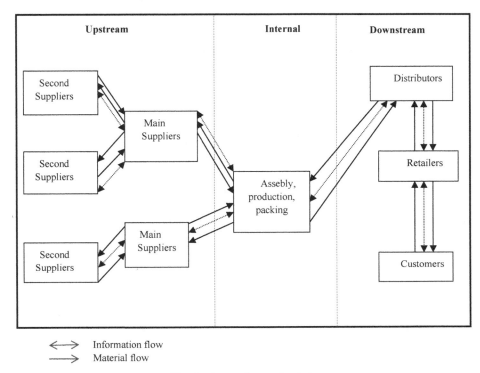

Fig. 1. A sample supply chain

Supplier based optimization is a continuous process in which future and present suppliers who cannot satisfy customer needs are replaced with appropriate number of suppliers. Optimization steps are as follows [6]:

- Eliminate those suppliers who have little trades with manufacturers so that the number of direct suppliers will be primarily reduced.
- Provide an appropriate performance measurement system to recognize suitable suppliers.
- Develop stronger commercial relationships with selected suppliers.

According to Fig. 1, the supply chain can be divided into three main parts:

- **Upstream chain**: includes suppliers, contractors and their primary suppliers. The main part of the upstream supply chain, includes sources and suppliers which need to be suitably evaluated and selected. Therefore, in this part supplier relationship management (SRM) is considered. Furthermore, the rate of interactions and exchanges in this part are of high importance. The systematic attitude towards the supply chain which streams from top to down shows the main role of the upstream part of the supply chain.

- **Internal chain**: includes all processes used by an organization that transform all inputs in the form of materials and data provided from suppliers into desired output.
- **Downstream chain**: includes all processes related to distribution and delivery of products and services to the final customers. In this part, customer relationship management (CRM) is considered.

Apart from CRM, considered in the downstream part of the supply chain, SRM, considered in the upstream part of the supply chain, plays an important role in improving the supply chain performance. SRM is a managerial concept, which has a close relationship with objectives of the organization and also with the approaches of managers. Through SRM organizations can control and lead the customers, market, suppliers and contractors to better relations and performance. In this paper, the upstream part of the supply chain is considered and the goal is to improve the performance of the supply chain with respect to effective factors in sourcing and selecting suitable suppliers. Therefore, we aim at presenting a systematic model and a decision support system (DSS) for evaluation and selection of suppliers in the upstream area. The proposed model includes well-known multi criteria decision making methods such as Analytic Hierarchy Process (AHP) and Simple Additive Weighting method (SAW) and also a Total Quality Management (TQM) tool that is Quality Function Deployment (QFD).

The rest of this paper is organized as follows. In Sect. 2, literature related to supplier selection is reviewed. In Sect. 3 materials and methods are presented. In Sect. 4 the proposed model is implemented to a case study taken from Mapna group and a sensitivity analysis is carried out. Finally, in Sect. 5 conclusions are provided along with some future research directions.

2   Literature Review

In today's competitive market, suitable supplier selection which meets manufacturer requirements is of interest.

According to the history, most of the existing methods for evaluation and selection of suppliers don't consider business objectives and requirements of stakeholders. Ignoring the 'voice' of company's stakeholders leads to inappropriate supplier selection. Supplier selection problem includes several factors, models, and decision making methods considering multiple criteria and alternatives [7]. Review of the literature (See Table 1) shows that MCDM methods along with QFD have been used for supplier selection in different areas. In this research, we make use of the hybrid MCDM and QFD method to select suitable suppliers for an industry in the field of energy and power plant. Through QFD the importance weights of the quality features (HOWs) are calculated, which themselves are derived by the importance ratings of customer requirements (WHATs) together with the relationship between customer requirements and quality characteristics. Apart from the basic applications of QFD, namely product development, quality management, and customer-needs analysis, QFD has been employed to other fields such as strategy development, planning, design, engineering, management, and so forth [8, 9]. Ho et al. [10] suggest QFD as a useful technique for supplier evaluation and selection with respect to the customer requirements. However, very few articles have used QFD method for supplier evaluation and selection.

Table 1. Related literature to supplier selection

Author(s)	Year	MCDM/OR method	TQM tool	Base selection criteria (requirements)	Case study
Bhattacharya et al. [11]	2005	AHP	QFD	Payload, Accuracy, Life-expectancy, Velocity of robot, Programming flexibility, Total cost	Robotic system in a manufacturing firm
Bevilacqua et al. [12]	2006	–	Fuzzy-QFD	Conformity, Cost, Punctuality, Efficacy, Programming, Availability	Clutch plate suppliers
Bhattacharya et al. [13]	2010	AHP with cost factor measure (CFM)	QFD	Delivery, Quality, Responsiveness, Management, Discipline, Financial position, Facility, Technical capabilities	Gibraltar Federation of Small Businesses (GFSB)
Dai and Blackhurst [14]	2012	AHP	QFD	Energy and climate, Material efficiency, Nature and resources, People and community	Large retail company
Dey et al. [15]	2012	DEMATEL	QFD	Reliability, IT Technology, JIT system, Social responsibilities	Laboratory
Bevilacqua et al. [16]	2012	–	Fuzzy-QFD	Taste, Smell, Appearance, Density	Food products
Dursun and karsak [17]	2013	Fuzzy weighted average method (FWA)	QFD	Conformity, Cost, Punctuality, Efficacy, Programming, Availability	Clutch coupling Company

(continued)

Table 1. (*continued*)

Author(s)	Year	MCDM/OR method	TQM tool	Base selection criteria (requirements)	Case study
Servert et al. [18]	2014	-	QFD	Technology, Social, Risk, Resource, Market, Economy, Environment	Solar energy applications for the medium and large size mining industry
Karsak and Dursun [19]	2014	DEA- FWA	QFD	Cost, Quality, Product Conformity, Availability, Customer support, Efficacy of corrective action	Hospital
Quoc Dat et al. [20]	2015	Fuzzy TOPSIS	QFD	Segment growth rate, Expected profit, Competitive intensity, Capital required, Level of technology utilization	Market segment selection for business company
Dat et al. [21]	2015	Fuzzy TOPSIS	Fuzzy-QFD	Conformity, Cost, Punctuality, Efficacy, Programming, Availability	Trading service and transportation company
Pramanik et al. [22]	2016	AHP-TOPSIS	QFD	Quality, Delivery time, Reliability, Processing time, Profit margin	Computer manufacturer

(*continued*)

Table 1. (*continued*)

Author(s)	Year	MCDM/OR method	TQM tool	Base selection criteria (requirements)	Case study
Yazdani et al [23]	2016	SWARA-WASPAS	QFD	Energy Consumption, Green Design, Reuse, Recycle rates, Quality Adoption, Price, Delivery Speed, Production planning	Manufacturing system
Yazdani et al. [24]	2016	DEMATEL	QFD	Green Design, Reuse, Recycle rate, Product and Process Quality, Energy and Natural resource consumption	Dairy company
Tavana et al. [25]	2016	ANP - MODM	QFD	Financial stability, Quality, Delivery time, Pollution control, Environmental management system, Reverse logistics, Energy management, Corporate social responsibility, Employment Practices	Dairy company-Kalleh
Lima et al. [26]	2016	MCDM	Fuzzy-QFD	Competitive price, On time delivery, Product conformance, Quality certification	First tier manufacturer in the automobile supply chain
Babbar and Hassanzadeh Amin [27]	2017	MODM	Fuzzy-QFD	Price, Quality, Service, Product & Packaging, Green Criteria	Beverages industry

(*continued*)

Table 1. (*continued*)

Author(s)	Year	MCDM/OR method	TQM tool	Base selection criteria (requirements)	Case study
Rajabi Asadabadi [28]	2017	ANP, Markov Chain	QFD	performance, Reliability, Price, Serviceability, Noise, Cost of maintenance	Manufacturer of water based air coolers-Absal
Bottani et al. [29]	2018	ANP	QFD	Delivery, Product quality, Responsiveness, Management Capabilities, Technical Capabilities, Facility, Supplier risk, Financial Stability	Italian company-food machinery industry
Abdel-Basset et al. [30]	2018	Neutrosophic set-AHP	QFD	Years of experience, Certification system of quality, Geographical location, Constancy of financial status, Raw materials safety, Organizational behavior, Discounts	Pharmaceutical company
Akkawuttiwanich and Yenradee [31]	2018	-	Fuzzy-QFD	Reliability, Responsiveness, Agility, Costs, Assets	Bottled water manufacturing
Devnath et al. [32]	2020	TOPSIS	QFD	Inventory, Over production, Transportation, Waiting, Motion, Defects, Over processing	Manufacturing industries

(*continued*)

Table 1. (*continued*)

Author(s)	Year	MCDM/OR method	TQM tool	Base selection criteria (requirements)	Case study
Current study		AHP- SAW	QFD	Quality and Standards, Programming and Time- scheduling, Sales and Support Services, Performance and Efficiency, Conditions and Facilities	Mapna Group

In this paper, a hybrid method is proposed to evaluate the suppliers of Meco company which is active in the field of energy and powerplant. The proposed method is made up of a combination of three tools from total management and decision making areas, i.e. Quality Function Deployment (QFD), Analytic Hierarchy Process (AHP), and Simple Additive Weighing (SAW) method.

3 Materials and Methods

In this paper, to evaluate and rank suppliers a qualitative-quantitative method is proposed in which the relationships amongst requirements expected from suppliers are specified through QFD. The relative weights of requirements are calculated by AHP method. At the end, to rank the suppliers the SAW method is employed to calculate final scores.

3.1 Criteria Considered for the Suppliers' Evaluation

In the proposed case study of this paper, some criteria are considered for evaluation of suppliers based on the following researches:

- Sasa et al. [33]: quality, accountability, delivery, orders, financial status, technical capability, facilities and supplier management.
- Choy et al. [34]: quality organizational culture, product development method, services provided to the customer, transportation quality.
- Buffa and Jackson [35]: supplier's experience in providing services.

In other researches price, delivery time, cost, flexibility and customer satisfaction were considered as well.

It is worth noting that various criteria can be considered for the purpose of evaluation considering the organizational requirements, information and the corresponding process flow. However, choosing the right criteria depends on sourcing strategies and long term programs of organizations. Therefore, it is not reasonable to consider limited numbers of criteria for the evaluation of the companies' suppliers.

3.2 QFD Process for Identifying Supplier Criteria

QFD is a tool for interpreting customers' needs and requirements expected from product/services, to design characteristics. QFD was initially developed by Mitsubishi Company in 1972. It is also employed by Akao (1972) to design ship tanks for Kobe ship Construction Company [36]. QFD provides a great insight into identifying customers' needs and requirements and the way of satisfying these requirements. In this paper, QFD is employed to evaluate the suppliers of a company. Customer requirements are considered as suppliers' measureable standards in the QFD matrix. By customer we mean the supply unit of the company which has some requirements based on which suppliers are evaluated and selected. Through QFD, the features needed for suppliers evaluation (i.e. "WHATs") are specified. Then, by conducting polls amongst the experts of the supply unit who deal with suppliers, these requirements are converted into measurable engineering criteria and indices (i.e. "HOWs").

The QFD process, proposed in this paper to evaluate and select suitable suppliers, is illustrated in Fig. 2, the steps of which are explained bellow:

Step 1: Extract a list of requirements expected from suppliers by the company's supply unit (i.e. organization's customer) (WHATs).
Step 2: Extract a list of measurable criteria through which the specified requirements are evaluated (HOWs).
Step 3: Calculate the relative weight or importance of customers' requirements through AHP pairwise comparisons (right hand side column in the QFD table).
Step 4: Form a central relationship matrix which shows the relation between requirements (in the row) and suppliers' evaluation criteria (in the column) and their interactions. This matrix and the total scores show the magnitude of correlation of the relationship between a certain evaluation criterion and a specific requirement, if this relationship exists.
Step 5: Calculate the final significance rate and weight of suppliers' evaluation criteria (row at the end of the QFD table).

3.3 HP Method Proposed for Estimation of Requirements Weights

Analytic Hierarchy Process (AHP) is a popular decision making method which calculates the weights of criteria, sub-criteria and alternatives in a hierarchy manner. It is based on pairwise comparisons valued by experts' opinions which may be consistent or inconsistent. In case of inconsistency, when the inconsistency ratio is less than 0.1 the comparison matrix is acceptable, otherwise it must be revised. The final ranks of alternatives are obtained by multiplying the weights of alternatives, regarding criteria/sub-criteria, by weights of criteria/sub-criteria. In this paper, AHP is employed through which first the preferences of customers' requirements comparing with each other are specified by an expert and then relative weights of requirements are calculated. In these comparisons,

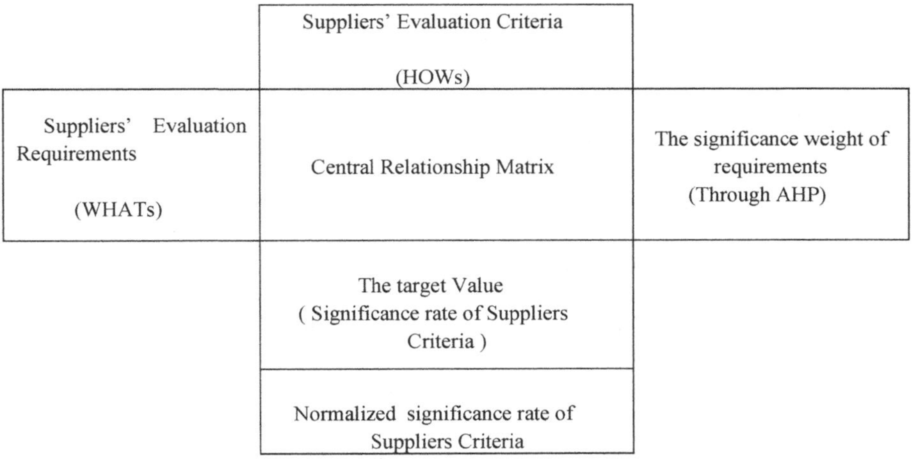

Suppliers' Evaluation Requirements (WHATs)	Suppliers' Evaluation Criteria (HOWs)	
	Central Relationship Matrix	The significance weight of requirements (Through AHP)
	The target Value (Significance rate of Suppliers Criteria)	
	Normalized significance rate of Suppliers Criteria	

Fig. 2. QFD Chart proposed for supplier evaluation

the experts may use verbal judgments (shown in Table 2) which are quantified by the corresponding numbers from 1 to 9.

Table 2. Preference values used in AHP pairwise comparisons [37]

Numerical	Preference (oral judgments)
9	Extremely preferred
7	Very strongly preferred
5	Strongly preferred
3	Moderately preferred
1	Equally preferred
2, 4, 6, and 8	Preferences between odd values

The steps of the proposed AHP are summarized as follows:

Step 1: Carry out pairwise comparisons: In this step pairwise comparisons between each pair of customers' requirements (i.e. the requirements which the supply unit expects from their suppliers) are carried out using the preference values presented in Table 2. The values located on the main diameter of the comparison matrix equals to 1 which means equal preferences. The values of elements in the top and bottom of the main diameter of the pairwise matrix are receproical. For this pairwise comparison matrix (Matrix A), the weights and the significances of suppliers' requirements are calculated (Vector W).

Step 2: Calculate the normalized matrix: To calculate the normalized matrix, first the values of each column are summed up and then each column's element is divided by this total value. It is worth mentioning that the total sum of values in each column of the normalized matrix equals to unity.

Step 3: Calculate the average (mean) of the elements in each row of the normalized matrix. These values show the relative preference or weights of requirements in the QFD Matrix.

Step 4: Check the consistency of pairwise comparison matrixes: To evaluate the consistency of pairwise comparison matrixes, the following steps are implemented:

I- Multiply pairwise comparison Matrix by its weight vector (A.W).
II- Calculate the maximum eigenvalue (λ).
III- Calculate inconsistency index according to $I.I = \frac{\lambda - m}{m-1}$.
IV- Compare the inconsistency index (I.I) with the random index (R.I) which is specified according to dimension of the pairwise matrix (m).
V- Calculate the inconsistency ratio (I.R) according to $I.R = \frac{I.I}{R.I}$. If $I.R \leq 0.1$, the consistency rate is acceptable. Else if $I.R > 0.1$, the pairwise matrix is highly inconsistent. Therefore, the expert is asked to revise their comparisons in order to achieve more consistency.

3.4 The Proposed Hybrid MCDM-QFD Method for Supplier Evaluation

In the followings the steps of the proposed method are presented.

Step 1. Pre-qualify the suppliers. In this step an initial screening of suppliers is conducted based on vital criteria considered by the company. As a matter of fact, the suppliers who can't meet the vital criteria are set aside and can't enter the next steps of evaluation. Selection of criteria depends on the organization and its supplier management strategy.

Step 2. Identify requirements (WHATs): This step identifies specified features of suppliers to be selected. In fact, the requirements (WHATs) which supply and purchase unit of the company expect their suppliers to have are specified through interviews..

Step 3. Specify evaluation criteria: In this step, the criteria related to supplier evaluation are extracted using the information obtained from the supply unit and the evaluation team of the company. Evaluation criteria are derived directly from requirements and features proposed in the previous step and are supposed to have relationships with those requirements.

Step 4. Calculate the weights of requirements: In this step AHP method is used to calculate the weights of requirements in the QFD table. For this purpose, the supply unit of the company and the evaluation team apply pairwise comparisons between each two elements (requirements).

Step 5. Create the central relationship matrix and calculating the weights of technical criteria: In this step, in order to recognize the strength of the relationships amongst criteria and requirements, a central relationship matrix is created. For this purpose, the information available in the supply unit of the company is used. In the QFD matrix, three numbers 1, 3 and 9 are used, where 1 represents weak relationship, 3 shows average relationship and 9 is used for strong relationship. These relationships and the corresponding numbers show to what extent each criterion is related to the proposed requirements. Of course, each criterion can have relationship with more than one requirement and vice-versa. If there is no relationship between any of the cases, zero number is assigned.

Step 6: Calculate the significance of each supplier's criterion: The significance of each supplier's criterion is obtained from total sum of each requirement's significant rate multiplied by the strength of its relationship with the related criterion in the central matrix of the QFD table. In fact, if the supplier's evaluation criterion has a relationship with the proposed requirements, the significance of each supplier's criterion is calculated according to Eq. (1).

$$W_j = \sum_{i-1}^{m} R_{ij}.C_i \tag{1}$$

W_j: The significance of the $j th$ supplier's criterion.

R_{ij}: The value of the relationship between the ith requirement and jth supplier evaluation criterion in the central relationship matrix.

C_i: The significance or the weight of $i th$ requirement (calculated by AHP).

Step 7: Normalize the significance of each criterion: The significance of each criterion is normalized and converted into a total weight according to Eq. (2).

$$W_j^N = \frac{W_j}{\sum_{j=1}^{n} W_j} \times 100 \tag{2}$$

Step 8: Calculate the final scores of suppliers: In this step, SAW which is an MCDM technique based on arithmetic mean or weighted average of value of each supplier regarding each criterion is employed to calculate the final scores of suppliers. Through SAW, the final weights of the evaluation criteria (obtained from the QFD Matrix) is multiplied by the score of each criterion for each supplier to calculate the final scores and rank the suppliers so that those suppliers with higher scores are put into the desired list of the company. The supplier with highest score (B^*) is specified by Eq. (3) in which the value of b_{ij} is the score of criterion j for supplier i, which can be obtained through methods such as visiting suppliers' site, the assessment of the documents and other qualitative or quantitative methods.

$$B^* = \left\{ B_i \middle| \max_i \sum W_i \cdot b_{ij} \right\} \tag{3}$$

Step 9: Select the high ranked suppliers: In case there are no limitations, the suppliers with the highest scores can be selected and inserted in the final list. If there are limitations in some areas such as selection of the suppliers, production capacity, quality, price and other issues, mathematical programming models can be used.

4   Case Study Implementation

In this section, the proposed model is implemented to a case study taken from Meco Power Construction, Supervision, and Engineering Company, which is a subordinate company of Mapna group.

The first step is the pre-qualification process through which suppliers are evaluated with respect to vital criteria of Meco company. In case a supplier fails to meet these vital criteria, it will be deleted from the evaluation list and does not proceed to the next step. To prequalify the suppliers of Meco Company, the following questions are posed:

1. Does the supplier exist in the vendor list of the Company?
2. Is the supplier approved by the licensor?
3. Does the selected supplier have a valid construction and operation license?

From the suppliers' list which are mostly foreigner, eleven suppliers passed the first stage and could enter the next stages of evaluation.

In the second step, the requirements that the supply unit of the company expect from their suppliers (i.e. WHATs) are determined. For this purpose, after interviewing and conducting polls among members and experts of the supply unit the following items were extracted as WHATs:

1. Quality and Standard
2. Scheduling and planning
3. Support and post sales services
4. Performance and efficiency
5. Terms and Facilities

It is worth mentioning that the above mentioned criteria and requirements have been provided according to experts' opinions as well as the information available in the company. Therefore, if the members and experts or even the products change, these issues may also change.

The third step is the selection of the evaluation criteria. After holding meetings with the company's supply chain evaluation team and making use of experts' opinions, the following criteria for the final evaluation were extracted:

1. Existing Technical documents and Maps in accordance with standards
2. Product Support and Guarantee
3. Delivery time scheduling and planning
4. The range and diversity of Supplier's products
5. The possibility of visiting the supplier's site
6. Supplier reputation and background
7. The product Catalogue Quality
8. Providing required trainings by the supplier
9. Certificate, Standard and system quality documents
10. Flexibility and ability to respond to customer's requests
11. Innovation and capability of designing and Manufacturing
12. Conformity of quality and technical features of the product with standards
13. Conditions and the possibility to return non- confirmed commodities
14. Geographical status and accessibility to foreign supplier
15. Financial transactions method
16. Punctual delivery of foreign commodities' documents

In the fourth step, after forming the pairwise comparison matrix and carrying out the AHP process, the weights of requirements are obtained. In the fifth step, a central relationship matrix is created to recognize the strength of the relationships between criteria and requirements. Table 3 shows the proposed QFD Table. The central relationship matrix is located at the middle of this table. In the sixth step, the significance of each supplier's criterion is calculated and shown in the bottom of the QFD Table. Then in the seventh step these values are normalized and entered to the eighth step for carrying out the SAW method through which the final rank and score of each supplier is obtained by multiplying the weight of each criterion to the related score of the same criterion and calculating the total sum as weight of each supplier.

In Fig. 3, the score and the status of the suppliers are shown.

Finally in the ninth step, each supplier will get a score out of 1000. This score will determine the supplier's rank. The ranks of considered suppliers are as follows: 4 > 7 > 6 > 10 > 8 > 1 >5 > 9 > 3 > 2 > 11.

This means that among 11 suppliers, the supplier number 4 with the total score of 816.7 is located in the first rank and the rest are placed in the subsequent ranks according to their scores.

Finally, the suppliers are classified according to their scores. Regarding the range of scores, suppliers who have the score of 800 and above are in "grade A", suppliers with scores between 600 to 800 are in "grade B", suppliers with scores between 400 and 600 are classified as "Grade C" and those with less than 400 are in "Grade D". The type of classification and placing the suppliers in each category depends on the company's requirements and its functional strategies which is changeable. The company can adopt either single-source or multiple-source strategy, and select only one supplier or more according to its requirements.

4.1 Sensitivity Analysis

In this section, a sensitivity analysis on the final weights of criteria is carried out in which the rank of supplier 4 is preserved in the first position.. This means to what extent the weight of each criterion can decrease or increase such that supplier 4 would be kept in the first rank. In fact, sensitivity analysis is carried out through changing the range of each criterion and ranges for which by changing the selected criteria the rank of a supplier would stay the same, are determined. As a result, a range for sensitive criteria is obtained for which increasing and decreasing the weights won't affect the rank of the superior supplier. The results show that the sensitivity of criteria for the first and second suppliers, which have close final scores, is more than other suppliers. Accordingly, the other sensitive criteria are obtained as follows:

From 16 available criteria, 5 criteria were sensitive to changing weights. These sensitive criteria are: delivery time scheduling and planning (criterion 3), the possibility of visiting the supplier's site (criterion 5), the product Catalogue Quality (criterion 7), providing required trainings by the supplier (criterion 8), and financial transactions method (criterion 15).

Each of these criteria can respectively bear percent change of 31%, 50%, 48%, 18% and 75% more than their weight without any change in the superior supplier's rank. As an instance, Fig. 4 shows a sensitivity analysis for criterion 3 and its range has been estimated. The same analysis has been carried out for other criteria.

The results indicate that by making use of systematic tools and appropriate standard weights, the proposed method is not sensitive to the most criteria and if any sensitivity exists, it does not exceed the acceptable range.

Table 3. Final QFD table

	Criteria (HOWs) → / Requirements (WHATs) ↓	1 Existing Technical documents and Maps Synchronizing with standards	2 Product Support and Guarantee	3 Delivery time –schedule and planning	4 The range and diversity of Supplier product	5 Capability to visit the supplier site and supervision	6 Supplier reputation and experience	7 Product Catalogue Quality	8 Providing required trainings by the supplier	9 Certificate , Standard and systems quality documents	10 Flexibility and ability to respond to customer's requests	11 Innovation and designing capability and production	12 Conformity of quality and technical features of the product with standards	13 Qualifications and the ability to return non- confirmed commodities	14 Geographical status and how to access foreign supplier	15 Financial transactions method	16 Punctual delivery of foreign commodities papers	Total requirements weight (through AHP)
1	Quality and Standard	3	0	0	0	0	1	1	0	9	0	0	9	0	0	0	0	0.512
2	Programming and Time-scheduling	0	0	9	0	0	0	0	0	0	0	3	0	0	3	0	1	0.091
3	Sales and Support Services	0	9	0	0	3	1	3	9	0	9	0	0	3	0	3	3	0.191
4	Performance and Efficiency	9	1	3	9	0	9	0	3	3	1	9	1	0	0	0	0	0.152
5	Conditions and Facilities	0	0	0	0	9	1	0	0	0	0	0	0	3	9	9	3	0.055
	Standard significance rate	2.904	1.871	1.275	1.863	1.068	2.126	1.085	2.175	5.064	1.871	1.641	4.760	0.738	0.768	1.068	0.829	
	Normalized significance Rate	9.3	6	4.1	6	3.4	6.8	3.5	7	16.3	6	5.3	15.3	2.4	2.5	3.4	2.7	

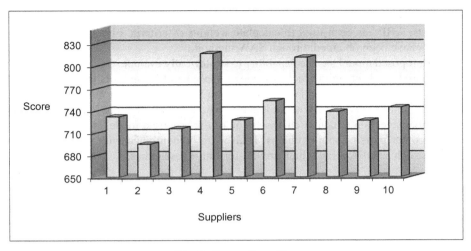

Fig. 3. Score of suppliers

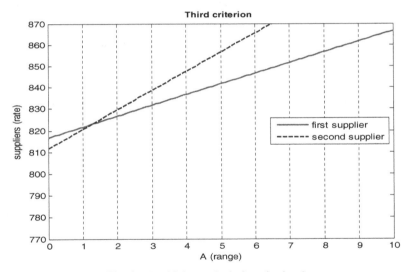

Fig. 4. Sensitivity analysis for criterion 3

5 Conclusion

In this research a hybrid MCDM-QFD method has been developed for supplier selection in Mapna group which is active in the field of energy and powerplant. For this purpose, a systematic decision making model has been proposed to evaluate and select suppliers based on experts' opinions of the company's supply unit. The QFD approach has been employed to suitably select suppliers which efficiently meet requirements of the supply unit of the studying company. In fact, through QFD technique suppliers' evaluation and selection criteria have been specified by taking thorough comments and opinions of experts. AHP has been used to calculate weights of criteria. Finally, final ranks of suppliers have been obtained through SAW method.

The proposed model has the capability to strategically evaluate the suppliers considering tangible and intangible factors through applying efficient tools taken from Total Quality Management and Multi Criteria Decision Making areas. For further research on this study, other decision- making tools can also be used and developed along with QFD for supplier selection. Another characteristic of the proposed model is its capability to evaluate other parts of supply chains. It also can be combined with other evaluation methods such as DEA all of which are left for future researches.

References

1. Handfield, R.B., Nicolas, E.L.: Introduction to Supply Chain Management. Prentice Hall, Newjersey (1999)
2. Monczka, R., Trent, R., Handfield, R.B.: Purchasing and Supply Chain Management. South Western College Publishing (1998)
3. Lummus, R.R., Krumwiede, D.W., Vokurka, R.J.: The relationship of logistics to supply chain management: developing a common industry definition. Ind. Manage. Data Syst. 101(8), 426–432 (2001)
4. Taylor, B.W., Russel, R.S.: Supply Chain Management and Enterprise Resourcing. Printice Hall, pp. 240–261 (2002)
5. Lummus, R.R., Vourka, R.J., Alber, K.L.: Strategic supply chain planning. Prod. Inventory Manage. J. Third Quarter 94–95 (1998)
6. Ghodsypour, S.H., O'brien C.: A decision support system for supplier selection using an integrated analytic hierarchy process and linear programming. International Journal of Production Economics, 199–212 (1998).
7. Govindan, K., Rajendran, S., Sarkis, J., Murugesan, P.: Multi criteria decision making approaches for green supplier evaluation and selection: a literature review. J. Clean. Prod. 98, 66–83 (2015)
8. Chan, L.K., Wu, M.L.: Quality function deployment: a literature review. Eur. J. Oper. Res. 143(3), 463–497 (2002)
9. Carnevalli, J.A., Miguel, C.: Review, analysis and classification of the literature on QFD Types of research, difficulties and benefits. Int. J. Prod. Econ. 114(2), 737–754 (2008)
10. Ho, W., Xu, X.W., Dey, K.: Multi-criteria decision making approaches for supplier evaluation and selection: a literature review. Eur. J. Oper. Res. 202(1), 16–24 (2010)
11. Bhattacharya, A., Sarkar, B., Mukherjee, S.K.: Integrating AHP with QFD for robot selection under requirement perspective. Int. J. Prod. Res. 43(17), 3671–3685 (2005)
12. Bevilacqua, M., Ciarapica, F.E., Giacchetta, G.: A fuzzy-QFD approach to supplier selection. J. Purchasing Supply Manage. 12, 14–27 (2006)
13. Bhattacharya, A., Geraghty, J., Young, P.: Supplier selection paradigm: an integrated hierarchical QFD methodology under multiple-criteria environment. Appl. Soft Comput. 10, 1013–1027 (2010)
14. Dai, J., Blackhurst, J.: A four-phase AHP–QFD approach for supplier assessment: a sustainability perspective. Int. J. Prod. Res. 50(19), 5474–5490 (2012)
15. Dey, S., Kumar, A., Ray, A., Pradhan, B.B.: Supplier selection: integrated theory using DEMATEL and quality function deployment methodology. Procedia Eng. 38, 3560–3565 (2012)

16. Bevilacqua, M., Ciarapica, F.E., Marchetti, B.: Development and test of a new fuzzy-QFD approach for characterizing customers rating of extra virgin olive oil. Food Qual. Prefer. **24**, 75–84 (2012)
17. Dursun, M., Karsak, E.: A QFD-based fuzzy MCDM approach for supplier selection. Appl. Math. Model. **37**, 5864–5875 (2013)
18. Serverta, J., Labandab, A., Fuentealbac, E., Cortesd, M., Pérez, R.: Quality function deployment analysis for the selection of four utility-scale solar energy projects in northern Chile. Energy Procedia **49**, 1896–1905 (2014)
19. Karsak, E., Dursun, M.: An integrated supplier selection methodology incorporating QFD and DEA with imprecise data. Expert Syst. Appl. **41**, 6995–7004 (2014)
20. Quoc Dat, L., Thi Phuong, T., Kao, H., Chou, S.H., Nghia, Ph.: A new integrated fuzzy QFD approach for market segments evaluation and selection. Appl. Math. Model. **39**, 3653–3665 (2015)
21. Dat, L.Q., Phuong, T.T., Kao, H.P., Chou, S.Y., Nghia, P.V.: A new integrated fuzzy QFD approach for market segments evaluation and selection. Appl. Math. Model. **39**, 3653–3665 (2015)
22. Pramanik, D., Haldar, A., Chandra Mondal, S., Kumar Naskar, S., Ray, A.: Resilient supplier selection using AHP-TOPSIS-QFD under a fuzzy environment. Int. J. Manage. Sci. Eng. Manage. (2016). https://doi.org/10.1080/17509653.2015.1101719
23. Yazdani, M., Hashemkhani Zolfani, S., Kazimieras Zavadskas, E.: New integration of MCDM methods and QFD in the selection of green suppliers. J. Bus. Econ. Manage. **17**(6), 1097–1113 (2016)
24. Yazdani, M., Chatterjee, P., Zavadskas, E.K., Hashemkhani Zolfani, S.: Integrated QFD-MCDM framework for green supplier selection. J. Clean. Prod. **142**(4), 3728–3740 (2016)
25. Tavana, M., Yazdani, M., Caprio, C.: An application of an integrated ANP–QFD framework for sustainable supplier selection. Int. J. Logist. Res. Appl. **20**(3), 254–275 (2016)
26. Lima Jr., F.R., Carpinetti, L.C.R.: A multicriteria approach based on fuzzy QFD for choosing criteria for supplier selection. Comput. Ind. Eng. **101**, 269–285 (2016)
27. Babbar, C., Hassanzadeh Amin, S.: A multi-objective mathemati-cal model integrating environmental concerns for supplier selection and order allocation based on Fuzzy QFD in beverages industry. Expert Syst. Appl. **92**, 27–38 (2017)
28. Mehdi Rajabi Asadabadi, M.: A customer based supplier selection process that combines quality function deployment, the analytic network process and a Markov chain. Eur. J. Oper. Res. **263**(3), 1049–1062 (2017)
29. Bottani, E., Centobelli, P., Murino, T., Shekarian, E.: A QFD-ANP method for supplier selection with benefits, opportunities, costs and risks considerations. Int. J. Inf. Technol. Decis. Making **17**(3), 911–939 (2018)
30. Abdel-Basset, M., Manogaran, G., Mohameda, M., Chilamkurti, N.: Three-way decisions based on neutrosophic sets and AHP-QFD framework for supplier selection problem. Future Gener. Comput. Syst. **89**, 19–30 (2018)
31. Akkawuttiwanich, P., Yenradee, P.: Fuzzy QFD approach for managing SCOR performance indicators. Comput. Ind. Eng. **122**, 189–201 (2018)
32. Devnath, A., Islam, M., Rashid, S., Islam, E.: An integrated QFD-TOPSIS method for prioritization of major lean tools: a case study. Int. J. Res. Ind. Eng. **9**(1), 65–76 (2020)
33. Humphreysa, P., Melvorb, R., Chan, F.: Using case-based reasoning to evaluate supplier environmental management performance. Expert Syst. Appl. 141–153 (2003)
34. Choy, K.L., Lee, W.B., Victor, L.: Design of a case based intelligent supplier relationship management system-the integration of supplier rating system and product coding system. Expert Syst. Appl. 87–100 (2003)
35. Ghodsypour S.H., O'Brien C.:The total cost of logistics in supplier selection under conditions of multiple sourcing, multiple criteria and capacity constraint. Int. J. Prod. Econ. 15–27 (2001)

36. Moradi, M., Raissi, S.: A quality function deployment based approach in service quality analysis to improve customer satisfaction. Int. J. Appl. Oper. Res. **5**(1), 41–49 (2015)
37. Saaty, T.L.: How to make a decision: the analytic hierarchy process. Eur. J. Oper. Res. **48**, 9–26 (1990)

A Game-Theoretic Approach for Pricing Considering Sourcing, Andrecycling Decisions in a Closed-Loop Supply Chain Under Disruption

Hamed Rajabzadeh, Alireza Arshadi Khamseh$^{(\boxtimes)}$, and Mariam Ameli

Industrial Engineering Department, Faculty of Engineering, Kharazmi University, Tehran, Iran
{ar_arshadi,m.ameli}@khu.ac.ir

Abstract. Nowadays, Pricing and sourcing under supply disruption risk playing a prominent role in the efficiency and responsiveness of a supply chain. However, being environmentally responsible is a vital element in regard to reputation enhancement and passing legislative requirements. This paper investigates two competing closed-loop supply chains: 1) a retailer with an expensive but reliable domestic supplier and cheap but unreliable foreign supplier, 2) a competitor with its integrated supplier. The retailer and competitor compete based on the game theory approach to decide on prices and return policies and prices when the retailer confronts disruption risk by the foreign supplier. The reverse supply chain is addressed by asking the retailer and a competitor to collect used products for recycling purposes by suppliers. The suppliers decide on the value paid for each the used product collected by retailers. The results indicate that in the dual sourcing situation there is a direct correlation between supply from a foreign supplier, and the return rate of used products. Besides, the competitor and its integrated supplier have to use the return policy as leverage to revive the competitive market.

Keywords: Pricing and sourcing · Recycling · Closed-loop supply chain · Disruption risk · Game theory

1 Introduction

Today, supply disruptions have become a crisis for businesses around the world. Disruptions may have a significant impact on the supply chain members and can also have a negative effect through all the Supply Chain (SC) echelons. Nowadays, supplying resources has become a major strategic decision in order to reduce the disruption risk [1]. The reliability and flexibility of suppliers are essential criteria for the long-term success of retailers who compete in the final market. Supply disruptions often lead to lower returns on sales and assets, loss of competitive advantage, market share, and goodwill, which in turn affects profitability [2].

The structure of various suppliers from different geographies is commonly used in the supply chain disruption risk literature [3]. The assumption is that many companies use

© Springer Nature Switzerland AG 2021
Z. Molamohamadi et al. (Eds.): LSCM 2020, CCIS 1458, pp. 137–157, 2021.
https://doi.org/10.1007/978-3-030-89743-7_9

low-cost suppliers in developing countries. Supplying resources from a foreign supplier in a foreign country will generate cost benefits, but will lead to supply disruptions due to shipping delays, customs delays, quality problems, etc. [4]. Recently, companies have shifted the resource supply strategies to a hybrid, cheaper supplier in a developing country and a trusted supplier unit near the final market. This strategy helps companies protect themselves against the potential supply disruptions and, at the same time, control the costs. For example, the 2011 Tsunami in Japan disrupted global semiconductor production. The disaster led to a shortage of Nikon and Canon parts which increased the price of these products. This increase was due to the less or lack of supply from suppliers to the retailers [4].

In this study it is attempted to answer the following questions:

- How disruption and non-disruption prices differ in both retailer and competitor?
- What is the effect of dual sourcing on both retailer and competitor's profits?
- What is the impact of different wholesale prices on the retailer's return rate of used products?
- How the payment for each returned item by suppliers affected by their various wholesale prices?

To answer all questions mentioned above Given the importance of considering disruption risks in the Closed-Loop Supply Chain (CLSC) and the ways to deal with them, the present study investigates the pricing and sourcing issues using a game-theoretic approach in two competing CLSC consisting of a retailer with its domestic (reliable but expensive) and foreign (with a risk of disruption but cheap) suppliers as well as a competitor with its integrated supplier. In both supply chain, indirect collection effort was investigated where both retailer and competitor collect used products for recycling purposes and give them back to suppliers. The suppliers pay for each returned product.

The rest of the paper is organized as follows: In Sect. 2, a literature review is presented, a problem description, functions, and solving methods are presented in Sect. 3, In Sect. 4, a numerical and sensitivity analysis are presented. The managerial insights are provided in Sect. 5. Finally, in Sect. 6, conclusions and guidance for future researches are presented.

2 Literature Review

A reverse supply chain refers to a set of activities such as collecting, recycling, repairing, redistributing, etc. to manage products after manufacturing and final sale to the customer. In recent years, due to the importance of the reverse supply chain, the focus on this topic has been increasing.For instance, Taleizadeh et al. [5] investigated the pricing strategies in two types of CLSCs, including 1) a single-channel forward supply chain with a dual-recycling channel (SD model), and 2) a dual-channel forward supply chain with a dual-recycling channel (DD model). Karimabadi et al. [6] studied the pricing and remanufacturing problems in a fuzzy dual-channel supply chain including a retailer and a manufacturer produces a product and sells it to the end customers through a retailer or direct channel. Li et al. [7] investigated the strategic effect of return policies in a dual-channel supply chain, in which a manufacturer can sell products directly to end customers and indirectly via an independent retailer. Jian et al. [8] designed a green CLSC with one manufacturer and a retailer considering fairness in the profit-sharing contract coordination to investigate supply chain members' decision-making.

Today's supply chains are complex networks with the possibility of disruption in their communication links. Disruptions are defined as major failures in the production or distribution nodes that are members of a supply chain. Disruptions might have a considerable impact on the supply chain members and can also have a significant negative effect on the supply of raw materials until the product reaches the customer [1]. Several studies have also been conducted in this field. Huang et al. [9] developed a Stackelberg game model for a three-level food supply chain (consisting of one retailer, one vendor, and one supplier) with production disruption which aims to study the optimal pricing, inventory, and preservation decisions that maximize the individual profit. Chakraborty et al. [10] investigated pricing issues in a single supply chain considering two scenarios, which a retailer may have a backup supply source or not in dealing with disruption risk. Jabbarzadeh et al. [11] provided a stochastic robust optimization model to design a CLSC network that is flexible in dealing with disruptions. Gupta et al. [12] in a supply chain including two suppliers and one retailer studied the impact of supply capacity disruption timing on pricing by a game-theoretic approach.

The supply chain depends on its "supply base", a group of suppliers from which, the company purchases products and services. The success of supply chains in global markets is heavily dependent on choosing the best source of supply [13]. Therefore, with increasing factors causing disruptions in the supply, supplying products from several suppliers is essential to the survival of the supply chain. The studies of sourcing can be found in the literature. For example, Freeman et al. [14] developed the strategies for supplying resources to the limited capacity manufacturer considering supply and demand disruption. In this model, the manufacturer produces two types of products, including high and low-quality versions of a product being only different in one vital component. In another study, Jain et al. [15] considered a simultaneous game in which one buyer assigns his or her supply needs to two suppliers in order to deal with uncertainty. Li et al. [16] studied a loss-averse company that sources from two suppliers with random demand. One of the suppliers is cheaper but faces a risk of disruption, while the other is trusted but more expensive to buy.

3 Description of the Model

This model is presented to identify the optimal price and sourcing in a CLSC with three suppliers and two retailers considering the risk of disruption and recycling requirements. As shown in Fig. 1, the competitor supplies the product from an integrated supplier. However, a retailer can order products from a reliable domestic supplier with a higher price or from a foreign one at a lower price but with the possibility of disruption. The retailer also has the possibility to order from both domestic and foreign suppliers simultaneously.

In the following, after obtaining optimal values for the decision variables of the first level, the Stackelberg game takes place between domestic/competitor's supplier and retailer/competitor. Here, the retailer and competitor are the leaders while the domestic and competitor's suppliers are considered as the followers (2–1) and (2–2), respectively. The goal is to determine the optimal return rate of the used products and the suppliers' optimal payment for each returned product.

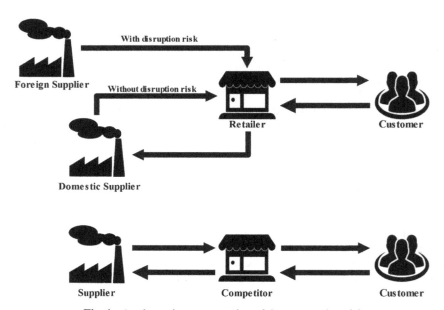

Fig. 1. A schematic representation of the proposed model

3.1 Notations

The following notations are used (Table 1):

Table 1. The notations of the proposed model

Indices	
$m1$	Foreign supplier
$m2$	Domestic supplier
mc	Competitor's supplier
$j = \begin{cases} R \\ C \end{cases}$	Retailer Competitor
$k = \begin{cases} N \\ D \end{cases}$	Non-Disruption scenario Disruption scenario

(continued)

Table 1. (*continued*)

Parameters	
c_{m1}	Production cost of raw materials for a foreign supplier
c_{m2}	Production cost of raw materials for a domestic supplier
c_{mc}	Cost of production from the raw materials for the competitor's supplier
c'_{m2}	Cost of production from the recycled materials for a domestic supplier
c'_{mc}	Cost of production from the recycled materials for the competitor's supplier
w_1	The cost of purchasing a unit product from the foreign supplier
w_2	The cost of purchasing a unit product from the domestic supplier
w_c	The cost of purchasing a unit product from the competitor's supplier
a_1	The market potential of the retailer
a_2	The market potential of the competitor
b_1	Own price elasticity of retailer
b_2	Own price elasticity of competitor
β	Scale parameter
q	The disruption probability for the foreign supplier
Decision variables	
b_j	The supplier's payment for a unit of returned used product by player j
y_j	The return rate of used products by player j $0 \leq y_j \leq 1$
$p_{j,k}$	The selling price of player j in mode k
α	The percentage of foreign supplier in dual sourcing $0 \leq \alpha \leq 1$
μ	KKT coefficient
Functions	
D_1	The retailer's demand function
D_2	The competitor's demand function
I	Rate of the return cost function
π_j	The value of profit for player j
π_{m1}	The profit value for the foreign supplier
π_{m2}	The profit value for the domestic supplier
π_{mc}	The profit value for the competitor's supplier
$\varnothing$	KKT function

At the first level, two competing retailers decide on the prices of their own products, which are substitutable. The demand for each product is a function of its price and that of the substitutable one. At the second level, the retailer/competitor collect used products and return them to the suppliers for recycling purposes. At this level, as shown in Figs. 2 and 3, the Stackelberg game is set between retailer and domestic supplier.

(2–1) and also a competitor with competitor's supplier (2–2) where the retailer/competitor is the Stackelberg leader. The retailer/competitor's decision variable is the target rate of used products return while the supplier's one is the payment for each returned product.

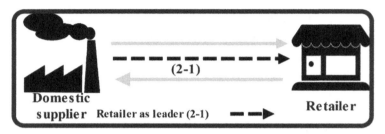

Fig. 2. Second level games part (A)

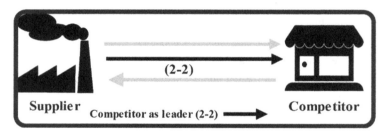

Fig. 3. Second level games part (B)

3.2 Assumptions

The following assumptions are used:

1. Information asymmetry is not allowed.
2. If $q < q^* < 1$ retailer supplies from both domestic and foreign suppliers, therefore, to reduce the complexity of the model in each mode we assumed $q = \frac{q^*+1}{2}$ to study only dual sourcing strategy.
3. The recycling procedure is done only by the domestic supplier due to the long-distance and high transportation cost of the foreign supplier.
4. The cross-price elasticity is equal to 1, and own-price elasticity is greater than cross-price elasticity $b_1, b_2 > 1$ [17].
5. The cost of production from recycled materials is less than from raw materials [18].

3.3 Demand Functions

A linear price-dependent demand structure is assumed as follows [19]:

$$D_1 = a_1 - b_1 p_1 + p_2 \tag{1}$$

$$D_2 = a_2 - b_2 p_2 + p_1 \tag{2}$$

Where p_1 and p_2 denote the prices set by the retailer and competitor, respectively.

3.4 Rate of the Return Cost Function

The cost function associated with the retailer or competitor spend on notifying and advertising activities in order to raise the return rate of the used product units is as follows [20]:

$$I = \beta (y_j)^2 \; 0 \le y_j \le 1 \tag{3}$$

Which y_j stands for the return rate of the used product from the customers and β is a scaling parameter.

3.5 The Expected Profit Functions

In this model, there are two retailers (namely retailer and competitor) and three suppliers as players, with the following value of profits:

The Profit Function of the Retailer
The retailer supplies from both the domestic and the foreign suppliers, therefore, there are two different types of prices: disruption mode price and non-disruption mode price. So, the expected value of profit for the retailer is as follows:

$$\pi_R = (1 - q)(p_{R,N} - \alpha w_1 - (1 - \alpha)w_2 + b_R y_R)(a_1 - b_1 p_{R,N} + p_{C,N})$$
$$+ q(p_{R,D} - w_2 + b_R y_R)(a_1 - b_1 p_{R,D} + p_{C,D}) - \beta y_R^2$$
$$\text{s.t:} (1 - \alpha)(a_1 - b_1 p_{R,N} + p_{C,N}) \ge (a_1 - b_1 p_{R,D} + p_{C,D}) \tag{4}$$

The Profit Function of the Competitor
The profit function of the competitor is as follows:

$$\pi_C = (1 - q)(p_{C,N} - w_C + b_C y_C)(a_2 - b_2 p_{C,N} + p_{R,N})$$
$$+ q(p_{C,D} - w_C + b_C y_C)(a_2 - b_2 p_{C,D} + p_{R,D}) - \beta y_C^2 \tag{5}$$

The Profit Function of the Foreign Supplier
The profit function of the foreign supplier is as follows:

$$\pi_{m1} = \alpha(1 - q)(w_1 - c_{m1})(a_1 - b_1 p_{R,N} + p_{C,N}) \tag{6}$$

The Profit Function of the Domestic Supplier
Here is the domestic supplier's profit function:

$$\pi_{m2} = (1 - q)(1 - \alpha)(w_2 - c_{m2}(1 - y_R) - c'_{m2}y_R - b_R y_R)(a_1 - b_1 p_{R,N} + p_{C,N})$$
$$+q(w_2 - c_{m2}(1 - y_R) - c'_{m2}y_R - b_R y_R)(a_1 - b_1 p_{R,D} + p_{C,D}) \tag{7}$$

The Profit Function of the Competitor's Supplier
Finally, the profit function of the competitor's supplier is as follows:

$$\pi_{mc} = (1 - q)(w_C - c_{mc}(1 - y_C) - c'_{mc}y_C - b_C y_C)(a_2 - b_2 p_{C,N} + p_{R,N})$$
$$+q(w_C - c_{mc}(1 - y_C) - c'_{mc}y_C - b_C y_C)(a_2 - b_2 p_{C,D} + p_{R,D}) \tag{8}$$

3.6 Games Between the Retailer and the Competitor

First of all, for determining optimal prices under disruption and non-disruption modes, as well as finding the optimal supply ratio of the retailer from the foreign supplier, the game is designed between retailer and competitor. In a Stackelberg moving game between retailer and competitor, cost efficiency determines the leader and the follower. Whoever has a cost advantage (minimum procurement cost), benefits from being a leader [21].

Lemma 1: The expected profit functions in (4) and (5) are concave in retailer's prices for a given value of competitor's prices and vice-versa.

Proof. See Appendix 1.

It is assumed that the purchasing cost for the competitor is greater than the retailer's one for both types of supply. Hence, the retailer and competitor act as the leader and follower, respectively based on the following equations:

$$w_1 < w_2 < w_c \tag{9}$$

$$w_2 = w_1 + \gamma_1 \tag{10}$$

$$w_c = w_1 + \gamma_1 + \gamma_2 \tag{11}$$

$$\gamma_1, \gamma_2 > 0$$

Owing to the fact that the competitor is treated as the follower, one can write:

$$\pi_{m1} = \alpha_0(1 - q)(w_1 - c_{m1})(a_1 - b_1 p_{R,N} + p_{C,N}) \tag{12}$$

$$p^*_{C,N} = \frac{a_2 + p_{R,N} + b_2 w_c - b_2 b_C y_C}{2b_2} \tag{13}$$

$$p^*_{C,D} = \frac{a_2 + p_{R,D} + b_2 w_C - b_2 b_C y_C}{2b_2} \tag{14}$$

$$p^*_{R,N} = \frac{a_2 - w_1 + 2b_2 a_1 + b_2 w_C + 2b_1 b_2 w_1 + b_R y_R - 2b_1 b_2 b_R y_R - b_2 b_C y_C}{2(2b_1 b_2 - 1)} \tag{15}$$

$$p^*_{R,D} = p^*_{R,N} + \frac{\gamma_1}{2q} \tag{16}$$

$$\alpha^* = \frac{\gamma_1(2b_1 b_2 - 1)}{q(a_2 + w_1 + 2b_2 a_1 + b_2 w_C - 2b_1 b_2 w_1 - b_R y_R + 2b_1 b_2 b_R y_R - b_2 b_C y_C)} \tag{17}$$

Since $\alpha^* < 1$,

$$q < \frac{\gamma_1(2b_1 b_2 - 1)}{a_2 + w_1 + 2b_2 a_1 + b_2 w_C - 2b_1 b_2 w_1 - b_R y_R + 2b_1 b_2 b_R y_R - b_2 b_C y_C} = q^* \tag{18}$$

The optimal value of the dual variable (KKT coefficient) for the constraint is given as:

$$\mu^* = (1 - q)\gamma_1 \tag{19}$$

The corresponding proofs are provided in Appendix 2.

3.7 Games Between Retailers and Their Suppliers

In the second level (Figs. 2 and 3), two Stackelberg games (2–1 and 2–2) are examined. Solving these games results in the optimal return rate of the used products as well as the purchase price of the returned used products for the suppliers. Since the optimal values obtained from solving the game between retailer and competitor has a direct impact on the retailer/competitor and domestic supplier/competitor's supplier games, these games are investigated considering the above-mentioned game.

In the games (2–1) and (2–2), the optimal values, which are achieved via game between retailer and competitor, are substituted in Eqs. (7) and (8). Since the suppliers (domestic or competitor) are always followers, the partial derivatives are first calculated for the domestic supplier/competitor's supplier, and the results are obtained as well.

In the second step, after putting the results obtained from the first step in Eqs. (4) and (5), the optimal return rate of used products is achieved. The results are illustrated in Tables 2 and 3, respectively.

Lemma 2. The expected profit functions in Eqs. (7) and (8) are concave in the purchase price of the domestic and competitor's suppliers for the used products returned by the retailer and competitor.

Proof. See Appendix 3.

Lemma 3. The expected profit functions in Eqs. (4) and (5) are concave in the return rate of the retailer and competitor for a given purchase price value of the domestic and competitor's suppliers for the used products returned by the retailer and competitor.

Proof. See Appendix 4.

Table 2. Results obtained from the game (2–1)

Domestic supplier	$b_R^* = -\frac{1}{2q(4b_1b_2-3)}(A_a + \frac{A_b-qbc y_C}{y_R})$	(20)
Retailer	$y_R^* = \frac{A_a(A_c-qb_2bc y_C)}{A_a^2-\beta q^2(64b_2^2b_1+32b_2)}$	(21)

Where:

$$A_a = q(c_{m2} - c'_{cm2}) + 2qb_1b_2(c'_{m2} - c_{m2}) \tag{22}$$

$$A_b = q(w_2 + w_1 - c_{m2} + a_2 - 2b_1b_2w_2 + w_Cb_2 + 2a_1b_2 - 2b_1b_2w_1 + 2b_1b_2c_{m2}) + \gamma_1 - 2b_1b_2\gamma_1 \tag{23}$$

$$A_c = q(w_2 - w_1 + c_{m2} + a_2 + w_Cb_2) + 2qb_2(b_1w_1 - b_1w_2 + a_1 + b_1c_{m2}) - \gamma_1 + 2b_1b_2\gamma_1 \tag{24}$$

Table 3. Results obtained from the game (2–2)

Competitor's supplier	$b_C^* = -\frac{1}{2b_2(4b_1b_2-3)}(B_a + \frac{b_Ry_R-2b_1b_2b_Ry_R+B_b}{y_C})$	(25)
Competitor	$y_C^* = \frac{B_a(b_Ry_R-2qb_1b_2b_Ry_R+B_c)}{B_a^2-\beta b_2(64+256b_1^2b_2^2-256b_1b_2)}$	(26)

Where:

$$B_a = 4b_1b_2^2(c'_{mc} - c_{mc}) - 3b_2(c'_{mc} - c_{mc}) \tag{27}$$

$$B_b = 2a_1b_2 - 3b_2c_{mc} - w_1 - \gamma_1 + 6w_Cb_2 - a_2 + 2b_1b_2(2c_{mc}b_2 + \gamma_1 + w_1 + 2a_2 - 4w_Cb_2) \tag{28}$$

$$B_c = 2a_1b_2 + 3b_2c_{mc} - w_1 - \gamma_1 - a_2 + 2b_1b_2(2c_{mc}b_2 + \gamma_1 + w_1 + 2a_2) \tag{29}$$

As presented in Tables 2 and 3, both of the return rate (y_r) and domestic supplier's payment for the returned products (b_r) are related to the return rate of the competitor (y_c) and its supplier's payment for the returned products (b_c).

For this reason, a system of linear equations is needed. The values obtained from the game (2–2), are plugged into the values obtained from the game (2–1). Then, the linear equation system is solved to obtain the optimal values of the retailer's rate of returned used products and purchase price of the domestic supplier for the returned used products. The summary of the results is illustrated in Table 4.

Table 4. Optimal values of b_r and y_r

Domestic supplier	$b_R^* = -\dfrac{\begin{aligned}&qE_aE_eB_g^2 + qB_g^2A_g^2E_a + 8b_1b_2A_cE_bB_g^2 - 6A_bE_bE_c\\&+8b_1b_2A_bE_bE_c + qE_bB_bA_g^2 + qB_bE_bE_c - 6A_cE_bA_g^2\end{aligned}}{\begin{aligned}&qA_g(2b_1b_2 - 1)(16b_1b_2A_cE_b + 2qB_bE_b - A_cB_g^2\\&+A_bB_g^2 + 2qE_aB_g^2 - 13A_cE_b + A_bE_b)\end{aligned}}$	(30)

Retailer	$y_R^* = \dfrac{\begin{aligned}&A_g(16b_1b_2A_cE_b + 2qB_bE_b - A_cB_g^2 + A_bB_g^2 + 2qE_aB_g^2\\&-13A_cE_b + A_bE_b)\end{aligned}}{-16b_1b_2E_cE_b + E_cB_g^2 + 13E_cE_b + B_g^2A_g^2 + E_bA_g^2}$	(31)

Where:

$$E_a = -4b_1c_{mc}b_2^2 + 2b_1b_2(\gamma_1 + 2a_2 + 2w_1) + 3b_2c_{mc} + 2b_2a_1 - w_1 - \gamma_1 - a_2 \quad (32)$$

$$E_b = B_g^2 - \beta b_2\left(64 + 256b_1^2b_2^2 - 256b_1b_2\right) \quad (33)$$

$$E_c = A_g^2 - \beta q^2\left(64b_2^2b_1 + 32b_2\right) \quad (34)$$

4 Numerical and Sensitivity Analysis

Investigating the background of pricing in a CLSC and considering the undeniable impact of the recycling process on the supply chain sustainability and adopting optimal strategies, it is important to examine the effects of parameter variations on the decision variables and total profit as well.

Table 5. The values of parameters

c_{m1}	c_{m2}	c_{mc}	c'_{m2}	c'_{mc}	w_1	w_2	w_c	a_1	a_2	b_1	b_2	β	q
70	90	100	70	50	90	130	140	400	200	3	3	800	0.5

According to Table 5, the optimal values of decision variables obtained as below:

Table 6. Decision variables' values

Decision variable	Value	Decision variable	Value
y_R	0.58	μ	20
b_R	44	α	0.75
y_C	0.32	π_{m1}	1133.3
b_C	130.6	π_{m2}	961.5
$p_{R,N}$	117.5	π_{mc}	292.5
$p_{R,D}$	157.5	π_R	3948.6
$p_{C,N}$	102.7	π_C	82.86
$p_{C,D}$	107.3		

4.1 The Effect of w_1 on Decision Variables and Profit Functions

In this section, the effects of foreign supplier's wholesale price (w_1) changes on the optimal values of decision variables have been considered.

An increase in the wholesale price (w_1) leads the retailer to decrease the supply from the foreign supplier (α) and consequently encourages to buy more products from a domestic supplier. Accordingly, with an increase in the supply from the domestic supplier, to stay in the competition, the retailer's price in disruption mode ($p_{R,D}$) and the competitor's price ($p_{C,D}$) decrease.

It is depicted from Fig. 4 in the case of normal condition where there is no disruption in the system, by decreasing supply from the external source, the price of the retailer ($p_{R,N}$) increases sharply in comparison with competitor's price ($p_{C,N}$).

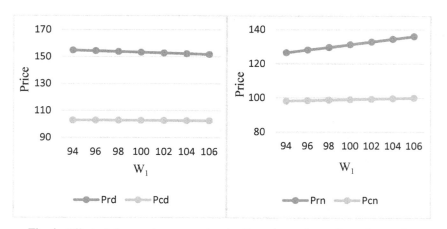

Fig. 4. Effect of changes in w_1 on prices in disruption and non-disruption modes

As shown in Fig. 5, by increasing w_1, the return rate of used products (y_R), and the payment for each returned product by supplier (b_R) decrease. This is because of that, rise in w_1 leads retailer to supply from internal supplier, which is expensive and drops retailer's profit and reduces retailer's willingness to collect used products and as a result, the payment for each returned product falls as well.

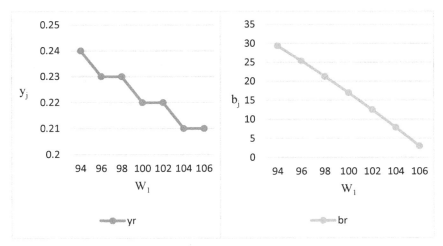

Fig. 5. Effect of changes in w_1 on the return rate and the supplier's payment for the returned product

It can be seen in Fig. 6, Increase in purchasing from domestic supplier brings more profit for this supplier. However, external supplier's and retailer's profits drop as the retailer has to purchase products at a higher price. In this case, the finished product price is high and consequently, the demand decreases. However, on the other hand, some of the retailer's demand shifts toward the competitor and rises competitor's and its supplier's profit.

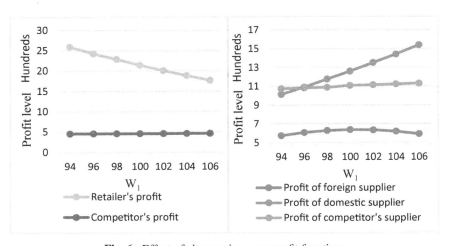

Fig. 6. Effect of changes in w_1 on profit functions

4.2 The Effect of w_2 on Decision Variables and Profit Functions

As shown in Fig. 7, by increasing the domestic supplier's wholesale price (w_2), the retailer finds a more tendency to supply from the foreign supplier, which is cheaper but has a risk of disruptions. This means that the percentage of supply from the foreign supplier increases (α).

Accordingly, the retailing price of the retailer in non-disruption mode ($p_{R,N}$) decreases. Besides, to stay in the competition, the competitor's price in non-disruption mode ($p_{C,N}$) declines gradually. In the case of disruption owing to the increasing desire of the retailer to supply from the foreign supplier, the price of the retailer in disruption mode increases ($p_{R,D}$), just like the competitor's one ($p_{C,D}$).

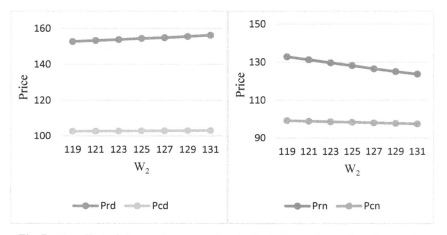

Fig. 7. The effect of changes in w_2 on prices in disruption and non-disruption modes

Also, the sensitivity analysis shows by increasing w_2, the return rate of used products (y_R), as well as the supplier's payment per returned product (b_R) increases. It is suggested that the retailer pays more for the collection by increasing the supply from a foreign supplier, which results in a rising percentage of return rate of used products and consequently, payments per unit of returned products by a domestic supplier.

Based on Fig. 9, by accepting this risk of supplying from a foreign supplier, a retailer receives cheaper products, which increases the profitability of both retailer and foreign supplier. Consequently, the profit of the domestic supplier shrink. On the other hand, competitor and its supplier lose their profits because of the higher demand for retailer's products and lower demand for competitor's ones.

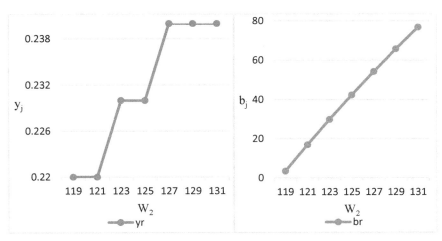

Fig. 8. The effect of changes in w_2 on the return rate and the supplier's payment for the returned product

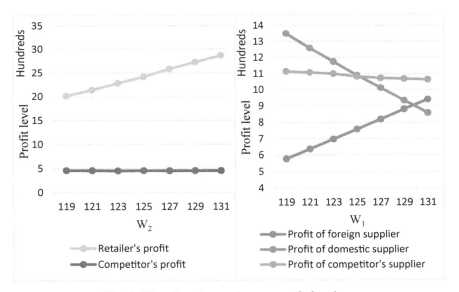

Fig. 9. The effect changes in w_2 on profit functions

4.3 The Effect of w_C on Decision Variables and Profit Functions

As can be seen in Fig. 10, by increasing the wholesale price of the competitor's supplier (w_c), the competitor's price rises in both disruption ($p_{C,D}$), and non-disruption ($p_{C,N}$) modes. Also retailer's prices in both disruption ($p_{R,D}$) and non-disruption ($p_{R,N}$) modes increase slightly.

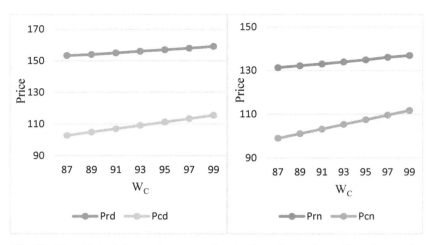

Fig. 10. The effect of changes in w_c on prices in disruption and non-disruption modes

Based on Fig. 11, in contrast to the retailer and domestic supplier by increasing the wholesale price (w_c), To revitalize the supply chain, the competitor increases the return rate of used products (y_C), which increases the payment of the competitor's supplier for each unit of returned products (b_C).

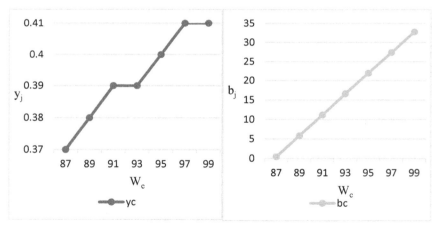

Fig. 11. The effect of changes in w_C on the return rate and the supplier's payment for the returned product

The rise in w_C falls the profit of competitor and its supplier (Fig. 12), which is reasonable since the increasing wholesale price has pushed competitor's retail prices up, and consequently drops the demand. As a result proportion of competitor's demands shift toward the retailer and the profit of the domestic supplier increases.

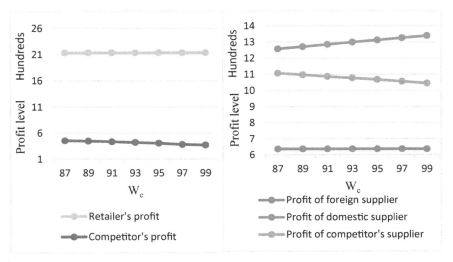

Fig. 12. The effect changes in w_C on profit functions

5   Managerial Insights

5.1   Managerial Insight 1

Considering Figs. 5 and 8, it can be inferred that as the percentage of supply from the foreign supplier increases, the return rate of the used products, or the level of retailer investment on the return policy increases and vice versa.

5.2   Managerial Insight 2

Supply from a cheap but risky foreign supplier provides a great opportunity for the retailer to attract more demand by low price products. In this regard, based on Fig. 7, the competitor tries to lower its prices to hold the demand for its products.

5.3   Managerial Insight 3

Based on Fig. 12, the retailer prefers to supply products from the domestic supplier which is reliable to respond to a demand shifted from a competitor due to increasing procurement cost of the competitor's supplier.

5.4   Managerial Insight 4

An increase in procurement cost of competitor's supplier and eventually the rise in retail prices and lack of alternative suppliers lead the competitor to lose demands, therefore the competitor's profit shrinks (Fig. 12). In response to these losses, the competitor increases the return rate of used products due to the fact that it helps the supplier to produce new products by recycled materials, which are cheaper than raw ones (Fig. 11).

6 Conclusion

In this study, two separate supply chains are proposed to compete with each other, wherein the first one there are domestic and foreign suppliers. The non-domestic one is cheaper but risky regarding disruption issues in supply and the domestic one is expensive and trustable. In the second SC, there is only one supplier and one retailer as a competitor. The reverse logistics are studied in both supply chains as an environmental strategy that affects supplier's wholesale prices as manufacturing from recycled materials is cheaper than raw materials.

Two levels of competition are proposed for the illustrated model, where in the first level there is a competition between retailer and competitor with retailer's leadership. In the next level owing to the environmental concerns, the competition is conducted between retailer/competitor and domestic/competitor's supplier to find the optimal values for the return rate of used products and supplier's payment for each returned ones. In these games, suppliers are considered as Stackelberg leaders.

The results show that based on the optimal value of the α gained in Sect. 4, the retailer with both internal and external suppliers inclines to supply from the cheaper one considering disruption risk probability. But as Fig. 12 shows in the case of a rise in the competitor's price, there would be a shift of demand toward the retailer. In this particular case, the retailer prefers to supply from the internal and expensive source to be ensured about satisfying shifted demands. Besides, it is found that collecting and recycling could play a prominent role in loss reduction and supply chain survival especially in the absence of an alternative supply chain.

This proposed model is suitable for studying and finding prices and reworks in several cases such as the food and automotive industries. For future researches, products with different green degrees could be considered as long as considering governmental subside for returned products.

Appendix

Appendix 1

Lemma 1. The expected profit functions in (4) and (5) are concave in the retailer's prices for a given value of competitor's prices and vice-versa.

Proof: In case (4), we can see that the π_R is concave in $p_{R,N}$ and $p_{R,D}$. With considering $g\left(p_{R,N}, p_{R,D}\right) = (1 - \alpha)\left(a_1 - b_1 p_{R,N} + p_{C,N}\right) - \left(a_1 - b_1 p_{R,D} + p_{C,D}\right)$, we find that $g\left(p_{R,N}, p_{R,D}\right)$ is convex. Hence, the Retailer's profit function is concave for in $p_{R,N}$ and $p_{R,D}$ for the given constraint. We can prove the concavity of the competitor's profit function (5).

Appendix 2

In this case, we assumed $w_1 < w_2 < w_c$, $w_2 = w_1 + \gamma_1$, and $w_c = w_1 + \gamma_1 + \gamma_2$, where $\gamma_1, \gamma_2 > 0$.

Considering Eqs. 4 and 5, and using K.K.T as below, we optimize the retailer's problem:

$$\emptyset\left(\alpha, \mu, p_{R,N}, p_{R,D}, p_{C,N}, p_{C,D}\right) = \pi_R + \mu[(1 - \alpha)\left(a_1 - b_1 p_{R,N} + p_{C,N}\right)$$

$$-\left(a_1 - b_1 p_{R,D} + p_{C,D}\right)] \tag{35}$$

The partial derivatives will be:

$$\frac{\partial \pi_C}{\partial p_{C,N}} = (1 - q)\left(a_2 - 2b_2 p_{C,N} + p_{R,N} + b_2 w_C - b_2 b_C y_C\right) \tag{36}$$

$$\frac{\partial \pi_C}{\partial p_{C,D}} = q\left(a_2 - 2b_2 p_{C,D} + p_{R,D} + b_2 w_C - b_2 b_C y_C\right) \tag{37}$$

$$\frac{\partial \emptyset}{\partial \mu} = (1 - \alpha)\left(a_1 - b_1 p_{R,N} + \frac{a_2 + p_{R,N} + b_2 w_C - b_2 b_C y_C}{2b_2}\right)$$

$$-\left(a_1 - b_1 p_{R,D} + \frac{a_2 + p_{R,D} + b_2 w_C - b_2 b_C y_C}{2b_2}\right) \tag{38}$$

$$\frac{\partial \emptyset}{\partial \alpha} = (1 - q)(w_2 - w_1)\left(a_1 - b_1 p_{R,N} + \frac{a_2 + p_{R,N} + b_2 w_C - b_2 b_C y_C}{2b_2}\right)$$

$$-\mu \, (a_1 - b_1 p_{R,D} + \frac{a_2 + p_{R,D} + b_2 w_C - b_2 b_C y_C}{2b_2}) \tag{39}$$

$$\frac{\partial \emptyset}{\partial p_{R,N}} = (1 - q)\left(a_1 + 2p_{R,N}\left(-b_1 + \frac{1}{2b_2}\right) + \frac{a_2 + b_2 w_C - b_2 b_C y_C}{2b_2}\right.$$

$$\left.-\left(-b_1 + \frac{1}{2b_2}\right)(\alpha w_1 - (1 - \alpha)w_2 + b_R y_R)\right) + \mu(1 - \alpha)\left(-b_1 + \frac{1}{2b_2}\right) \tag{40}$$

$$\frac{\partial \emptyset}{\partial p_{R,D}} = q\left(a_1 + \left(-b_1 + \frac{1}{2b_2}\right)(2p_{R,D} - w_2 + b_R y_R) + \frac{a_2 + b_2 w_C - b_2 b_C y_C}{2b_2}\right) - \mu\left(-b_1 + \frac{1}{2b_2}\right) \tag{41}$$

Appendix 3

Lemma 2. The expected profit functions in Eqs. (7) and (8) are concave in the purchase price of the domestic supplier and competitor's supplier for the products returned by the retailer and the competitor.

Proof: Considering the profit functions of domestic and competitor's suppliers (Eqs. 7 and 8), we can see that the π_{m2}, and π_{mc} are concave in b_R, and b_C, respectively.

Appendix 4

Lemma 3. The expected profit functions in Eqs. (4) and (5) are concave in the return rate of used products for the retailer and the competitor for the given value of purchase price of the domestic supplier and competitor's supplier for the returned products.

Proof: Considering the profit functions of the retailer and competitor (Eqs. 4 and 5), we can see that the π_R, and π_C are concave in y_R, and y_C, respectively.

References

1. Katsaliaki, K., Galetsi, P., Kumar, S.: Supply chain disruptions and resilience: a major review and future research agenda. Ann. Oper. Res. **1**, 38 (2021). https://doi.org/10.1007/s10479-020-03912-1
2. Kumar, M., Basu, P., Avittathur, B.: Pricing and sourcing strategies for competing retailers in supply chains under disruption risk. Eur. J. Oper. Res. **265**(2), 533–543 (2018)
3. Golan, M.S., Jernegan, L.H., Linkov, I.: Trends and applications of resilience analytics in supply chain modeling: systematic literature review in the context of the COVID-19 pandemic. Environ. Syst. Decis. **40**(2), 222–243 (2020). https://doi.org/10.1007/s10669-020-09777-w
4. Fang, Y., Shou, B.: Managing supply uncertainty under supply chain cournot competition. Eur. J. Oper. Res. **243**(1), 156–176 (2015)
5. Taleizadeh, A.A., Alizadeh-Basban, N., Niaki, S.T.A.: A closed-loop supply chain considering carbon reduction, quality improvement effort, and return policy under two remanufacturing scenarios. J. Clean. Prod. **232**, 1230–1250 (2019)
6. Karimabadi, K., Arshadi-khamseh, A., Naderi, B.: Optimal pricing and remanufacturing decisions for a fuzzy dual-channel supply chain. Int. J. Syst. Sci. Operat. Logist. **7**, 1–14 (2019)
7. Li, G., et al.: Return strategy and pricing in a dual-channel supply chain. Int. J. Prod. Econ. **215**, 153–164 (2019)
8. Jian, J., et al.: Decision-making and coordination of green closed-loop supply chain with fairness concern. J. Clean. Prod. **298**, 126779 (2021)
9. Huang, H., He, Y., Li, D.: Pricing and inventory decisions in the food supply chain with production disruption and controllable deterioration. J. Clean. Prod. **180**, 280–296 (2018)
10. Chakraborty, T., Chauhan, S.S., Ouhimmou, M.: Mitigating supply disruption with a backup supplier under uncertain demand: competition vs. cooperation. Int. J. Prod. Res. **58**(12), 3618–3649 (2020)
11. Jabbarzadeh, A., Haughton, M., Khosrojerdi, A.: Closed-loop supply chain network design under disruption risks: a robust approach with real world application. Comput. Ind. Eng. **116**, 178–191 (2018)
12. Gupta, V., Ivanov, D., Choi, T.-M.: Competitive pricing of substitute products under supply disruption. Omega **101**, 102279 (2021)
13. Torabi, S., Baghersad, M., Mansouri, S.: Resilient supplier selection and order allocation under operational and disruption risks. Transp. Res. Part E Logist. Transp. Rev. **79**, 22–48 (2015)
14. Freeman, N., et al.: Sourcing strategies for a capacitated firm subject to supply and demand uncertainty. Omega **77**, 127–142 (2018)
15. Jain, T., Hazra, J.: Dual sourcing under suppliers' capacity investments. Int. J. Prod. Econ. **183**, 103–115 (2017)

16. Li, X., Li, Y.: On the loss-averse dual-sourcing problem under supply disruption. Comput. Oper. Res. **100**, 301–313 (2018)
17. Biswas, I., Avittathur, B., Chatterjee, A.K.: Impact of structure, market share and information asymmetry on supply contracts for a single supplier multiple buyer network. Eur. J. Oper. Res. **253**(3), 593–601 (2016)
18. Savaskan, R.C., Bhattacharya, S., Van Wassenhove, L.N.: Closed-loop supply chain models with product remanufacturing. Manage. Sci. **50**(2), 239–252 (2004)
19. Sinayi, M., Rasti-Barzoki, M.: A game theoretic approach for pricing, greening, and social welfare policies in a supply chain with government intervention. J. Clean. Prod. **196**, 1443–1458 (2018)
20. Savaskan, R.C., Van Wassenhove, L.N.: Reverse channel design: the case of competing retailers. Manag. Sci. **52**(1), 1–14 (2006)
21. Amir, R., Stepanova, A.: Second-mover advantage and price leadership in Bertrand duopoly. Games Econom. Behav. **55**(1), 1–20 (2006)

Production/Scheduling
and Transportation in Supply Chain
Management

A Multi-product EPQ Model for Defective Production and Inspection with Single Machine, and Operational Constraints: Stochastic Programming Approach

Reza Askari, Mohammad Vahid Sebt[(✉)], and Alireza Amjadian

Department of Industrial Engineering, Faculty of Engineering, Kharazmi University, Tehran, Iran
{std_reza.askari,sebt,std_amjadian}@khu.ac.ir

Abstract. In the current study, a mixed-integer nonlinear programming model is presented for the development of an economic production quantity (EPQ) model that encompasses warehouse space and budget constraints while it does not consider shortage. To improve the simulation of reality, the model consists of haphazard factors with differing scenarios. The proposed model aims at maximizing the revenue by the calculation of the optimal level of production of every item in individual scenarios. Moreover, we concentrated on inspection, various inspection errors, defective items and their costs. The defective items are categorized regarding to the action taken on them. All related costs are also considered in the model. The model is solved using two metaheuristic algorithms, and the effectiveness of the approaches is demonstrated by generating various instances in a variety of dimensions. Finally, the results are calculated accurately by GAMS software to determine the deviation between the optimal solution and the output of the algorithms.

Keywords: Inventory model · EPQ model · Inspection error · Meta-heuristic algorithm · TOPSIS

1 Introduction and Research Literature

The technological advancements and developing plants worldwide have inspired production executives in the modernization of the inventory system and paying attention to both production and demand quantities. Consequently, this has necessitated the development of novel inventory models that account for production capability. The internationalization of the economy and liberalization of markets at a progressively fast rate have strengthened the necessity for the incorporation of the operating uncertainties and financial risks into the production and inventory control decision-making in companies. The prime economic production quantity (EPQ) inventory model for a single product-single step fabrication system was presented by Taft (1918). Eilon (1957) and Rogers (1958)

© Springer Nature Switzerland AG 2021
Z. Molamohamadi et al. (Eds.): LSCM 2020, CCIS 1458, pp. 161–193, 2021.
https://doi.org/10.1007/978-3-030-89743-7_10

were possibly the pioneers who examined the multi products-single fabrication system. Amongst studies performed on optimum lot-sizing of the EPQ models, the recently published investigations are denotable chronologically. Pasandideh (2008) designed a model aiming at using the EPQ model for production in reality by the assumption that the delivery of orders might occur in the form of discrete manifold pallets. Additionally, over a single product might be present alongside storehouse space restriction. An EOQ model for a two-level supply chain system was designed by Pasandideh et al. (2011) who proposed a non-linear integer-programming model with a restriction on the storehouse volume for the retail merchant. This model took account of the backordering shortages in manifold product conditions to discover the order numbers and maximal backorder levels, thereby minimizing the overall inventory cost of the supply chain. Taleizadeh et al. (2014) invented an EPQ model with manifold products and a machine that considers interruptions in processes, scraps, and reworks. The model allows shortages in production and are taken as a stockpile, aiming at minimizing the expectable overall costs. Although the EPQ model is popular, it suffers from limits and it is not possible to consider this model as an all-inclusive inventory model. A limit is that the production system responsible for the classic EPQ inventory model will never fabricate faulty items throughout the production process. Nonetheless, defective products might be produced in every production cycle in the majority of realistic conditions. Thus, it has been interesting to study the existence of faulty items on inventory models. For this reason, it is greatly important to deal with this subject in the EPQ model.

Generally, all produced items, are considered as perfect items in manufacturing process. As a viewpoint, it is not always true to consider all the completed products as perfect products. There are several reasons for occurring defection during the production process. A production process which randomly shifts from in-control state to out-of-control state, is described in Sarkar and Saren (2016). Considering the inspection of the defects in this model, the product inspection policy is profitable for making a diminution in the cost enforced by manufacturing imperfect products. Sarkar and Saren (2016) integrated Type 1 and type 2 errors in this model in order to form a more realistic model rather than the existing ones. A warranty policy is also included in this model for some fixed time periods. Hsu (2013) expanded two EPQ models and considered defective fabrication processes, faults in examination processes, prearranged backorders, and sales returns. They obtained a solution for optimum production lot size and maximal shortage degree for the two models. Faulty products found throughout the examination procedure are dividable into two discrete groups: reworkable and non-reworkable. An economic production quantity, introduced by Barzoki et al. (2011), can consider manufacturing faulty items (both reworkable and non-reworkable) and work-in-process inventory. The non-reworkable faulty items are vended at a less cost. This researcher achieved a solution for this problem. Rahim and Ben-Daya (2001) investigated the impacts on economic production quantity, examination timetables, and economically designing control charts resulting from both imperfect products and deterioration of production procedures concurrently.

In today's environment, 100% inspection plays a crucial role. Not only in industrial processes but also it is applied to decision-making processes where the consequences of inordinate deviations from target values are very high. Feng (2006) developed an inspection strategy considering the whole system costs including both the costumer and the producer costs. A general optimization model was also formulated by Feng (2006) to investigate the necessity of performing a 100% inspection. Alamri et al. (2016) developed a general EOQ model for those items that are likely to be inspected for their imperfect quality. The lots received by the sorting facility undergoes a 100% screen. According to a learning curve, the percentage of defective items per lot decreases. Rezaei and Salimi (2012) investigated the relationship between buyer and supplier regarding the inspection and the fact that it could lead to a change in economic order quantity and purchasing price under two conditions: (1) there are no associations between purchaser's retailing fee, purchaser's buying fee, and customer demand, and (2) there are interrelations between purchaser's retailing fee, purchaser's buying fee, and customer demand. Chiu et al. (2007) proposed a method that determines an optimum runtime for EPQ models by considering the scraps, rework, and haphazard machine failures. Producing the imperfect items and random breakdowns of machines are unavoidable in real life production systems. For several decades, the inventory control has been studied as a means of cost saving for enterprises that have intended to maintain appropriate inventory levels to deal with stochastic customer demands and to increase customer satisfaction (Axsäter (2000); Moinzadeh (2002); Zipkin (2000)). Khalilpourazari and Pasandideh (2017) introduced a multiple constraints model that featured a nonlinear maintenance cost. The primary objective of this model was to bring the Economic Order Quantity (EOQ) model closer to the real world. Furthermore, the presented model considers the re-order rate during the shortage period of a decision variable. Cárdenas-Barrón et al. (2020) presented the EOQ model from the point of view of a retailer with the purpose of maximizing profit for each unit time. Demand and maintenance cost was considered to be non-linear during this study. In the same year, Khalilpourazari and Pasandideh (2019) presented the multiple products EOQ model intended for growing items for the very first time with the objective of maximizing total profit. They also utilized multiple constraints in order to bring the model closer to real world conditions. Eventually, they applied metaheuristic algorithms to solve the model. Recently, Shaikh et al. assessed (2019) the EOQ model in two means: 1-Zero-ending case, and 2-Inventory model for shortages case; in which the demand in these models depended on the price. Liao and LI (2021) also examined the EOQ model in the Closed Loop Supply Chain (CLSC). They have applied random analysis to calculate the influence of market uncertainty on production and operations.

The EPQ classic model mainly assumes the certainty and stability of the demand. For this reason, the EPQ model is recommendable for use in a setting with a stable and expectable demand; however, the demand is really random in realistic conditions. Efficient management of the inventory level of individual participants inside supply chains is vital to improve the service levels of the initiatives (Kwon et al. 2009). Sana et al. (2007) designed an EPQ/EOQ model where the demand for faultless products was either ensued by distributing uniformly or constantly, and the defective products or those with no rework demand rate were dependent on the lowered vending point. Pal et al.

(2013) could specify the optimum buffer inventory for random demand in the marketplace throughout preventive maintenance by designing an EPQ inventory model. Chang (2004) established two fuzzy models for an inventory problem having faulty quality stuff. In the 1st model, a fuzzy number characterizes the faulty rate, whereas the yearly demand level is regarded as an unchanged constant. In the 2nd model, a fuzzy number presents both the faulty level and the yearly demand level. Pal et al. (2013) invented an EPQ inventory model for determining the optimum buffer inventory for random demand in the marketplace throughout the preventive maintenance or repairing fabrication installation with an EPQ model in a defective production system. The model includes a minor likelihood that the system could shift from in-control situation to out-of-control situation following a definite time that ensues a probability density function. In both in-control and out-of-control situations, faulty stuff undergo reworking in price issues exactly following a typical production time. Moreover, Khan et al. (2017) introduced the Economic Production Quantity (EPQ) model for supply chains that have a random lead time. On the other hand, bearing in mind that some of the goods that are provided by the seller might include defective items, screening costs have also been taken into consideration. In the same year, Nobil et al. (2017) introduced a multi-product single machine EPQ model for the vendor-buyer system with the purpose of determining the optimal cycle length and number of delivered batches. In this model, capacity has been regarded as one of the limitations of the model due to the fact that multiple products are produced using a single piece of machinery. The optimization and numerical examples presented in this model have been performed by means of metaheuristic algorithms. After that, Cunha et al. proposed an EPQ model in which shortages were recognized. Meanwhile, revenue from selling low quality items (imperfect quality batches) was also calculated. Recently, Gharaei et al. (2019) presented a dual-purpose EPQ mode with the purpose of minimizing the total cost of inventory, and maximizing profits. They considered multiple random constraints for the model closest to the real conditions. Finally, the Mixed-Integer Non-Linear Programming (MINLP) model has been optimized using the Generalized Cross Decomposition (GCD) method. Gharaei et al. (2019) also presented an integrated dual-objective model that has random constraints, while taking into account quality control and green production policies. The cost of emissions was also taken into account in this model, and the ultimate objective is to optimize the total amount of inventory, and maximize profit. Table 1 demonstrates the summary of previous investigations conducted in the field of inventory models.

The present article represents the development of an EPQ model for multi-product single-machine case/condition, aiming at specifying the economic production level for maximizing the revenue. The model considers the net demand as a function of several internal and external systematic elements, allowing the model to approximate to the realistic circumstances. Moreover, all parameters in the suggested model are stochastic and have various probabilities. The model's usage in real world situations got validated considering errors type 1 and 2 and their combination in the model and also considering the classic model constraints. The formulation of the EPQ problem for a defective production system is as a mixed integer non-linear programming (MINLP). To solve this problem, we used genetic and invasive weed optimization algorithm and investigated their efficiency for the exact answer. Eventually, the outputs of these two algorithms were compared in two ways in order to identify the more efficient solution. The most important contributions of this paper are as follows:

- Developing and optimizing an EPQ model in multi-product machine-single condition to determine lot sizing and maximize profits
- Determining quality control (QC) policies and considering its cost
- Determining penalty cost for customer dissatisfaction
- Considering the stochastic parameters in order to cope with uncertainty and approaching reality
- Using GA and IWO algorithms to optimize the model
- Solving various numeric examples in different cycles and analyzing the results to evaluate the performance of algorithms
- Comparing the algorithms with different methods to determine the more efficient algorithm
- Employing a variety of stochastic constraints to agree and approximate with the reality

The definition of the problem and its presumptions are given in Sect. 2. Mathematic formulations of this article are given in Sect. 3. Section 4 provides the two metaheuristics as solution approaches. Section 5 represents the findings numerically to illustrate that the introduced model is applicable alongside evaluating the functionality of the solution approaches. Following managerial implications in Sect. 7, Sects. 8 concludes the article.

Table 1. A summary of some studies of the Inventory control system

Study	Inventory control system		Product		Shortage	Model Type					Solution Method		Type of Constraints			Penalty policy
	EOQ	EPQ	single	multi		LP	NLP	MILP	MIP	MINLP	Metaheuristic	Exact	NO Constraints	Deterministic Constraints	Stochastic Constraints	
Chiu et al (2007)	-	✓	✓	-	-	-	✓	-	-	-	-	✓	✓	-	-	-
Pasandide et al (2008)	-	✓	-	✓	-	-	-	-	-	✓	✓	-	-	✓	-	-
Pasandide et al(2011)	✓	-	-	✓	✓	-	-	-	-	✓	✓	-	-	✓	-	-
Barzoki (2011)	-	✓	✓	-	-	-	✓	-	-	-	-	✓	✓	-	-	-
Yoo et al(2012)	-	✓	✓	-	-	-	✓	-	-	-	-	✓	✓	-	-	-
Hsu et al(2013)	-	✓	✓	-	✓	-	✓	-	-	-	-	✓	-	✓	-	-
Pal et al(2013)	-	✓	✓	-	✓	-	✓	-	-	-	-	✓	✓	-	-	-
Brojeswar et al. (2013)	-	✓	✓	-	-	-	✓	-	-	-	-	✓	-	✓	-	-
Taleizadeh et al. (2014)	-	✓	-	✓	✓	-	✓	-	-	-	-	✓	✓	-	✓	-
Zhou and et al. (2015)	-	✓	✓	-	-	-	✓	-	-	-	-	✓	-	✓	-	-
Alamri et al. (2016)	✓	-	-	✓	-	-	✓	-	-	-	-	✓	-	✓	-	-
Khalilpourazari and Pasandideh (2017)	✓	-	✓	-	-	-	✓	-	-	-	✓	-	-	✓	-	-
Khalilpourazari and Pasandideh (2019)	✓	-	-	✓	-	-	✓	-	-	-	✓	-	-	✓	-	-
Nobil et al. (2020)	-	✓	-	✓	-	-	-	-	-	✓	✓	-	✓	-	-	-
Current Work	-	✓	-	✓	-	-	-	-	-	✓	✓	-	-	-	✓	✓

2   Problem Definition and Assumptions

In this section, we first look into a closed-loop supply chain where manufacturers production and inspection processes are firm and non-deteriorating but not completely reliable;
therefore, some errors lead to delivering imperfect items to costumers, and subsequently
cause the return of defective items and finally, the company destroys them. Contrary to
the assumptions of the classic EPQ model, in real world situations, effective parameters
on the production quantity are not certain. Some of these parameters are demand, selling
price, space limitation, budget restrictions on varied conditions, and so forth. Therefore,
managers ought to answer two questions considering the stochastic situation with the
aim of increasing the profit: (1) how much is the amount of net demand and (2) how
much is the amount of optimal production?

Yoo et al. (2012) worked on a variety of presumptions for a mathematical model
to establish an EPQ system with faulty products. By enriching and developing a few
of their presumptions, we tried to have more closeness to realistic circumstances. By
redefining the novel presumptions, therefore, the problem was further clarified as shown
below:

- Several products are manufactured by a single machine.
- All parameters are Stochastic.
- Products are not spoilable.
- There is a budgeting limit for different products.
- Warehouse space is limited.
- A penalty is considered for costumer's dissatisfaction in QC policy and its total cost
  has a maximum limit.

3   Problem Definition and Assumptions

3.1   Notation

Besides the terminology of Yoo et al. (2012), the definition of few novel symbols here
clarifies the problem. Moreover, few symbols are altered in our research to widen the
understanding of this investigation. The undermentioned symbolizations, consisting of
the factors and the decision variables, are applied to utilize the mathematical formulation
of the present problem:

Indices:

i *Index of a product* $i=1.2.3,...,n$

s *index of a scenario* $s=1,2,3,...,m$

Parameters:

P_i^s *The probability of each scenario for product i.*

D_i^s *demand rate of product i in scenario s (unit/unit time)*

$(\acute{D}_i^s)$ *net depletion rate of product i in scenario s in production line(unit/unit time)*

M_i^s *production rate f product i in scenario s (unit/unit time) $M_i^s > (D_i^s)$*

pr_i^s *The selling price of a unit of product i in scenario s($/unit)*

θ_i^s *sampling proportion for inspection of product i in scenario s $0 \le \theta_i^s \le 1$*

u_i^s *production cost per unit of product i in scenario s($/unit)*

w_i^s *rework cost per unit of product i in scenario s($/unit)*

ρ_i^s *proportion of inspection failure during inspectionof product i in scenario s $0 \le \rho_i^s \le 1$*

f_i^s *Volume of product i in scenario s(volume/unit)*

K_i^s *setup cost per production run of product i in scenario s($/cycle)*

G_i^s *depletion rate of defective items by inspection Of product i in scenario s (unit /unit time)*

E_i^s *exchange rate of product i in scenario s (unit/unit time)*

R_i^s *accumulation rate of items being reworked or depletion rate of items reworked of product i in scenario s*

V_i^s *accumulation and depletion rate of salvage items of product i in scenario s (unit / unit time)*

Kr_i^s *rework setup cost per rework run of product i in scenario s ($/rework cycle)*

W_i^s *rework rate of product i in scenario s (unit/unit time) $W_i^s > R_i^s$*

π_i^s *proportion of defective items during productionof product i in scenario s $0 \le \pi_i^s \le 1$*

h_i^s *inventory holding cost rate fraction of product i in scenario s (fraction/unit /unit time)*

H_i^s *inventory holding cost rate of product i in scenario s ($/unit/unit time)*

n_i^s *rework (setup)frequency per cycle time of product i in scenario s*

j_i^s *inspection cost per unitof product i in scenario s ($/unit)*

r_i^s *return cost per unit of product i in scenario s ($/unit)*

$\alpha_{i_{1.2.3}}^s$ *sampling proportion of rework. salvage. scrap among defectives. respectively of product i in scenario s $0 \le \alpha_{i_{1.2.3}}^s \le 1$*

T_i^s *cycle time of a production lot of product i in scenario s $T_i^s = Q_i^s / (\acute{D}_i^s)$*

$T_{i_p}^s$ *production run time of product i in scenario s $T_{i_p}^s = Q_i^s / M_i^s$*

$T_{i_r}^s$ *cycle time of rework items of product i in scenario s $T_{i_r}^s = T_i^s / n_i^s = Q_i^s / n_i^s (\acute{D}_i^s)$*

$T_{i_w}^s$ *rework run time of product i in scenario s $T_{i_w}^s = Q_{i_r}^s / W_i^s = \alpha_{i_1}^s \pi_i^s Q_i^s / n_i^s W_i^s$*

v_i^s *salvage price per unit of product i in scenario s ($/unit)*

g_i^s *scrap cost per unit of product i in scenario s ($/unit)*

l_i^s *penalty cost per unit due to customers' quality dissatisfaction of product i in scenario s (unit)*

β_i^s *proportion of exchange among returns of product i in scenario s $0 \le \beta \le 1$*

F *The available storage space.*

$B_{1.2.3}$ *The budget limitation of product. inspection process. penalty cost due to customers quality dissatisfaction.*

Decision variables:

Q_i^s *lot size of product i in scenario s(unit/cycle)*

3.2   Mathematical Formulation

In this research, our intention for modelling is to determine the production quantity by considering the constraints of the profit maximization. However, supply chain cycle and the parameters affecting it, should be studied before the process. Here, a closed-loop supply chain is investigated that considers the stability of the production and examination procedure, but it is not of perfect reliability; hence, clients receive a part of defective stuff, which dissatisfies them and ultimately, imposing some additional costs to the system.

In each cycle of every scenario, each product is generated at the amount of Q_i^s and at a rate of M_i^s which includes the defective items in the manufacturing process as well. Moreover, products are defective with the ratio of π_i^s; thus total defective products of a production cycle equals $\pi_i^s Q_i^s$ to get inspected. (See Fig. 1).

Defective items are spotted and put aside with a probability of $1 - \rho_i^s$ in the inspection process; so the amount of detected defective items in the inspection process equals $\theta_i^s (1 - \rho_i^s) Q_i^s$. Accordingly, the amount of the detected defective items from the total defective items of production process cycle in inspection process equals $G_i^s = \theta_i^s (1 - \rho_i^s) \pi_i^s Q_i^s$.

On the other hand, inspection process is not completely perfect and its failure probability equals ρ_i^s, thus the amount of defective items that are not discovered in inspection process, for each product in a specific scenario equals $\theta_i^s \rho_i^s \pi_i^s Q_i^s$.

The amount of product produced and not inspected equals $(1 - \theta_i^s) Q_i^s$ which contains some defective items. Therefore, in every cycle, a specific amount of defective items is delivered to the customers that are classified in two different groups; first, items that are not discovered in the inspection process and second, defective items that are not inspected. The total amount is equal $(1 - \theta_i^s + \theta_i^s \rho_i^s) \pi_i^s Q_i^s$.

Costumers dealing with defective items are divided into two separate categories regarding their choices; the first group of costumers demand exchange for defective products with the ratio of β_i^s; thus, the amount of products to be exchanged equals $\beta_i^s (1 - \theta_i^s + \theta_i^s \rho_i^s) \pi_i^s Q_i^s [E_i^s]$.

The second group of costumers demand refund for defective products; therefore, the amount of products which their cost should be refunded equals $(1 - \beta_i^s)(1 - \theta_i^s + \theta_i^s \rho_i^s) \pi_i^s Q_i^s$.

As we mentioned earlier, the total defective items of a circle for a single product in a specific scenario equals $\pi_i^s Q_i^s$ which includes two categories; items detected in the inspection and items spotted by the costumers which equal $\pi_i^s Q_i^s = G_i^s + E_i^s$.

The collected defective products are split into three groups; the first group is the defective items that are reworked and their ratio for each product in a specific scenario equals $\alpha_{i_1}^s$, so the amount of defective items that are reworked equals $W_i^s = \alpha_{i_1}^s \pi_i^s Q_i^s$. The reworked items either get sold or are used for exchanging the defective items delivered to the costumers.

The second group is the defective items that are salvaged and their ratio equals $\alpha_{i_2}^s$; thus, the amount of products that can be used after the salvage equals $V_i^s = \alpha_{i_2}^s \pi_i^s Q_i^s$. These items will be sold in a second market and with a different price from the normal products.

The third group of the defective items is the scraps which their amount equals $\alpha_{i_3}^s \pi_i^s Q_i^s$. It also worthy of considering that $\alpha_{i_1}^s + \alpha_{i_2}^s + \alpha_{i_3}^s = 1$.

As demand is one of the important parameters in inventory systems, in this model there is also a demand by the amount of D_i^s for each product in a specific scenario; however, this amount is not the exact amount that must be produced.

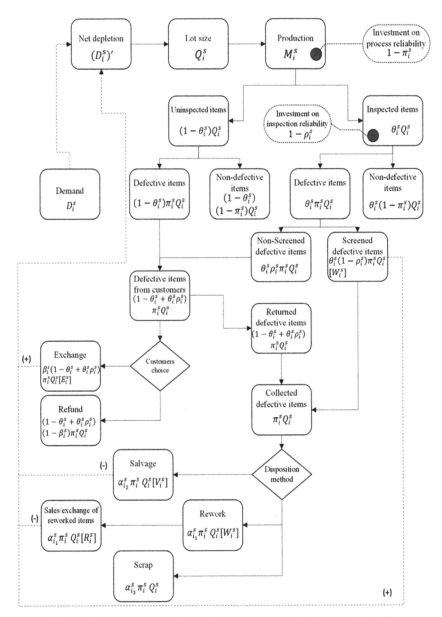

Fig. 1. Inventory status of imperfect products considering the QC policy

As shown in the flow chart number 1, we calculate net demand for each product; considering previously mentioned facts for each product, it equals $\left(D_i^s\right)' = D_i^s + G_i^s + E_i^s - R_i^s - V_i^s$.

As we can see, Fig. 2 shows the inventory behavior in a classic EPQ model in which the products are manufactured in the period of $T_{i_p}^s$ with the rate of M_i^s, and are used with the rate of D_i^s. The production is not completely perfect and it leads to a reduction in the inventory level. Now to rewrite it, the inspection, rework and salvage process should be considered.

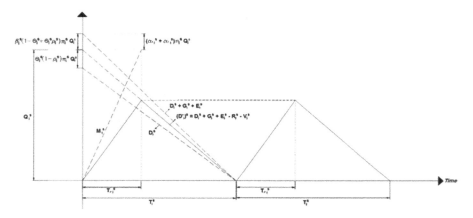

Fig. 2. The behavior of in producer's inventory

After inspection process, some defective items are spotted. The amount of these items will be added to the demand quantity. The remaining defective items that are not spotted by the inspection process, will be delivered to the costumers and will be returned by them afterwards and these defective items will be added to the demand as well. It leads to an increase in demand rate.

As shown, some of the defective items are usable after the rework. To answer E_i^s and D_i^s, we assumed that reworked items are usable like the normal products. Figure 3 shows these items which are compiled by the rate of R_i^s and are used constantly in $T_{i_r}^s$ cycles to prevent the maintaining of the unnecessary inventory.

Moreover, we showed that some items are also usable after the salvage and are sold in a certain price. The behavior of these savaged items is depicted in the Fig. 4. They are compiled with the rate of V_i^s in T_i^s cycles and are depleted with the same rate. Lastly, the imperfect items cannot be suitably reworked or salvaged, and are disposed of promptly at certain scrapping costs.

After explaining the supply chain cycle, we can discuss the model. As we mentioned in the first paragraph of this section, we model the problem with the intention of determining the lot size of each product in a specific scenario to maximize the profit. To achieve this purpose, we identify two groups of factors;

1- Factors that make profit for the system
2- Factors that impose cost to the system.

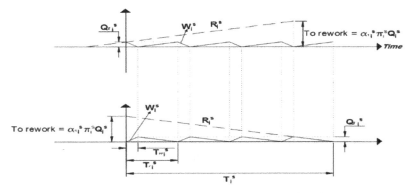

Fig. 3. The behavior of Inventory for reworked products

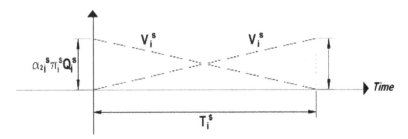

Fig. 4. The behavior of Inventory for salvage products

In this production cycle, selling the perfect products is the first factor that makes profit for the system. The price of each product in every scenario is pr_i^s and the probability of each scenario is P_i^s. The products that we can sell include perfect manufactured products and products recovered by rework; therefore, the income from selling the products can be formulated as Eq. (1):

$$\sum_{s=1}^{s}\sum_{i=1}^{i} P_i^s\left(pr_i^s\left(1 - \pi_i^s + \alpha_{1_i}^s \pi_i^s\right)Q_i^s\right) \tag{1}$$

The Second factor making income for this system is selling the salvaged items in a different price from the perfect ones; thus, if the price of each item in a specific scenario is v_i^s, the income from selling the salvaged products can be formulated as Eq. (2):

$$\sum_{s=1}^{s}\sum_{i=1}^{i} P_i^s v_i^s \alpha_{2_i}^s \pi_i^s Q_i^s \tag{2}$$

Now we take a look at the factors that impose cost for the system. The first factor is the production cost. In each cycle, the amount of the production equals Q_i^s and if the production cost of each product in a specific scenario is u_i^s, the production cost of the all items in a cycle will be given by Eq. (3):

$$\sum_{s=1}^{s}\sum_{i=1}^{i} P_i^s u_i^s Q_i^s \tag{3}$$

The inspection process is another cost making factor. If the inspection cost of each product for a specific scenario is shown by j_i^s, then the products inspection cost in a cycle is configured as follow:

$$\sum_{s=1}^{s}\sum_{i=1}^{i} P_i^s j_i^s \theta_i^s Q_i^s \tag{4}$$

When the defective items are returned by the costumers, a returning cost is imposed to the system. As shown, the amount of defective items delivered to the costumers equals $\left(1 - \theta_i^s + \theta_i^s \rho_i^s\right) Q_i^s \pi_i^s$, so if r_i^s is considered as the returning cost of each item, then the total returning cost will be given by Eq. (5):

$$\sum_{s=1}^{s}\sum_{i=1}^{i} P_i^s r_i^s \left(1 - \theta_i^s + \theta_i^s \rho_i^s\right) \pi_i^s Q_i^s \tag{5}$$

As mentioned before and shown in the flowchart, a specific number of costumers dealing with the defective items, refund their products which imposes a fine to the system due to the costumers' dissatisfaction. If this cost is considered l_i^s for each product, the total cost hitting the system because of the costumers' dissatisfaction can be formulated as Eq. (6):

$$\sum_{s=1}^{s}\sum_{i=1}^{i} P_i^s l_i^s \left(1 - \beta_i^s\right)\left(1 - \theta_i^s + \theta_i^s . \rho_i^s\right) \pi_i^s Q_i^s \tag{6}$$

After detecting and collecting the defective items, some of them need to be reworked $(w_i^s \alpha_{1_i}^s \pi_i^s)$. If rework cost of each item is w_i^s, then total cost is configured as follow:

$$\sum_{s=1}^{s}\sum_{i=1}^{i} P_i^s w_i^s \alpha_{1_i}^s \pi_i^s Q_i^s \tag{7}$$

According to the flowchart 1, some other parts of the defectives are the scraps. If the cost of each scrap is g_i^s, then the total cost of the scraps will be given by Eq. (8):

$$\sum_{s=1}^{s}\sum_{i=1}^{i} P_i^s g_i^s \alpha_{3_i}^s \pi_i^s Q_i^s \tag{8}$$

Another cost making factor is the system set up cost. It contains two kinds of costs: set up cost for production and set up cost for rework. The constraints are formulated mathematically in the following Eq. (9):

$$\sum_{s=1}^{s} \sum_{i=1}^{i} P_i^s \frac{\left(K_i^s + Kr_i^s n_i^s\right)\left(D_i'\right)^s}{Q_i^s} \tag{9}$$

The final cost making factor in this cycle is the storing cost. If the rate of storing cost for each item in a specific scenario is h_i^s, then the Eq. (10) represents the storing cost of the manufactured products and Eq. (11) depicts the storing cost of the reworked products.

$$\sum_{s=1}^{s} \sum_{i=1}^{i} P_i^s \frac{h_i^s u_i^s}{2} \left(1 - \frac{\left(D_i'\right)^s}{M_i^s}\right) Q_i^s \tag{10}$$

$$\sum_{s=1}^{s} \sum_{i=1}^{i} P_i^s \frac{h_i^s \left(2u_i^s + 2j_i^s + 2r_i^s(1 - \theta_i^s + \theta_i^s \cdot \rho_i^s) + w_i^s\right)}{2} \left(1 - \frac{\alpha_{1_i}^s \pi_i^s \left(D_i'\right)^s}{W_i^s}\right) \frac{\alpha_{1_i}^s \pi_i^s}{n_i^s} Q_i^s \tag{11}$$

According to Eq. (1) to (11), we obtain total profit per unit time (TPU) as a function of Q_i^s:

$$
\begin{aligned}
TPU = & \sum_{s=1}^{s} \sum_{i=1}^{i} P_i^s \left(pr_i^s\left(1 - \pi_i^s + \alpha_{1_i}^s \pi_i^s\right)Q_i^s\right) + \sum_{s=1}^{s} \sum_{i=1}^{i} P_i^s \, v_i^s \alpha_{2_i}^s \pi_i^s Q_i^s \\
& - \sum_{s=1}^{s} \sum_{i=1}^{i} P_i^s u_i^s \, Q_i^s - \sum_{s=1}^{s} \sum_{i=1}^{i} P_i^s j_i^s \theta_i^s Q_i^s \\
& - \sum_{s=1}^{s} \sum_{i=1}^{i} P_i^s r_i^s(1 - \theta_i^s + \theta_i^s \rho_i^s)\pi_i^s \, Q_i^s \\
& - \sum_{s=1}^{s} \sum_{i=1}^{i} P_i^s l_i^s(1 - \beta_i^s)(1 - \theta_i^s + \theta_i^s \cdot \rho_i^s) \, \pi_i^s \, Q_i^s \\
& - \sum_{s=1}^{s} \sum_{i=1}^{i} P_i^s w_i^s \alpha_{1_i}^s \pi_i^s Q_i^s - \sum_{s=1}^{s} \sum_{i=1}^{i} P_i^s g_i^s \alpha_{3_i}^s \pi_i^s Q_i^s \\
& - \sum_{s=1}^{s} \sum_{i=1}^{i} P_i^s \frac{\left(K_i^s + Kr^s n_i^s\right)(D_i')^s}{Q_i^s} \\
& - \sum_{s=1}^{s} \sum_{i=1}^{i} P_i^s \frac{h_i^s u_i^s}{2} \left(1 - \frac{(D_i')^s}{M_i^s}\right) Q_i^s \\
& - \sum_{s=1}^{s} \sum_{i=1}^{i} P_i^s \frac{h_i^s\left(2u_i^s + 2j_i^s + 2r_i^s(1 - \theta_i^s + \theta_i^s \cdot \rho_i^s) + w_i^s\right)}{2} \Big(1 \\
& \quad - \frac{\alpha_{1_i}^s \pi_i^s(D_i')^s}{W_i^s}\Big) \frac{\alpha_{1_i}^s \pi_i^s}{n_i^s} Q_i^s
\end{aligned}
$$

Now we look into the restrictions mentioned earlier in the second section of this research. The first constraint is the amount of free space in the warehouse. At first, the maximum inventory of each item should be calculated that equals $I_{m_i}^s = Q_i^s \left(1 - \frac{\left(D_i'\right)^s}{M_i^s}\right)$.

Then, the required space equals $\sum_{s=1}^{s} \sum_{i=1}^{i} P_i^s f_i^s I_{m_i}^s$, if the volume of each item is considered f_i^s. Therefore, if the available space of the warehouse is F, the available space constraint is configured as follow:

$$\sum_{s=1}^{s} \sum_{i=1}^{i} P_i^s f_i^s \sqrt{\frac{2K_i^s M_i^s \left(D_i'\right)^s}{h_i^s u_i^s \left(M_i^s - \left(D_i'\right)^s\right)}} \left(1 - \frac{\left(D_i'\right)^s}{M_i^s}\right) \leq F \tag{12}$$

Another constraint is the allocated budget to the production. If the allocated budget is adequate for production, then the constraint of the production cost will be given by Eq. (13):

$$\sum_{s=1}^{s} \sum_{i=1}^{i} P_i^s u_i^s \sqrt{\frac{2K_i^s M_i^s \left(D_i'\right)^s}{h_i^s u_i^s \left(M_i^s - \left(D_i'\right)^s\right)}} \leq B_1 \tag{13}$$

Inspection cost is also a constraint. In real world situations, budget is allocated to the inspection process according to the conditions. This issue is investigated in this research. If the budget allocated to inspection process is B_2, then inspection cost constraint is as follows:

$$\sum_{s=1}^{s} \sum_{i=1}^{i} P_i^s j_i^s \theta_i^s Q_i^s \leq B_2 \tag{14}$$

In real world situations, costumers' satisfaction is a crucial factor in completion, so it has a direct impact on production. Therefore, producers cannot disregard this factor and due to its importance, we managed to consider it as a parameter in our model and calculated the fine imposed to the system caused by costumer dissatisfaction:

$$\sum_{s=1}^{s} \sum_{i=1}^{i} P_i^s l_i^s \left(1 - \beta_i^s\right)\left(1 - \theta_i^s + \theta_i^s \cdot \rho_i^s\right) \pi_i^s Q_i^s \tag{15}$$

In order to make the model more real, a limitation is placed on the fine imposed to the system by costumer dissatisfaction. If the acceptable limitation for the fine imposed by costumer dissatisfaction for total manufactured products is B_3, then this constraint will be given by Eq. (16):

$$\sum_{s=1}^{s} \sum_{i=1}^{i} P_i^s l_i^s \left(1 - \beta_i^s\right)\left(1 - \theta_i^s + \theta_i^s \cdot \rho_i^s\right) \pi_i^s Q_i^s \leq B_3 \tag{16}$$

This constraint helps us control one of the most important costs of the system and also gives us the opportunity to calculate the quality of products by measuring the costumers' satisfaction.

Now the optimal production quantity is determined to maximize the profit in different ways.

4 Solution Method

In this paper, a MINLP model is presented. Using accurate methods for solving this kind of problems is sophisticated, therefore two metaheuristic algorithms are developed to find the nearest solution to the optimal one. Optimization method is developed on the basis of invasive weed optimization algorithm and a genetic algorithm which their frameworks are discussed in the following paragraphs.

4.1 Invasive Weed Optimization Algorithm

The successfully utilized IWO led to high differences in differing optimization problems, including the attuning of a robust controller (Mehrabian and Lucas 2006), the design of an E-formed MIMO antenna (Zhang et al. 2009), optimally positioned piezoelectric actuators (Mehrabian and Yousefi-Koma 2007), the study of electricity marketplace dynamics (Sahraei-Ardakani et al. 2008), and solving different electromagnetic problems (Karimkashi and Kishk 2010). Accordingly, linear array antenna synthesis – a standard antenna engineering problem – was represented to exemplify the utilization of IWO by Karimkashi and Kishk (2010). Lastly, the application of IWO to a U-slot patch antenna aimed at achieving the desirable dual-band features. Ojha and RamuNaidu (2015) used IWO to solve six nonlinear constrained optimization problems. Furthermore, they introduced the integration of particle swarm optimization (PSO) and invasive weed optimization (IWO). By consolidating the stochastic ranking technique, they handled the restraints, named PSO-IWO-SR. Compared to other evolution-based algorithms, the distinct specifications of IWO include the methods of multiplication, spatial dispersion, and competitory elimination (Abu-Al-Nadi et al. 2013).

The algorithm has a continued procedure by the time it reaches a maximal number of reiterations. In IWO, only the superiorly fitted plants are survivable and have seed production, whereas the rest undergoes elimination. Then, the superiorly fitted plants are closest to the optimum solution. The detailed procedures are presented in the section below.

Initialization. Invasive weed optimization algorithm starts with a primary population which equals to the number of seeds that are produced and dispersed. The primary answers are stochastically placed in the problem space. The evaluation of the fittingness of the primed weeds depends on the fitness function or the objective function selected for the optimization problem.

Reproduction. In this phase, each participant in the primary population is allowed to reproduce in accordance with its fitness level. Therefore, the better is the population growth rate, the higher number of seeds are produced.

According to the Fig. 5, the number of seeds starts from the value of S_{min} for the worst participant and increases linearly to the value of S_{max} for the best one as follows:

$$Number\ of\ seeds\ around\ weed\ i = \frac{F_i - F_{worst}}{F_{best} - F_{worst}}(S_{max} - S_{min}) + S_{min} \qquad (17)$$

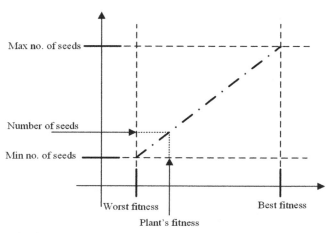

Fig. 5. Seed production procedure in a colony of weeds (reproduced from Pourjafari and Mojallali (2012), with permission).

Spatial Dispersal. The seed distribution is done with a random distance from the parental plant after the completion of the multiplication stage. The distance is on a normally distributed basis with an average of 0 and a variance of σ^2.

In this case, standard deviation is a variable which varies in accordance with time. The initial value of standard deviation is predetermined (σ_{init}) which decreases in every generation to a final value (σ_{final}), so that the algorithm gradually shifts from exploration behavior to exploitation behavior. The value N is the controller of the standard deviation's reduction pace which will be linear and uniform if it equals 1.

$$\sigma_{cur} = \frac{(iter_{max} - iter)^n}{(iter_{max})^n}\left(\sigma_{init} - \sigma_{final}\right) + \sigma_{final} \qquad (18)$$

Where $iter_{max}$ is the maximum number of iterations, σ_{cur} is the standard deviation at the current generation.

Competitive Exclusion. Following the passage of few iterations, the count of generated plants is maximized colonially (P_{max}). Thus, the plants should compete necessarily to restrict the maximization of plants in a population. Individual weeds have the possibility of producing seeds and spreading throughout the space. The seeds are ranked together with their parents right after they find their position (Mehrabian and Lucas (2006)). In this phase, plants are ranked regarding their fitness level. The worst ones are eliminated and the survived ones are permitted to grow and reproduce.

4.2 Genetic Algorithm

Recently, the development of the genetic algorithm (GA) has led to solving MINLP problems. Amongst metaheuristic algorithms, GA has been accepted as a robust one to solve MINLP models (Yokota et al. 1996). Genetic algorithm has been very useful when it has come to solving the optimization problems. Mondal and Maiti (2003) used genetic algorithm to optimize an EOQ model in FNLP condition. Roozbeh Nia (2015) used genetic algorithm to find a close answer to the optimized one while trying to expand an EOQ model for the order management in a green supply chain. Moreover, Pasandideh and Niaki (2008) used genetic algorithm to solve a MINLP which had been made by expanding an EPQ model. The main rules of GA introduced by Holland, are:

Initially, a starter population of chromosomes is created to assess their fittingness. Next, novel chromosomes are created using genetic operators, including crossover, mutation, and reproduction. This is followed by evaluating the fittingness of the novel population of the prepared chromosomes. The procedure is terminated by satisfying the ending situation, thereby returning the superior chromosome; if this is not the case, novel chromosomes are created by repeating the procedure with the use of genetic operators, including crossover, mutation, and reproduction.

Chromosome Representation. In the genetic algorithm, the chromosome is a string or a sequence of genes that is represented as an encoded form of a response (reasonable or unreasonable). The most important phase of running GA in the process of solving a problem is to design a proper chromosome. In this research each chromosome is composed of produced values for each scenario.

Evaluation and Initial Population. In this algorithm, we need an initial population (like what was discussed in IWO) which are the chromosomes. In this phase, all the values produced go under the evaluation. In order to avoid producing unreasonable values, enormous fines are added to the objective function.

Selection. Selection gives us the opportunity to transfer the good genes of a solution to the later generations. In this research each chromosome is selected by a roulette wheel. Based on this method, the probability of selecting each chromosome depends on its fitness level.

Crossover Operator. Few genes of chromosomes are exchanged by the crossover operator to produce e few offspring via breaking and rejoining two chosen chromosomes. A prearranged factor Pc use is representative of the probable application of the crossover operator in a population. The arithmetical crossover is applied to do this. By considering that $X_1 = (x_{11}.x_{12}, \ldots, x_{1n})$ and $X_2 = (x_{21}.x_{22}, \ldots, x_{2n})$ are chosen as two parents, if $\alpha = (\alpha_1.\alpha_2, \ldots, \alpha_n)$, in which $0 \leq \alpha_i \leq 1$, the offspring are produced as below:

$$\begin{cases} y_{1i} = \alpha_i x_{1i} + (1 - \alpha_i)x_{2i} \ i = 1, 2, 3 \ldots, n \\ y_{2i} = \alpha_i x_{2i} + (1 - \alpha_i)x_{1i} \ i = 1, 2, 3 \ldots, n \end{cases} \tag{19}$$

It is noteworthy that Ramezanian et al. (2012) demonstrated proper functionality of the arithmetical crossover operator in continual space. It should be noted that this operator in the chromosome is only utilized for continuously encoded segments. A two-point crossover is selected to discover the solution space for binary-coded segments of the chromosome. Indeed, if a chromosome pair is chosen as the parental one from the chromosome generation, the action of the crossover operator will be as: initially, a part is chosen randomly from individual chromosomes. Next, the progenies are produced by the exchange of chosen parental chromosomes.

Mutation and Stopping Criterion. The mutation operator modifies the parental characteristics at random to make a progeny solution. This operator effectively generates a sensible degree of diversification in the population. Besides, it performs searching by skipping locally optimized solutions. An exchange mutation is adopted herein. This mutation exchanges the values of the two haphazardly chosen genes of the present solution with each other. The operator is executed in every segment of the chromosome. The searching procedure of the algorithm discontinues when the count of obtained generations exceeds a maximal count of generations or at a certain determined count of generations with no enhancement of the superior recognized solution.

5   Numerical Analyses

In this section we solved 30 different numeric examples of the problem using two mentioned algorithms. The solved problems have different dimensions. Problem parameters are selected stochastically in the specific intervals given in Table 2. Three scenarios are considered for each parameter: good, mediocre and bad. Accordingly, this section illustrates the numerical function of the introduced model. The model generic feature is demonstrated by providing and analyzing three groups of problems, namely small, medium, and large-sized problems. The three groups establish a wide range of problems capable of evaluating the functionality of the model and algorithms.

Important parameters in these two algorithms are selected stochastically from the specific intervals shown in Table 3.

The results are gathered in the Table 4 and the condition of the various examples' objective function in two algorithms are shown in the Fig. 6.

5.1   Parameter Tuning

In this section, we intend to calculate the best values for the parameters of each algorithm to ensure that the algorithms produce the best performance. Thus, three basic parameters of each algorithm are shown in the Table 3 with their intervals which has already been determined. The intervals are determined based on the experiments and experiences.

In this study, regression analysis is used to tune the parameters. Therefore, the values of the object function of numeric problems which are solved, are shown in Table 5 and are considered as the answer's variable. Moreover, the basic parameters of each algorithm are fitted based on them.

Table 2. Generation of parameter ranges.

Parameter	Good	Mediocre	Bad
P_i	0.1–0.3	0.1–0.4	0.1–0.3
$*(P_i^{good} + P_i^{Mediocre} + P_i^{Bad} = 1 \forall i)$			
Kr_i	1000–2000	2001–3000	3001–4000
v_i	25–45	46–60	61–75
K_i	2000–3000	3001–4000	4001–5000
r_i	60–80	81–100	101–120
ρ_i	0.03–0.079	0.08–0.14	0.141–0.2
D_i	1600–1800	1200–1599	1000–1199
θ_i	0.21–0.25	0.13–0.2	0.07–0.12
j_i	8–14	15–24	25–35
n_i	1–2	3–4	5–6
π_i	0.03–0.07	0.08–0.19	0.2–0.3
u_i	40–60	61–80	81–100
g_i	50–69	70–99	100–120
w_i	10–19	20–49	50–70
l_i	100–149	150–249	250–300
h_i	0.05–0.09	0.1–0.19	0.2–0.25
β_i	0.61–0.8	0.31–0.6	0.1–0.3
α_{1_i}	0.35–0.4	0.25–0.35	0.2–0.3
α_{2_i}	0.3–0.4	0.2–0.3	0.1–0.25
α_{3_i}	$1 - (\alpha_{1_i} + \alpha_{2_i})$	$1 - (\alpha_{1_i} + \alpha_{2_i})$	$1 - (\alpha_{1_i} + \alpha_{2_i})$
f_i	0.5–1.99	2–3.99	4–5.5
M_i	1701–1900	1301–1700	1200–1300
W_i	0.05–0.1	0.11–0.2	0.21–0.3
Pr_i	551–600	401–550	300–400
F	10000		
B_1	500000000		
B_2	50000000		
B_3	30000000		

Table 3. Generation of parameter ranges

IWO		GA	
Parameter	Range	Parameter	Range
S_{max}	4–7	n_{pop}	50–100
σ_{final}	0.001–0.0099	P_m	0.1–0.4
n	1–3	P_c	0.4–0.8

Table 4. Results before tuning the parameters

Problem	Size	The number of products	Objective function		Problem	Size	The number of products	Objective function	
			GA	IWO				GA	IWO
1	Small	3	1623954	1617847	16	Medium	5	2433952	2319924
2			1694720	1685619	17			2570863	2551777
3			1475205	1485753	18			2313715	2264127
4			1314473	1314457	19			2510341	2489083
5			1551725	1551725	20			2421682	2286879
6	Medium	4	2037419	2017035	21			2753138	2680663
7			1985191	1919719	22			2894364	2964717
8			1637010	1657078	23			2793858	2729797
9			2066686	2046967	24			2742153	2646409
10			1964463	1948825	25			2509862	2436320
11			2026191	1961523	26	Large	7	3016151	2900441
12			2300386	2270107	27			3040923	2984002
13			1956230	1916285	28			3247652	3103285
14			2132085	2109238	29			3095523	3074645
15			2016221	1978921	30			3012999	2916975

For each algorithm, a nonlinear multivariate regression equation is calculated by R software and statistical analysis. The IWO algorithm is fitted by a nonlinear multivariate simple regression.

$$IWO = 2603358 + 4333330\sigma_{final} + 5533400000(\sigma_{final})^2 + 70812S_{max}$$
$$+ 4468(S_{max})^2 + 63482n + 4559(n)^2$$

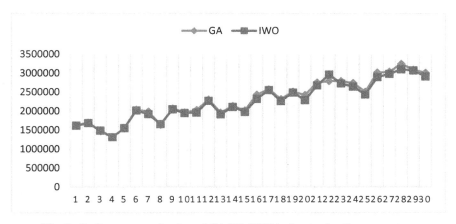

Fig. 6. Performance evaluation of GA VS. IWO before tuning the parameters.

s.t.

$$\begin{cases} 4 \leq S_{max} \leq 7 \\ 0.001 \leq \sigma_{final} \leq 0.0099 \\ 1 \leq n \leq 3 \\ n, \sigma_{final} \geq 0 \quad S_{max} \geq 0 \; and \; integer \end{cases} \tag{20}$$

Since determining a simple nonlinear regression equation for GA algorithm was not possible, we determined a nonlinear regression equation for GA algorithm using the exponential distribution after multiple experiments.

$$GA = 2084580 + 3006030e^{55.8P_c} + 333333e^{0.884P_m} + 644393e^{1.83n_{pop}}$$

s.t.

$$\begin{cases} 0.1 \leq P_m \leq 0.4 \\ 0.4 \leq P_c \leq 0.8 \\ 50 \leq n_{pop} \leq 100 \\ P_m, P_c \geq 0 \, n_{pop} \geq 0 \, and \, integer \end{cases} \tag{21}$$

The equations are considered as objective functions, and intervals are considered as constraints, therefore we ought to maximize the objective function regarding to the constraints. We calculate the Eqs. (20) and (21) in GAMS and the optimized value of parameters are determined as Eqs. (22) and (23).

$$S^*_{max} = 4 \, \sigma^*_{final} = 0.004 \, n^* = 1.08 \tag{22}$$

$$P^*_m = 0.4 \, P^*_c = 0.4 \, n^*_{pop} = 86 \tag{23}$$

Table 5. Performance evaluation of GA VS. IWO after tuning the parameters.

Problem	The number of products	Objective function							
		IWO	Time	Gams	$Deviation_{IWO}$	GA	Time	Gams	$Deviation_{GA}$
1	3	1623528	6.6	1624195	0.000410	1623954	2.97	1624195	0.000148
2		1694720	6.39	1695732	0.000597	1694720	3.57	1695732	0.000597
3		1492737	6.59	1493326	0.000394	1492781	3.29	1493326	0.000365
4		1688016	6.17	1700726	0.007529	1699273	3.22	1700726	0.000855
5		1550378	6.7	1551768	0.000896	1551725	2.75	1551768	0.000027
6	4	2034555	6.08	2114225	0.039158	2037412	3.49	2114225	0.037701
7		1976022	6.07	1985407	0.004749	1985388	3.22	1985407	0.000009
8		1801591	6.54	1802591	0.000555	1801971	4.19	1802591	0.000344
9		2170389	6.62	2418885	0.114493	2184942	3.21	2418885	0.107070
10		1946276	6.45	2042901	0.049646	1964457	2.9	2042901	0.039931
11		2025018	6.48	2026723	0.000841	2026189	3.48	2026723	0.000263
12		2297481	6.2	2301163	0.001602	2300378	3.36	2301163	0.000341
13		1944089	6.77	2172108	0.117288	1956223	3.23	2172108	0.110358
14		2119343	6.84	2158512	0.018481	2132073	3.24	2158512	0.012400
15		2005853	6.68	2077194	0.035566	2016212	2.89	2077194	0.030245
16	5	2407155	6.32	2440742	0.013952	2433799	3.71	2440742	0.002852
17		2543566	6.31	2576242	0.012846	2569784	3.33	2576242	0.002513
18		2304063	6.25	2323871	0.008596	2313715	4.03	2323871	0.004389
19		2480463	8.37	2533075	0.021210	2510270	3.12	2533075	0.009084
20		2385092	6.42	2428500	0.018199	2421471	4.01	2428500	0.002902
21		2720558	6.31	2754430	0.012450	2752955	3.06	2754430	0.000535
22		2970248	6.33	3015510	0.015238	3014483	4.5	3015510	0.000340
23		2757200	6.39	2800679	0.015769	2793750	4.14	2800679	0.002480
24		2719224	6.15	2747518	0.010405	2742063	3.66	2747518	0.001989
25		2483507	6.67	2586701	0.041551	2509666	3.52	2586701	0.030695
26	7	2967810	6.23	3016618	0.016445	3015347	2.63	3016618	0.000421
27		2936332	6.69	3043809	0.036602	3039644	3.15	3043809	0.001370
28		3176588	6.36	3293011	0.036650	3245281	3.36	3293011	0.014707
29		3011382	6.91	3091574	0.026629	3091249	3.08	3091574	0.000105
30		2945244	6.34	3068020	0.041686	3011532	3.65	3068020	0.018757

6   Comparison of the Method

6.1   Results

In this section, the solved examples of Sect. 5 are calculated considering the optimal parameters calculated in Sect. 5 in order to evaluate the capability of the model; the calculating time is also recorded. The GAMS/BARONR 23.5 software on an Intel R core™ i7, 2.59 GHz laptop with 12 GB RAM is used to solve the problems. Note that the branch-and-reduce optimization navigator (BARON) proposed by Tawarmalani and Sahinidis (2005) is the algorithm for the universal solution of NLP and MINLP. A comparison is made between the findings of applying the algorithms and those of BARON for validation. These are compared using a quality scale named the relative percentage deviation of the objective function, the definition of which for individual solutions is given below:

$$Deviation_{Obj} = \frac{z_{algorithm} - z_{BARON}}{z_{BARON}} \times 100\% \qquad (24)$$

Results shown in Table 5 depicts that the suggested algorithms have had good performance and have produced answers close to the optimal answer. The deviation mean of the GA algorithm's answers from precise answers is 0.01446. This number is 0.02401 for IWO. The example 3 shows the minimum deviation for IWO which is 0.000394. These numbers are 5 and 0.000027 for GA algorithm respectively. Figure 7 shows the comparison between the two algorithms regarding to their deviation from the optimal precise answer.

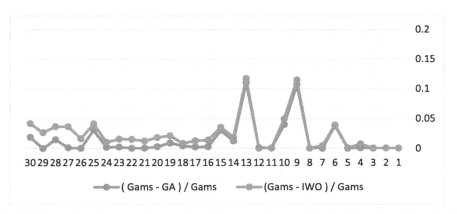

Fig. 7. Comparison between the two algorithms regarding to their deviation from the optimal precise answer

6.2   Comparison

In this section, the comparison between two algorithms is made according to the results produced in the Sect. 5.1 and analysis of the outputs, in order to find the more efficient algorithm. There are two ways to make this comparison:

Statistical Approaches. This subdivision evaluates the functionality of IWO and compares it with the GA. The statistical analyses for the findings of the two algorithms using paired samples t-tests are shown in Table 6. It is worth mentioning that this statistical test can usefully test the average of the values of the expectable overall revenue.

Through this hypothesis test we can examine whether there is a difference between the averages of observations that are tested in pairs and under the same conditions or not. The important part is that the data distribution for each pair should be normal. The way to do this test is to gather n samples. According to the central limit theorem, at least 30 samples must be tested to ensure the normality of the data. Once more the values of the solved examples in the Sect. 4-3 are recorded under GA's number 1 test and also are recorded under IWO's number 2 test. Then the average of the difference between two tests $\left(\overline{d} \right)$ and the standard deviation of the difference between two tests (s_d) are calculated in order to calculate Test statistic (t_0). The Test statistic is calculated by this formula:

$$t_0 = \frac{\overline{d} - 0}{s_d / \sqrt{n}} \tag{25}$$

The significance level of the test is 95% according to experiences.

$$\begin{cases} H_0: \mu_{Z_{iwo}} - \mu_{Z_{ga}} = 0 \\ H_1: \mu_{Z_{iwo}} - \mu_{Z_{ga}} \neq 0 \end{cases} \tag{26}$$

According to formula 25 and the recorded data in Fig. 7, the test statistic for comparison of objective function means equal $t_0^Z = 5.526$, therefore $P_{value} \cong 0.001$ and the null hypothesis gets rejected. The test is done for one more time regarding to data recorded in Table 7 about solving time for the two algorithms so that the comparison of two algorithms with two criterion would be possible.

The test statistic for the comparison of the mean time was $t_0^{Time} = 5.526$ and $P\_value \cong 0.0001$, therefore, the null hypothesis is rejected.

$$\begin{cases} H_0: \mu_{time_{iwo}} - \mu_{time_{GA}} = 0 \\ H_1: \mu_{time_{iwo}} - \mu_{time_{GA}} \neq 0 \end{cases} \tag{27}$$

The result of our experiments in this section showed that at 95% confidence level, the two algorithms do not differ significantly in the mean value of the target function and the mean solution time.

Multiple Attribute Decision Making (MADM). The diverse methods of MADM can usefully be applied to make comparisons. One of its components is the TOPSIS procedure employed in our study. TOPSIS is the shortened from of Technique for Order Preference by Similarity to Ideal.

Solution mean values are preferable approaches in terms of being similar to the perfect solution. Hoang and Yong (1981) initially introduced this technique, where m options are assessed by n indices. The ideal solution (positive) and the negative ideal solution are defined by the logistic rule of this model. The ideal solution (positive) raises

Table 6. Calculated values for paired-samples T Test - objective functions value

Problem	Z_{IWO}	Z_{GA}	$D_i = Z_{IWO} - Z_{GA}$	$(D_i - \overline{D})$	$(D_i - \overline{D})^2$
1	1623528	1623954	426	-24716.63333	610911963.2
2	1694720	1694720	0	-25142.63333	632152010.8
3	1492737	1492781	44	-25098.63333	629941395
4	1688016	1699273	11257	-13885.63333	192810813
5	1550378	1551725	1347	-23795.63333	566232165.6
6	2034555	2037412	2857	-22285.63333	496649452.9
7	1976022	1985388	9366	-15776.63333	248902159.2
8	1801591	1801971	380	-24762.63333	613188009.4
9	2170389	2184942	14553	-10589.63333	112140334.1
10	1946276	1964457	18181	-6961.63333	48464338.62
11	2025018	2026189	1171	-23971.63333	574639204.5
12.	2297481	2300378	2897	-22245.63333	494868202.3
13	1944089	1956223	12134	-13008.63333	169224541.1
14	2119343	2132073	12730	-12412.63333	154073466.2
15	2005853	2016212	10359	-14783.63333	218555814.4
16	2407155	2433799	26644	1501.36667	2254101.878
17	2543566	2569784	26218	1075.36667	1156413.475
18	2304063	2313715	9652	-15490.63333	239959721
19	2480463	2510270	29807	4664.36667	21756316.43
20	2385092	2421471	36379	11236.36667	126255935.9
21	2720558	2752955	32397	7254.36667	52625835.78
22	2970248	3014483	44235	19092.36667	364518465.1
23	2757200	2793750	36550	11407.36667	130128014.3
24	2719224	2742063	22839	-2303.63333	5306726.519
25	2483507	2509666	26159	1016.36667	1033001.208
26	2967810	3015347	47537	22394.36667	501507658.6
27	2936332	3039644	103312	78169.36667	6110449886
28	3176588	3245281	68693	43550.36667	1896634437
29	3011382	3091249	79867	54724.36667	2994756307
30	2945244	3011532	66288	41145.36667	1692941198
μ	2305947.6	2331090.233	25142.634		

Table 7. Calculated values for paired-samples T Test – CPU time value

Problem	$Time_{IWO}$	$Time_{GA}$	$D_i = Time_{IWO} - Time_{GA}$	$(D_i - \overline{D})$	$(D_i - \overline{D})^2$
1	2.97	6.6	−3.63	−0.53	0.2809
2	3.57	6.39	−2.82	0.28	0.0784
3	3.29	6.59	−3.3	−0.2	0.04
4	3.22	6.17	−2.95	0.15	0.0225
5	2.75	6.7	−3.95	−0.85	0.7225
6	3.49	6.08	−2.59	0.51	0.2601
7	3.22	6.07	−2.85	0.25	0.0625
8	4.19	6.54	−2.35	0.75	0.5625
9	3.21	6.62	−3.41	−0.31	0.0961
10	2.9	6.45	−3.55	−0.45	0.2025
11	3.48	6.48	−3	0.1	0.01
12	3.36	6.2	−2.84	0.26	0.0676
13	3.23	6.77	−3.54	−0.44	0.1936
14	3.24	6.84	−3.6	−0.5	0.25
15	2.89	6.68	−3.79	−0.69	0.4761
16	3.71	6.32	−2.61	0.49	0.2401
17	3.33	6.31	−2.98	0.12	0.0144
18	4.03	6.25	−2.22	0.88	0.7744
19	3.12	8.37	−5.25	−2.15	4.6225
20	4.01	6.42	−2.41	0.69	0.4761
21	3.06	6.31	−3.25	−0.15	0.0225
22	4.5	6.33	−1.83	1.27	1.6129
23	4.14	6.39	−2.25	0.85	0.7225
24	3.66	6.15	−2.49	0.61	0.3721
25	3.52	6.67	−3.15	−0.05	0.0025
26	2.63	6.23	−3.6	−0.5	0.25
27	3.15	6.69	−3.54	−0.44	0.1936
28	3.36	6.36	−3	0.1	0.01
29	3.08	6.91	−3.83	−0.73	0.5329
30	3.65	6.34	−2.69	0.41	0.1681

the profit benchmark and reduces the cost benchmark. The optimum choice is described as the one with the shortest distance to the ideal solution and the longest distance to the negative ideal solution. Otherwise stated, in choices ranked by the TOPSIS technique, those are more properly ranked that have the highest proximity to the ideal solution. In the objective space between two criteria (Fig. 8). A^+ and A^- are respectively the ideal and negative ideal solutions. The distance of A_1 is lower from the ideal solution and is farther to the negative ideal solution in comparison to A_2.

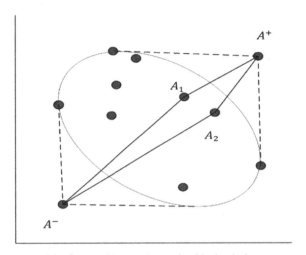

Fig. 8. The ideal and negative ideal solutions

In this research the optimal answer is the one calculated in GAMS software. The TOPSIS technique algorithm is as shown below:

Formation and Normalization of the Decision Matrix
In this phase indicators of decision making are chosen and the decision matrix is formed. Then the matrix gets normalized by using the norm method.

In this research, the objective function value's mean and the solution time mean (seconds) are considered as decision indicators for comparing the two algorithms. On the other hand, the answers obtained in GAMS software are considered as optimal answers. The decision matrix is as shown below (Table 8):

Table 8. The decision matrix.

	OBJ	CPU (s)
GA	2331090	3.398
IWO	2305947	6.5
GAMS	2362858	1.02

If r_{ij} is a component of the decision matrix, and n_{ij} is a component of the normalized decision matrix, there will be:

$$n_{ij} = \frac{r_{ij}}{\sqrt[2]{\sum_{i=1}^{m} r_{ij}^2}} \tag{28}$$

Therefore, the normalized matrix would be (Table 9):

Table 9. The normalized decision matrix.

	OBJ	CPU (s)
GA	0.5767	0.458
IWO	0.877	0.877
GAMS	0.137	0.137

Criteria Weighting
In this phase the importance degree of each indicator is considered in the normalized decision matrix. If w_j is the weight of each indicator and v_{ij} is an element of the weighted normalized matrix then we will have:

$$v_{ij} = w_j \times n_{ij} \tag{29}$$

Importance degree of each indicator equals ½ and the weighted normalized decision matrix is:

$$\frac{1}{2} \times \begin{pmatrix} 0.458 & 0.5767 \\ 0.877 & 0.5705 \\ 0.137 & 0.5846 \end{pmatrix} = \begin{pmatrix} 0.229 & 0.288 \\ 0.438 & 0.285 \\ 0.068 & 0.292 \end{pmatrix} \tag{30}$$

Determining the Positive and Negative Ideal Answers
In this phase, first the positive and negative indicators are determined. In this problem the objective function's mean is the positive indicator and the higher the better it is in this problem. On the other hand, the higher solution time would be worse in this problem.

$$\begin{cases} A^+ = \begin{pmatrix} 0.068 & 0.292 \end{pmatrix} \\ A^- = \begin{pmatrix} 0.438 & 0.285 \end{pmatrix} \end{cases} \tag{31}$$

Determining the Coefficient of Closeness of Each Option
In this phase the option's distance to the negative and positive (33) ideals are calculated as shown below:

$$\begin{cases} d_i^+ = \sqrt{\sum_{j=1}^{n} \left(v_{ij} - v_j^+ \right)^2} \\ d_i^- = \sqrt{\sum_{j=1}^{n} \left(v_{ij} - v_j^- \right)^2} \end{cases} \tag{32}$$

The numbers for two algorithms are shown as below:

$$GA = \begin{cases} d_i^+ = 0.161 \\ d_i^- = 0.209 \end{cases} \tag{33}$$

$$IWO = \begin{cases} d_i^+ = 0.37 \\ d_i^- = 0 \end{cases} \tag{34}$$

Ranking Options Using Relative Proximity Criteria.
In the last phase, the options are ranked using cl_i criteria in which $0 \leq cl_i \leq 1$ and is calculated as below:

$$cl_i = \frac{d_i^-}{d_i^+ + d_i^-} \tag{35}$$

The closer cl_i to 1 the more ideal answer it is. The criteria are as shown below for the two algorithms:

$$\begin{cases} cl_{GA} = 0.564 \\ cl_{Iwo} = 0 \end{cases} \tag{36}$$

The result numbers show that the GA algorithm is more useful for this problem and produces the closest answers to the optimal one in less time.

7   Managerial Implication

According to what is analyzed here, our suggested administrative intimations and visions are helpful for inventory directors in taking decisions for product lot-sizing. The presented model can be really useful for QC and production managers regarding to its applications in calculating lot size for various items in single machine-multi product systems and always covers the demand. The net demand is a function of various factors. In this model for instance, QC policy has direct effects on the net demand. On the other hand, in real world situations, factors like demand, cost, and inspection errors are not deterministic which could affect cost and income directly. Therefore, the parameters are considered stochastic in order to make the model more flexible dealing with different scenarios. More about this model's application is its constraints which make the model much more applicable. Warehouse space is one of the factors affecting manufacturing system's production quantity in real world situations if which be used with stochastic demand in the model, it will give us the opportunity to calculate the maximum production quantity for each product. Nowadays customers' satisfaction is one of the important comparing factors. Therefore, a dissatisfaction fine has been considered in this model because of the probable occurrence of defective item delivery. To control this factor, an upper limit is determined for total dissatisfaction fine in order to ignore customer satisfaction level to become lower than a specified level. The budget constraint is one of the constraints that makes the model more applicable. In real world situations the budget is never infinite; so allocating costs to different products is a key point in raising profits in multi-product systems which is considered in this model. Moreover, this model can help the managers' decision making to eliminate the defectives of their production system on time due to the classification of defective items in three categories or to improve QC policies due to these factors. Altogether, the introduced model has a high closeness to realistic circumstances and flexibility to tackle various situations, making it better applicability. At the same time, the details provided in this model helps the managers in decision making in different departments.

8   Conclusion and Future Research

This paper presents a nonlinear programming model to develop an EPQ model in which warehouse space limitation and budget constraints are included but shortage is not permitted. This model contains stochastic parameters with different scenarios which allow it to become more real. The present model aims at maximizing the revenue with the calculation of the optimal quantity of producing every product in individual scenarios. In this model, demand gets affected by a variety of factors which change its level. In this paper, we concentrated on inspection, various inspection errors, defective items and their costs. The defective items are categorized regarding to the action taken on them; all their costs are also considered in the model. The model is solved using two metaheuristic algorithms, the parameters of which are attuned by multivariate nonlinear regression. In order to show the capability of the models, a variety of examples in different dimensions are evaluated and later calculated accurately by GAMS software to determine the deviation between the optimal answer and the algorithms. Statistical analysis shows that the algorithms work efficiently and produce answers close to the optimal ones. In the interest of choosing a better algorithm, Paired T and TOPSIS technique are used. The value of the objective function and solving time are taken as the comparing indexes. The comparisons demonstrate th more effectiveness of the GA algorithm in this model. Lastly, our findings confirm that the introduced model and the procedures chosen for solving the problem are applicable. To extend to forthcoming studies, the introduced model will be developed by taking account of the working degree and machine reliability objectives under the multi-product – multi-machine circumstances. Additionally, the spoilable items will be considered under the shortage situation. With regard to solution approaches, a precise method will be utilized for solving the novel MINLP model. Moreover, the findings of the novel precise method will be compared with the employed metaheuristic methodologies in the areas of optimality criteria, including the count of considered reiterations, infeasibility, optimality error, and complementarity.

References

Abu-Al-Nadi, D.I., Alsmadi, O.M., Abo-Hammour, Z.S., Hawa, M.F., Rahhal, J.S.: Invasive weed optimization for model order reduction of linear MIMO systems. Appl. Math. Model. **37**(6), 4570–4577 (2013)

Alamri, A.A., Harris, I., Syntetos, A.A.: Efficient inventory control for imperfect quality items. Eur. J. Oper. Res. **254**(1), 92–104 (2016)

Axsäter, S.: Exact analysis of continuous review (R, Q) policies in two-echelon inventory systems with compound Poisson demand. Oper. Res. **48**(5), 686–696 (2000)

Barzoki, M.R., Jahanbazi, M., Bijari, M.: Effects of imperfect products on lot sizing with work in process inventory. Appl. Math. Comput. **217**(21), 8328–8336 (2011)

Cárdenas-Barrón, L.E., Shaikh, A.A., Tiwari, S., Trevino-Garza, G.: An EOQ inventory model with nonlinear stock dependent holding cost, nonlinear stock dependent demand and trade credit. Comput. Ind. Eng. **139**, 105557 (2020)

Chang, H.C.: An application of fuzzy sets theory to the EOQ model with imperfect quality items. Comput. Oper. Res. **31**(12), 2079–2092 (2004)

Chiu, S.W., Wang, S.L., Chiu, Y.S.P.: Determining the optimal run time for EPQ model with scrap, rework, and stochastic breakdowns. Eur. J. Oper. Res. **180**(2), 664–676 (2007)

Cunha, L.R.A., Delfino, A.P.S., dos Reis, K.A., Leiras, A.: Economic production quantity (EPQ) model with partial backordering and a discount for imperfect quality batches. Int. J. Prod. Res. **56**(18), 6279–6293 (2018)

Samuel, E.: Scheduling for batch production. Inst. Prod. Eng. J. **36**(9), 549–570 (1957)

Feng, Q., Kapur, K.C.: Economic design of specifications for 100% inspection with imperfect measurement systems. Qual. Technol. Quant. Manage. **3**(2), 127–144 (2006)

Gharaei, A., Hoseini Shekarabi, S.A., Karimi, M., Pourjavad, E., Amjadian, A.: An integrated stochastic EPQ model under quality and green policies: generalised cross decomposition under the separability approach. Int. J. Syst. Sci. Oper. Logist. **8**, 1–13 (2019)

Gharaei, A., Hoseini Shekarabi, S.A., Karimi, M.: Modelling and optimal lot-sizing of the replenishments in constrained, multi-product and bi-objective EPQ models with defective products: generalised cross decomposition. Int. J. Syst. Sci. Oper. Logist. **7**(3), 262–274 (2020)

Hsu, J.T., Hsu, L.F.: Two EPQ models with imperfect production processes, inspection errors, planned backorders, and sales returns. Comput. Ind. Eng. **64**(1), 389–402 (2013)

Karimi, M., Pasandideh, S.H.R., Niknamfar, A.H.: A newsboy problem for an inventory system under an emergency order: a modified invasive weed optimization algorithm. Int. J. Manage. Sci. Eng. Manage. **12**(2), 119–132 (2017)

Karimkashi, S., Kishk, A.A.: Invasive weed optimization and its features in electromagnetics. IEEE Trans. Antennas Propag. **58**(4), 1269 (2010)

Khalilpourazari, S., Pasandideh, S.H.R.: Multi-item EOQ model with nonlinear unit holding cost and partial backordering: moth-flame optimization algorithm. J. Ind. Prod. Eng. **34**(1), 42–51 (2017)

Khalilpourazari, S., Pasandideh, S.H.R.: Modeling and optimization of multi-item multi-constrained EOQ model for growing items. Knowl.-Based Syst. **164**, 150–162 (2019)

Khan, M., Hussain, M., Cárdenas-Barrón, L.E.: Learning and screening errors in an EPQ inventory model for supply chains with stochastic lead time demands. Int. J. Prod. Res. **55**(16), 4816–4832 (2017)

Kwon, D., Lippman, S.A., McCardle, K., Tang, C.S.: Time-based contracts with delayed payments. Working paper. UCLA Anderson School, Los Angeles (2009)

Liao, H., Li, L.: Environmental sustainability EOQ model for closed-loop supply chain under market uncertainty: a case study of printer remanufacturing. Comput. Ind. Eng. **151**, 106525 (2021)

Mehrabian, A.R., Lucas, C.: A novel numerical optimization algorithm inspired from weed colonization. Ecol. Inform. **1**(4), 355–366 (2006)

Mehrabian, A.R., Yousefi-Koma, A.: Optimal positioning of piezoelectric actuators on a smart fin using bio-inspired algorithms. Aerosp. Sci. Technol. **11**(2–3), 174–182 (2007)

Moinzadeh, K.: A multi-echelon inventory system with information exchange. Manage. Sci. **48**(3), 414–426 (2002)

Mondal, S., Maiti, M.: Multi-item fuzzy EOQ models using genetic algorithm. Comput. Ind. Eng. **44**(1), 105–117 (2003)

Nia, A.R., Far, M.H., Niaki, S.T.A.: A hybrid genetic and imperialist competitive algorithm for green vendor managed inventory of multi-item multi-constraint EOQ model under shortage. Appl. Soft Comput. **30**, 353–364 (2015)

Nobil, A.H., Sedigh, A.H.A., Cárdenas-Barrón, L.E.: A multiproduct single machine economic production quantity (EPQ) inventory model with discrete delivery order, joint production policy and budget constraints. Ann. Oper. Res. **286**(1–2), 265–301 (2020). https://doi.org/10.1007/s10 479-017-2650-9

Ojha, A.K., RamuNaidu, Y.: Hybridizing particle swarm optimization with invasive weed optimization for solving nonlinear constrained optimization problems. In: Das, K.N., Deep, K., Pant, M., Bansal, J.C., Nagar, A. (eds.) Proceedings of Fourth International Conference on Soft Computing for Problem Solving. AISC, vol. 336, pp. 599–610. Springer, New Delhi (2015). https://doi.org/10.1007/978-81-322-2220-0_49

Pal, B., Sana, S.S., Chaudhuri, K.: A mathematical model on EPQ for stochastic demand in an imperfect production system. J. Manuf. Syst. **32**(1), 260–270 (2013)

Pasandideh, S.H.R., Niaki, S.T.A.: A genetic algorithm approach to optimize a multi-products EPQ model with discrete delivery orders and constrained space. Appl. Math. Comput. **195**(2), 506–514 (2008)

Pasandideh, S.H.R., Niaki, S.T.A., Nia, A.R.: A genetic algorithm for vendor managed inventory control system of multi-product multi-constraint economic order quantity model. Expert Syst. Appl. **38**(3), 2708–2716 (2011)

Pourjafari, E., Mojallali, H.: Solving nonlinear equations systems with a new approach based on invasive weed optimization algorithm and clustering. Swarm Evol. Comput. **4**, 33–43 (2012)

Rahim, M.A., Ben-Daya, M.: Joint determination of production quantity, inspection schedule, and quality control for an imperfect process with deteriorating products. J. Oper. Res. Soc. **52**(12), 1370–1378 (2001)

Ramezanian, R., Rahmani, D., Barzinpour, F.: An aggregate production planning model for two phase production systems: Solving with genetic algorithm and tabu search. Expert Syst. Appl. **39**(1), 1256–1263 (2012)

Rezaei, J., Salimi, N.: Economic order quantity and purchasing price for items with imperfect quality when inspection shifts from buyer to supplier. Int. J. Prod. Econ. **137**(1), 11–18 (2012)

Rogers, J.: A computational approach to the economic lot scheduling problem. Manage. Sci. **4**(3), 264–291 (1958)

Sahinidis, N.V., Tawarmalani, M.: BARON 7.2. 5: Global optimization of mixed-integer nonlinear programs. User's manual (2005)

Sahraei-Ardakani, M., Roshanaei, M., Rahimi-Kian, A., Lucas, C.: A study of electricity market dynamics using invasive weed colonization optimization. In: IEEE Symposium on Computational Intelligence and Games, CIG 2008, pp. 276–282. IEEE, December 2008

Sana, S.S., Goyal, S.K., Chaudhuri, K.: An imperfect production process in a volume flexible inventory model. Int. J. Prod. Econ. **105**(2), 548–559 (2007)

Sarkar, B., Saren, S.: Product inspection policy for an imperfect production system with inspection errors and warranty cost. Eur. J. Oper. Res. **248**(1), 263–271 (2016)

Shaikh, A.A., Khan, M.A.A., Panda, G.C., Konstantaras, I.: Price discount facility in an EOQ model for deteriorating items with stock-dependent demand and partial backlogging. Int. Trans. Oper. Res. **26**(4), 1365–1395 (2019)

Taft, E.W.: The most economical production lot. Iron Age **101**(18), 1410–1412 (1918)

Taleizadeh, A.A., Cárdenas-Barrón, L.E., Mohammadi, B.: A deterministic multi product single machine EPQ model with backordering, scraped products, rework and interruption in manufacturing process. Int. J. Prod. Econ. **150**, 9–27 (2014)

Tawarmalani, M., Sahinidis, N. V.: A polyhedral branch-and-cut approach to global optimization. Math. Program. **103**(2), 225–249 (2005)

Yokota, T., Gen, M., Li, Y.X.: Genetic algorithm for non-linear mixed integer programming problems and its applications. Comput. Ind. Eng. **30**(4), 905–917 (1996)

Yoo, S.H., Kim, D., Park, M.S.: Inventory models for imperfect production and inspection processes with various inspection options under one-time and continuous improvement investment. Comput. Oper. Res. **39**(9), 2001–2015 (2012)

Yoon, K., Hwang, C.L.: TOPSIS (Technique for Order Preference by Similarity to Ideal Solution)– A Multiple Attribute Decision Making, W: Multiple Attribute Decision Making–Methods and Applications, a State-of-The-Art Survey. Springer, Berlin (1981)

Zhang, X., Wang, Y., Cui, G., Niu, Y., Xu, J.: Application of a novel IWO to the design of encoding sequences for DNA computing. Comput. Math. Appl. **57**(11–12), 2001–2008 (2009)

Zipkin, P.H.: Foundations of inventory management (2000)

Recursive Heuristic Algorithm for Balancing Mixed–Model Assembly Line Type–II in Stochastic Environment

Samah A. Aufy$^{(\boxtimes)}$ and AllaEldin H. Kassam

Production Engineering and Metallurgy Department, University of Technology, Baghdad, Iraq
{70221,70150}@uotechnology.edu.iq

Abstract. Straight assembly line and U-shaped assembly line usually required to be extended and updated in order to solve line balancing problems in the real-world based on computational intelligence. In the literature, most models are presented for solving assembly line balancing (ALBP) assume deterministic processing time. This paper is extended solving ALBP with the stochastic environment using fuzzy theory as one of the main pillars of computational intelligent, it's called "Worker–Task Stochastic Assigned to Workstation Heuristic" (W–TSAWH) is adopted. The framework can be structured by creating stochastic work environment (SWE) and assigning process. Firstly, SWE is adopted, with three fuzzy logic models as fuzzy skill level, fuzzy work stability and dynamic fuzzy processing time models. These models are used in order to represent uncertainty associated with task processing time in real assembly system. Secondly, a heuristic algorithm is developed to obtain best solution. The algorithm organized by sequence vector, and a mathematical model that assigns task and worker that subjected to some constraints into constant number of workstations to minimize cycle time. Finally, the performance validation of the methodology is proved using a numerical example.

Keywords: Fuzzy logic · Mixed-model assembly line · Heuristic algorithm · Cycle time

1 Introduction

The early production began by assembly of one unit of production. The assembly line which have the procedure of specific product produced on the line, which operate manually or with the aid of instruments for handling or completing the procedure of product assembly. A typical assembly line consists of a series of successive workstations, each of which contains some work elements known as tasks. Each performed in a crisp value called task processing time. In addition, the balancing problem refers to equality of output of each successive operation in the sequence of the assembly line. If they are all equal, then it is a condition of perfect balance can be considered to be smooth [1]. The assembly line balancing problem (ALBP) is one of a class problems which are known to be computationally difficult (NP-hard problems), basically that based on the assignment the set of tasks into given workstations in a way that not violating precedence constraint

© Springer Nature Switzerland AG 2021
Z. Molamohamadi et al. (Eds.): LSCM 2020, CCIS 1458, pp. 194–218, 2021.
https://doi.org/10.1007/978-3-030-89743-7_11

among tasks and some other constraints. The assignment of tasks to workstations is done to insure that the assembly can meet the demand rate. Thus each workstation is given a fixed amount of time to complete its work, max of them called cycle time [1–3]. Most common class of ALBP is based on objectives are usually classified into four categorize: Minimize the number of workstations for given cycle time is (Type–I), this type is utilized for installing a new assembly line; Minimize the cycle time of the assembly line for given number of workstations is given (Type–II), this is applied when changes occur for improving of an existing assembly line [multi-objective fuzzy assembly line balancing using genetic algorithm]; Minimizing both cycle time and number of workstation simultaneously is given (Type–E); Generating a feasible solution while both cycle time and number of workstation is given (Type–F) [4].

Today's increasingly competitive global market climate forces manufacturing companies to improve productivity plans with the goal of increasing efficiency and effectiveness. In this direction, an efficient assembly line considered as the most important in developing assembly line [3]. Therefore. assembly line also can categorized into other classification according to shape layout into two versions: straight assembly line balancing (SALB), that allocating assembly tasks only predecessors are already allocated to straight workstations, and U-shaped (UALB), which allocating assembly tasks whose predecessors and successors are already allocated to workstations along the assembly line with respect to the solution efficiency can be estimated to be better for the same number of workstations due to available more options of assignable tasks than straight version of the assembly line [3, 5, 6]. Figure 1 illustrates the two versions of the assembly line according to layout.

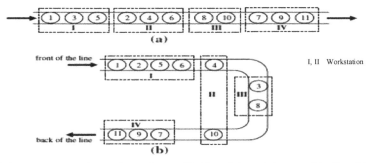

Fig. 1. Tasks arrangement in versions of assembly line ((a) straight assembly line (b) U-shaped assembly line) [7]

However, the task processing time of almost assembly line was considered a crisp, that mean task may be completed in standard time and this may lead to delay in specified cycle time and in product completion [8]. Whilst, in practical there is a high uncertainty, ambiguity, and vagueness in processing time of tasks are performed by worker. This case of worker, significant variation may result from worker fatigue, non skilful workers, motivations of the employees, lack of training, etc. also varying production rate may result from machine breakdowns. To incorporate process time uncertainty in ALBP, task processing time may be treated by estimating uncertain data [1].

Computational intelligent technique (CI) proved their abilities to reach more efficient solutions for real-world problems. The main aspect of real–world problem is imprecise and uncertain data, thus, the input data must be only estimated as within uncertainty due to both machine and worker factors. This uncertainty can be organized a fuzzy number to reduce errors of uncertainty. Therefore, stochastic nature of task processing time is to be considered [9].

2 Relevant Literature

There are numerous literatures have been reviewed studies for solving ALBP, and most of them related with solving ALPB assuming that the processing time as a crisp, but others reported with fuzzy processing time, some of them were summarized as the following: Samah and AllaEldin (2020) [10] presented a novel methodology for solving a mixed-model assembly line balancing problem using a worker-assigned heuristic workstation (W-TAWH) model to address both SALB and UALB versions. The proposal enhanced performance measures depending on the number of suitable workers and tasks that assigned to the given workstation. Finally, these measures are integrated and optimized by employing the desirability function approach for optimization. Salehi et al. (2017) [6] proposed a new hybrid fuzzy interactive approach to solve a new multi-objective ALB problem and it was formulated in a fuzzy environment. Two examples and case studies were adopted for experimental study to demonstrate the effectiveness of the solution and proved the notability of the proposed approach compared with benchmark approaches in the literature. Alavidoost et al. (2016) [3] introduced a novel bi-objective fuzzy mixed-integer linear programming model to represent uncertainty associated with task processing time, the proposed model considered to optimize two conflicting objectives (minimizing cycle time and workstations number) simultaneously. A numerical example, besides benchmark study was considered over some test problems to assess the performance of the proposed solution. The results show that the proposed model can be utilized not only in ALBP but also it would be helpful to handle any practical multi-objective linear programming. Anthony et al. (2016) [9] proposed fuzzy logic model for balancing a single model assembly line. The fuzzy toolbox was used in the analysis of the data, these data obtained from a tricycle assembly line. Results show that the efficiency of the assembly line increased from 88.1% to 92.4%, while idle time was reduced by 56.5% as well as reduction of the bottleneck. Yilmaz et al. (2016) [11] used a genetic algorithm and heuristic priority rule to solve stochastic two-sided U-shape assembly line balancing problem The proposed procedure aims to minimize the number of positions and minimize number of workstations for given cycle time. Finally, to validate the efficiency of the proposed algorithm a comparison study for test problems taken from the literature is conducted. The obtained results demonstrate that the proposed algorithm performs well. Zeqiang and wenming (2015) [12] improved a heuristic procedure based on traditional ranked positional weight method to solve mixed –model U-line balancing problem (MMULBP) with two parameters are task processing time and cycle time as fuzzy numbers. The results obtained of an experimental study show that the improved procedure is effective. Zacharia (2012) [13] proposed a fuzzy of the simple assembly line balancing problem type–II. The task processing times formulated as triangular fuzzy membership functions. A multi-objective genetic algorithm is presented for solving FALBP-II. Fitness function represented as a total fuzzy cost function with the weighted sum of multiple fuzzy objectives. These weights were studied for three different methods (fixed, random, adaptive weight). The aim of this work is to solve the assembly line balancing problem which consists of SALB problem and UALB problem in real-world. The organization of the study is described as follows. In the next section, formulate the W-TSAWH model, then proposed a solution approach to solve this problem. The next numerical example is used to demonstrate the effectiveness, validity, and reality of the developed approach. Finally, conclusion and future works are given.

3 Formulation of W–TSAWH Problem

In this study, the research effort towards exploitation the features of the ALBP to assign a proper task and best available worker that subjected to a set of constraints into suitable fixed workstation numbers as a consequence reducing the cycle time in stochastic environment. In another words, the aim is to increase the efficiency of overall throughput of the assembly line. The extension of assembly line balancing model type–II (W-TSAWH) developed to achieve a satisfactory best or near best solution with respect to both straight and U-shaped models. Generally, the framework is to solve W-TSAWH problem described by designing three–phases. Firstly, is devoted to convert deterministic task processing time into stochastic using fuzzy theory. Whereas the second and third are dedicated to developing an inclusive mathematical model for recursive algorithms for assigning processes for both tasks and workers, with the main objective to minimize workstations cycle time in both straight and U-shaped assembly line models. Figure 2 demonstrates the framework of W-TSAWH. The developed algorithm is subject to the following constraints:

1. Task assigned to workstation if the precedence relationship not violated.
2. Every task processing time is considered in stochastic work environment.
3. The time of set-up, loading, unloading material are involved in the processing time.
4. The number of task processing times allocated to each workstation must not exceed the cycle time specified.
5. Given workstations number must be identical with worker number (it guarantees that each worker will be allocated only to one the workstation).

3.1 Phase 1: Stochastic Work Environment

To ensure taking into account the uncertainty of processing time in real – assembly systems, this phase intended to response the need for a comprehensive capability for work conditions in which fuzzy parameters such as skill level and work stability. The traditional methods for solving assembly line balancing depend on deterministic time. This study presented approach to address balancing assembly line under stochastic work environment, it is devoted to estimating data according to human intuition to cover variability and uncertainty in task processing time. This phase strictly interferes with customers satisfied in the form of due date that the most distinctive quality of modern industry.

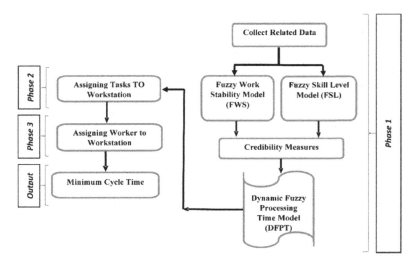

Fig. 2. The framework of the developed W–TSAWH approach.

Data required to be collected so as to structure the integral framework for the developed model are related data which includes product data, worker data, and machine data. Typically, at each workstation there are variables considered as a main because inherent stochastic in assembly process. In this study, the variables can be categories into worker skills that assigned to perform tasks and work stability as important curability measures associated with nature of the complete time of tasks for uncertain work environment. According to author's knowledge, such variables are not reported in existing literature. Thus, this study proposes stochastic work environment (SWE). The basic idea of this module is treating the related input data utilizing fuzzy logic theory, while, the output is representing process time as stochastic output. However, the outline of SWE organized by developing three fuzzy logic models. These are fuzzy skill level (FSL) model, fuzzy work stability (FWS) model, and dynamic fuzzy processing time (DFPT) model.

Fuzzy Skill Level (FSL) Model

FSL model is structured to estimate skill level for each worker (SLk) measure on an assembly line to satisfy the constraint (2) that says a given task in difference processing time due to different worker efficiency in the form of processing time the change amount of time due to the diversity in work accumulated experience according to employment period (EP) and training period (TP) variables, which are structured according to standard classification of occupation. The output variable (SL) is controlled using 25 fuzzy rules reasoning according to general formula (IF <condition> THEN <result>). Finally, the Mamdani interference method was used to fuzzy logic of FSL model in order to get crisp value (SLk) value. Figure 3 depicted the outline of the FSL model.

Fuzzy Work Stability (FWS) Model

FWS model allows taking into account the degree of available of worker (AWk) and available of machine (AM) as fuzzy input variables in order to estimate work stability

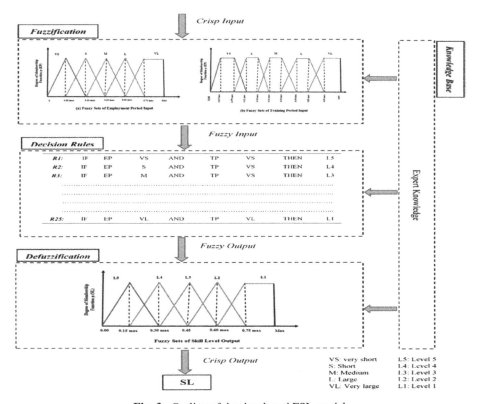

Fig. 3. Outline of the developed FSL model.

for each worker (WSk). These variables can be formulated with relative importance which defined according to expert knowledge, the use of FWS was proposed according to the following procedure:

Step-1: Calculate the working ratio (WRk) for each worker by using Eq. (1).

$$WRk = (WDk)/(TWD) \tag{1}$$

Where:

WD_k: Working days of worker (k)
TWD: Total working days

Step-2: Calculate the ascertain ratio (ARk) for each worker by using Eq. (2).

$$ARk = (ATk)/(TTL) \tag{2}$$

Where:

AT_k: *Assigned tasks of worker*
TTL: Total tasks on assembly line

Step-3: Calculate the availability of worker (AWk) which results from average of two percentages are working ratio, and ascertain ratio.

While the availability of machine (AM) is the actual time that the machine is capable of production as a percent of total planned production time and formulated in Eq. (3).

$$AM = RT/PPT \tag{3}$$

Where.

AM: *available of machine*
RT: run time
PPT: planned production time

The input/output variables are fuzzified trapezoidal membership function shaped. Finally, all computation procedures of the inference process is achieved by mamdani inference method. Figure 4 denote the outline of the FWS model.

Dynamic Fuzzy Processing Time (DFPT) Model
Dynamic Fuzzy Processing Time (DFPT) model is designed to treat the difference in max value of processing time (PT). When a max value in each fuzzy set turns, dynamic process is developed. The output of FSL model and FWS model (SL & WS) are used as input variables for DFPT model. Input/output variables are fuzzified into set of triangular and trapezoidal membership function shape in range from 0–1. Finally, a fuzzy inference procedure is similar to those used in FSL and FWS models in order to estimate PTik of task (i) needed by worker (k). Figure 5 denote the outline of the FWS model.

3.2 Phase 2: Assigning Tasks to Workstation

The assignment aims to minimize the cycle time for balancing assembly tasks along assembly line. Thus, task- heuristic recursive algorithm (T-HRA) is developed to achieve this aim. The evolved algorithm's search process is based on achieving maximum equality in total execution time along the assembly line, in such a manner that maximum equality in partition sequence vector (SV) data is achieved for all workstations. Under this consideration, there was lower variation in workstation time between workstations, as shown in Fig. 6. For further details, the following steps will summarize the procedure of the developed algorithm.

Step 1:- To address the imposed precedence relationship constraint between all of tasks, a positional weight priority rule of the form SV was used to rank the set of tasks. In the academic scene, a set of heuristic priority rules is used in forming SV to rank the set of tasks according to their priority function and precedence relationships among them. In

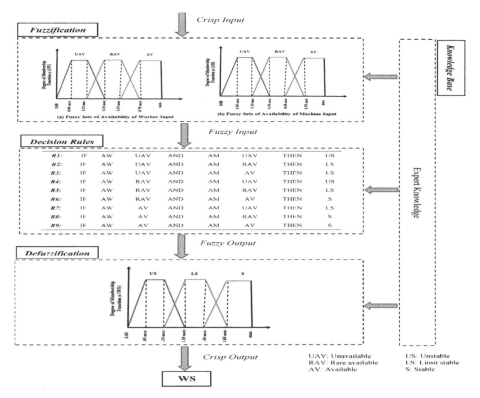

Fig. 4. Outline of the developed FWS model.

this study, the maximum total number priority rule for SALB and UALB was used and its corresponding equation is presented in Eq. (4) & (5), respectively.

$$P_{tp\,(max)} = \max \left\{ \sum_{i \in p} i \right. \tag{4}$$

$$P_{maxf\,(c)} = \max \left\{ \text{number of task} \in \mu_c^s, \text{number of tasks} \in \mu_c^p \right\} \tag{5}$$

Step 2:- Segmentation the sequence vector into A & B parts by dividing the given workstations by 2.

Step 3:- Calculate workstation ratio (WR), that display the ratio of the given workstations that allocated to each sub-vector, that subject to the impost condition, that say, WR ≤ 1.

Step 4:- Calculate time ratio (TR), displays the set of data ratio assigned to each part (A & B), and the idea is based on dividing the SV into two parts called sub-vector, each one can be represented by the left and right positions (PL & RP), as formulated in Eq. (6).

$$TR = \sum_{j=PL}^{i} APT_j / \sum_{j=i+1}^{Pr} APT_j \tag{6}$$

Step 5:- Checking of the condition that says (TR ≤ WR), if yes, a new position (i + 1) must be added, if not, continue to step (6), which ensures that the amount of time allotted for the sub-vector has the least variation.

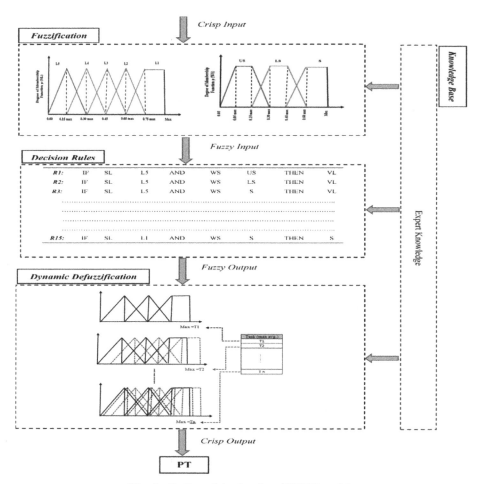

Fig. 5. Outline of the developed DFPT model.

Step 6:- The last position (i) must be deleted from the sub-vector (A), to ensure that the TR ≤ WR condition not violating.

Step 7:- All steps from (1–6) should be repeated until the rest of the given workstations become 1, in another words, each sub-vector which represents a workstation that has a number of tasks assigned to it.

3.3   Phase 3: Assigning Worker to Workstation and Evaluation

The assignment is aimed for minimizing the cycle time for worker assembly line balancing problem. Thus, worker - heuristic recursive algorithm (W-HRA) is developed to achieve this goal. Workers assigned to given workstations have been summarized in the following procedure and can be shown in Fig. 7.

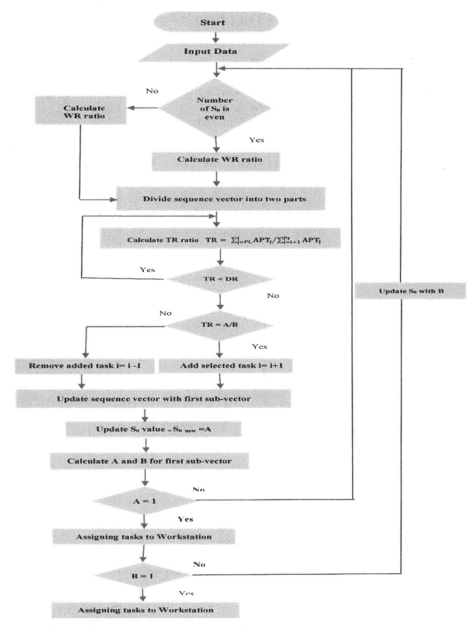

Fig. 6. Flowchart of the recursive algorithm for assigning tasks to workstations.

Step 1:- Compute workstation time (WT), as shown in Eq. (7), it shows the total time needed to finish the assigned tasks to the workstation.

$$T_{sw} = \sum_{i=1}^{n} \sum_{i \in s} TT_{ki} A_{si} \quad for \quad k = 1, \ldots, W \quad (7)$$

Step 2:- Repeat step (1) for available workers until given workers have been assigned to a workstation based on the minimum Tsw.
Step 3:- The steps above should be repeated for each workstation.
Step 4:- Finally, determine the minimum cycle time of assembly line using Eq. (8).

$$CT = \max (Tsw) \quad for \quad \forall s \in S \quad (8)$$

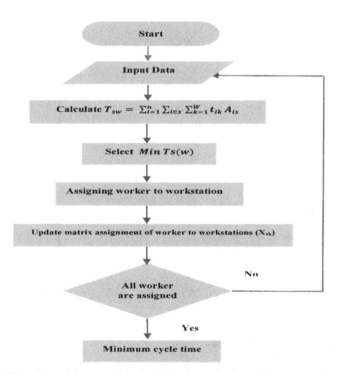

Fig. 7. Flowchart of the recursive algorithm for assigning workers to workstations.

4 Implementation Mechanism of the Developed W-TSAWH Approach

Generally, instead of using deterministic data, the evolved mechanism is experimentally tested using theoretical data to account for the uncertainty associated with task processing time. So, the reality, effectiveness, and validity of the developed approach can be highlighted. Two types of mixed models are critical for the criteria of the mechanism of two products (A&B) with data given in Table 1 and 2 respectively and the precedence relationship is given in Fig. 8. Generally, the two mentioned products (A&B) are required 12 tasks, each can be performed by any one of four workers with different capabilities, i.e. process time. The experiments testes through four points of parameters i.e. (EP, TP, AW, AM). EP and TP are dealt by two critical points (VL, VS), while the two others AW and AM are tested at (AV, UAV), which represents the extreme levels. The crisp value of the extreme points is belonging to fuzzy sets. Tables 3 and 4 denoted the extreme point range of each fuzzy set under study. Then, maximum values of input &output variables are specified by authors and listed in Table 5. Obviously, fuzzy inference will be executed using toolbox Graphical User Inference (GUI) in MATLAB.

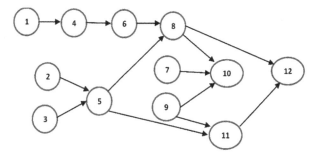

Fig. 8. Precedence graph of combined product.

In order to demonstrate the applicability of the proposed mathematical model and effective solution of the developed solution approach, two cases (case A & case B) covering diversity of the work condition are studied for both examined layout of mixed assembly line SAL and UAL. These cases were solved according to the above-mentioned parameter values and the final decision is focusing on estimated processing time and then the best solution of cycle time required for the analysis and comparative is approached.

Case A: Fixed Skill Level, Variable Work Stability
This case examined data for two test problems (A1, A2), each test will benchmark the performance of the developed approach through the two examined WS extreme levels. Table 6 presents the results from the developed approach for SAL & UAL, respectively.

i.   A1 ➡SL is VS, WS is AV & UAV

   A11 ⇒EP is VS AND TP is VS AND AW is AV AND AM is AV.
   A12 ⇒EP is VS AND TP is VS AND AW is UAV AND AM is UAV.

ii. A2 ⇒SL is VL, WS is AV & UAV

   A21 ➡EP is VL AND TP is VL AND AW is AV AND AM is AV.
   A22 ⇒EP is VL AND TP is VL AND AW is UAV AND AM is UAV.

Table 1. Data set of the product (A).

Task no.	Task time (per unit time)			
	W1	W2	W3	W4
1	17	23	17	13
2	15	14	17	13
3	22	15	27	25
4	10	17	13	20
5	21	25	16	32
6	28	18	20	21
7	42	28	23	34
8	17	23	40	25
9	19	18	17	34
10	16	27	35	26
11	27	23	14	19
12	13	15	10	11

Table 2. Data set of the product (B).

Task no.	Task time (per unit time)			
	W1	W2	W3	W4
1	18	22	19	13
2	21	22	16	20
3	12	25	17	15
4	29	21	19	16
5	31	25	26	22
6	25	14	22	15
7	11	20	23	12
8	27	33	40	25
9	19	13	17	34
10	26	27	35	16
11	9	14	10	13
12	7	9	12	12

Table 3. Fuzzy sets ranges of model 1

Employment period (EP)			Training period (TP)		
Fuzzy set	Boundary	Values	Fuzzy set	Boundary	Values
VS	0 × 35	0	VS	0 × 60	0
	0.05 × 35	1.75		0.037 × 60	2.22
	0.15 × 35	5.25		0.075 × 60	4.5
	/			0.11 × 60	6.6
⋮	⋮		⋮	⋮	
VL	0.50 × 35	17.5	VL	0.83 × 60	49.8
	0.75 × 35	26.25		0.90 × 60	54
	1 × 35	35		1 × 60	60
	/			1 × 60	60

Table 4. Fuzzy sets ranges of model 2

Availability of worker (AW)			Availability of machine (AM)		
Fuzzy set	Boundary	Values	Fuzzy set	Boundary	Values
UAV	0 × 0.9	0	AV	0 × 0.85	0
	0.05 × 0.9	0.04		0.05 × 0.85	0.04
	0.10 × 0.9	0.09		0.10 × 0.85	0.08
	0.30 × 0.9	0.27		0.30 × 0.85	0.25
⋮	⋮		⋮	⋮	
AV	0.50 × 0.9	0.45	UAV	0.50 × 0.85	0.72
	0.70 × 0.9	0.63		0.70 × 0.85	0.59
	1 × 0.9	0.9		1 × 0.85	0.85
	1 × 0.9	0.9		1 × 0.85	0.85

Table 5. Data for setting input/output fuzzy variables

Models	Description	Data
1	Maximum employment period	35 year
2	Maximum training period	60 day
3	Maximum skill level	0.95
4	Maximum available of worker	0.9
5	Maximum available of machine	0.85
6	Maximum work stability	0.8
7	Maximum processing time	Dynamic process

Table 6. Results of W – TSAWH approach for case A

Tests no.	Model (1) SL				Model (2) WS				Model (3) PT				Straight layout					U-shaped layout				
	SL1	SL2	SL3	SL4	WS1	WS2	WS3	WS4	W1	W2	W3	W4	Assigned tasks & worker				CT	Assigned tasks & worker				CT
													w3	w1	w2	w4		w3	w1	w2	w4	
A11	0.23	0.19	0.27	0.14	0.58	0.59	0.59	0.59	20.29	21.88	17.75	24.75	1	2	8	9	114.95	1	12	4	5	107.80
									17.59	18.96	15.38	21.45	4	3	7	10		2	7	10	6	
									23.45	25.29	20.51	28.6	6	5		11		3	9	11	8	
									22.10	23.83	19.33	26.95				12						
									28.41	30.64	24.85	34.65										
									23.90	25.77	20.91	29.15										
									29.31	31.61	25.64	35.75										
									36.08	38.91	31.56	44										
									30.67	33.07	26.83	37.4										
									31.57	34.04	27.62	38.5										
									18.49	19.94	16.17	22.55										
									13.53	14.59	11.83	16.5										
A12	0.23	0.19	0.27	0.14	0.25	0.24	0.25	0.25	29.21	30.97	26.28	34.50	1	2	8	9	160.23	1	12	4	5	150.26
									25.31	26.84	22.78	29.90	4	3	7	10		2	7	10	6	
									33.75	35.79	30.37	39.86	6	5		11		3	9	11	8	
									31.80	33.73	28.62	37.56				12						
									40.89	43.37	36.80	48.30										
									34.40	36.48	30.96	40.63										
									42.19	44.74	37.97	49.83										

(continued)

Table 6. (*continued*)

Tests no.	Model (1) SL				Model (2) WS				Model (3) PT				Straight layout Assigned tasks & worker				CT	U-shaped layout Assigned tasks & worker				CT
	SL1	SL2	SL3	SL4	WS1	WS2	WS3	WS4	W1	W2	W3	W4	w3	w2	w4	w1		w3	w2	w4	w1	
									51.93	55.07	46.73	61.33					38.26					35.88
									44.14	46.81	39.72	52.13										
									45.43	48.19	40.89	53.66										
									26.61	28.22	23.95	31.43										
									19.47	20.65	17.52	23										
A21	0.74	0.74	0.75	0.74	0.58	0.59	0.59	0.59	8.23	8.23	8.23	8.23	1	2	8	9		1	12	4	5	
									7.13	7.13	7.13	7.13	4	3	7	10		2	7	10	6	
									9.51	9.51	9.51	9.51	6	5		11		3	9	11	8	
									8.97	8.96	8.96	8.97				12						
									11.53	11.53	11.52	11.53										
									9.70	9.70	9.69	9.70										
									11.89	11.89	11.88	11.89										
									14.64	14.64	14.63	14.64										
									12.44	12.44	12.43	12.44										
									12.81	12.81	12.80	12.81										
									7.50	7.50	7.49	7.50										
									5.49	5.49	5.48	5.49										

(*continued*)

Table 6. (*continued*)

Tests no.	Model (1) SL				Model (2) WS				Model (3) PT				Straight layout Assigned tasks & worker				CT	U-shaped layout Assigned tasks & worker				CT
	SL1	SL2	SL3	SL4	WS1	WS2	WS3	WS4	W1	W2	W3	W4	w3	w2	w4	w1		w3	w2	w4	w1	
A22	0.74	0.74	0.75	0.74	0.25	0.24	0.25	0.25	8.23	8.23	8.23	8.23	1	2	8	9	38.26	1	12	4	5	35.88
									7.1	7.13	7.13	7.13	4	3	7	10		2	7	10	6	
									9.51	9.51	9.51	9.51	6	5		11		3	9	11	8	
									8.97	8.96	8.96	8.97				12						
									11.53	11.53	11.52	11.53										
									9.70	9.70	9.69	9.70										
									11.89	11.89	11.88	11.89										
									14.64	14.64	14.63	14.64										
									12.44	12.44	12.43	12.44										
									12.81	12.81	12.80	12.81										
									7.50	7.50	7.49	7.50										
									5.49	5.49	5.48	5.49										

Table 7. Results of W – TSAWH approach for case B

Tests no.	Model (1) SK				Model (2) WS				Model (3) SL				Straight layout					U-shaped layout				
													Assigned tasks & worker				CT	Assigned tasks & worker				CT
	SL1	SL2	SL3	SL4	WS1	WS2	WS3	WS4	W1	W2	W3	W4	w3	w1	w2	w4		w3	w1	w2	w4	
B11	0.23	0.19	0.27	0.14	0.58	0.59	0.59	0.59	20.29	21.88	17.75	24.75	1	2	8	9	114.95	1	12	4	5	107.8
									17.59	18.96	15.38	21.45	4	3	7	10		2	7	10	6	
									23.45	25.29	20.51	28.6	6	5		11		3	9	11	8	
									22.10	23.83	19.33	26.95				12						
									28.41	30.64	24.85	34.65										
									23.90	25.77	20.91	29.15										
									29.31	31.61	25.64	35.75										
									36.08	38.91	31.56	44										
									30.67	33.07	26.83	37.4										
									31.57	34.04	27.62	38.5										
									18.49	19.94	16.17	22.55										
									13.53	14.59	11.83	16.5										
B12	0.74	0.74	0.75	0.74	0.58	0.59	0.59	0.59	8.23	8.23	8.23	8.23	w3:1 w2:2 w4:8 w1:9				38.26	w3:1 w2:12 w4:4 w1:5				35.88
									7.1	7.13	7.13	7.13	4	3	7	10		2	7	10	6	
									9.51	9.51	9.51	9.51	6	5		11		3	9	11	8	
									8.97	8.96	8.96	8.97				12						
									11.53	11.53	11.52	11.53										
									9.70	9.70	9.69	9.70										
									11.89	11.89	11.88	11.89										

(continued)

Table 7. (*continued*)

Tests no.	Model (1) SK				Model (2) WS				Model (3) SL				Straight layout					U-shaped layout				
													Assigned tasks & worker				CT	Assigned tasks & worker				CT
	SL1	SL2	SL3	SL4	WS1	WS2	WS3	WS4	W1	W2	W3	W4	w3	w1	w2	w4		w3	w1	w2	w4	
									14.64	14.64	14.63	14.64										
									12.44	12.44	12.43	12.44										
									12.81	12.81	12.80	12.81					160.23					150.26
									7.50	7.50	7.49	7.50										
									5.49	5.49	5.48	5.49										
B21	0.23	0.19	0.27	0.14	0.25	0.24	0.25	0.25	29.21	30.97	26.28	34.50	1	2				1	12	4	5	
									25.31	26.84	22.78	29.90	4	3	8	9		2	7	10	6	
									33.75	35.79	30.37	39.86	6	5	7	10		3	9	11	8	
									31.80	33.73	28.62	37.56				11						
									40.89	43.37	36.80	48.30				12						
									34.40	36.48	30.96	40.63										
									42.19	44.74	37.97	49.83										
									51.93	55.07	46.73	61.33										
									44.14	46.81	39.72	52.13										
									45.43	48.19	40.89	53.66										
									26.61	28.22	23.95	31.43										
									19.47	20.65	17.52	23										

(*continued*)

Table 7. (continued)

Tests no.	Model (1) SK				Model (2) WS				Model (3) SL				Straight layout					U-shaped layout				
													Assigned tasks & worker				CT	Assigned tasks & worker				CT
	SL1	SL2	SL3	SL4	WS1	WS2	WS3	WS4	W1	W2	W3	W4	w3	w2	w4	w1		w3	w2	w4	w1	
B22	0.74	0.74	0.75	0.74	0.25	0.24	0.25	0.25	8.23	8.23	8.23	8.23	1	2	8	9	38.26	1	12	4	5	35.88
									7.1	7.13	7.13	7.13	4	3	7	10		2	7	10	6	
									9.51	9.51	9.51	9.51	6	5		11		3	9	11	8	
									8.97	8.96	8.96	8.97				12						
									11.53	11.53	11.52	11.53										
									9.70	9.70	9.69	9.70										
									11.89	11.89	11.88	11.89										
									14.64	14.64	14.63	14.64										
									12.44	12.44	12.43	12.44										
									12.81	12.81	12.80	12.81										
									7.50	7.50	7.49	7.50										
									5.49	5.49	5.48	5.49										

Case B: Variable Skill Level, Fixed Work Stability
This case treated the examined data for two test problems (B1, B2), as listed below, each test will benchmark the performance of the developed approach through the examined SL extreme levels. Table 7 presents the results from the developed approach for SAL & UAL, respectively.

i. B1 ⟹ SL is VS & VL, WS is AV

B11 ⟹ EP is VS AND TP is VS AND AW is AV AND AM is AV.
B12 ⟹ EP is VL AND TP is VL AND AW is AV AND AW is AV.

ii. B2 ⟹ SL is VS & VL, WS is UAV

B21 ⟹ EP is VS AND TP is VS AND AW is UAV AND AM is UAV.
B22 ⟹ EP is VL AND TP is VL AND AW is UAV AND AM is UAV.

5 Numerical Results and Discussion

The performance of W-TSAWH approach was evaluated over A & B cases. In the experiments, we included eight test, concerning the two examined versions of ALB problem. Hence, for each test, the # of tasks = 12 must be assigned to 4 workstations/workers. Comparing the results yielded by the two versions of W-TSAWH approach, one can observe that W-TSAWH with U-shaped outperforms the other one version. This is established from the fact that the increasing ratio of minimum cycle time about rang 0.06–0.07. It is cleared that the processing time of the tests are sensitive with respect to the change credibility measures. Generally, in comparison between all the eight tests, that the tests have best minimum cycle time because in these tests they take higher SL scores while WS take lower or higher score for the examined extreme points, that means SL was the more impact because of its related with manual and semi-automated assembly line. Figure 9 display the divergence of the minimum cycle time obtained by the two versions of the developed approach over the cases examined. It is clear from this figure that, in most of the tests the U-shape version perform better than other one (although the difference is not large), in other words, this study proved the U–shaped version was preferred for examined parameters basically based on worker. Figure 10 and 11 shows the comparison of the cycle time obtained by two versions over the straight layout, case A, case B and U- shaped layout, case A, case B. The diagram confirm that the tests represented A21, A22, B12, B22 related with high SL are given CT = 38.26 & 35.88 for SAL and UAL respectively, in the contrary, that the tests A11 and B11 have high score of WS are given CT = 114.95, 107.8 for SAL and UAL respectively.

As mentioned earlier, the most important stochastic parameters, EP, TP, AW, AM, are considered in the stochastic work environment of W–TSAWH approach. To obtain sufficient details of the effect of these parameters and their levels on experimental results as listed in Table 8, Taguchi method was used. This method has developed a special design of orthogonal array to study the entire problem parameters space with small number of experiments based on number of factors and their levels.

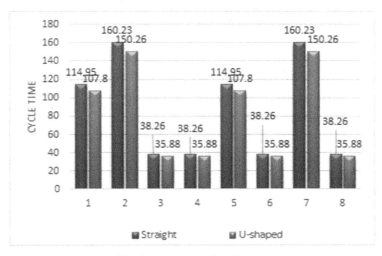

Fig. 9. Cycle time for all tests.

To determine the best parameters level, a robust design criterion entitled Signal–to–Noise (S/N) ratio, which establishes the relative importance of each parameter with respect to its main impacts on the objective function. Usually, the S/N ratio classifies objective function into three types: a nominal the better, smaller the better, larger the better, each type calculates the S/N ratio differently [14, 15]. The examined objective function are classified as smaller the better. MINITAB 17 was used for Taguchi method implementation. The parameter values with impact effect expected to have great improvement of objective function (minimum cycle time). The standard orthogonal array (L16) (24) is conducted for the SWE parameter combinations on an examined problem. Analysis of variance (ANOVA) is applied to investigate the effect of the parameters and their interactions, from Tables 9 and 10 EP, AW, EP*TP, EP*AW, TP*AW, have significant impact on the objective function because taking P–value less than 0.05, thus

Table 8. Input variables with their fuzzy values

Parameter	Levels symbol	Extreme range	Setting
EP	VS	0–5.25	w1: 4, w2: 3, w3: 5, w4: 1
	VL	17.5–35	w1: 18, w2: 29, w3: 23, w4: 25
TP	VS	0–6.6	w1: 2, w2: 5, w3: 4, w4: 6
	VL	49.8–60	w1: 50, w2: 53, w3: 60, w4: 57
AW	UAV	0–0.27	w1: 0.15, w2: 0.20, w3: 0.25,w4: 0.10
	AV	0.45–0.9	w1: 0.5, w2: 0.7, w3: 0.65, w4: 0.75
AM	UAV	0.69	/
	AV	0.20	/

these obtained results proved the main role of worker rather than machine that were considered in manual and semi-automation assembly systems.

Table 9. ANOVA results for straight assembly line

Parameters	Sum of Square	Mean of Square	DF	F value	P - value
EP	6.759	0.00	1	0.00	0.997
TP	20.437	20.437	1	6.13	0.132
AW	3.333	80.701	1	24.22	0.039
AM	3.333	4.999	1	1.50	0.345
EP * AW	108.402	182.358	1	54.73	0.018
EP * AM	1.666	0.00	1	0.00	1.000
TP * AW	80.542	80.542	1	24.17	0.039
TP * AM	4.918	4.918	1	1.50	0.345
AW * AM	3.332	3.332	1	1.000	0.423

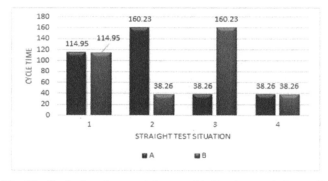

Fig. 10. Cycle time for a straight assembly line in case A & case B.

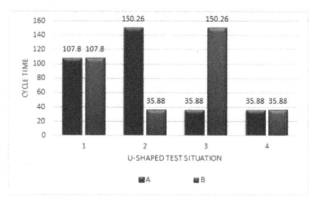

Fig. 11. Cycle time for U - shaped assembly line in case A & case B.

6 Conclusions and Future Works

In this study a developed W–TSAWH approach for balancing mixed-model straight and U–shaped is presented. We concern with the assigned of the suitable task/worker to the suitable workstation based on minimizing cycle time as the main objective that subjected to some constraints. The problem also was formulated in an uncertain environment with fuzzy parameters. SWE is used to estimate stochastic processing time, and then recursive algorithm is developed for tasks/worker allocation to proper workstation. Finally, the performance of the developed approach was validated through set of numerical experiments. The developed model is proven to be capable to address the two types of assembly line are straight and U–shaped balancing problem under uncertainty work environment. Virtually, it is proved their efficiency through specify the more effect element in uncertain work environment. The extensive computational study proved the superiority of the parameters regard to workers over the other examined because we are dealing with manual or semi-automation. Although, it is limited by two versions of assembly line balancing models are straight and U–shaped, future works can focus on solving more complicated problems such as two-sided or parallel layout of assembly line using the W–TSAWH model. In addition, the developed algorithm could solve by employing multi-objective genetic algorithm.

References

1. El Awady, A.E.: Solving assembly line balancing problems using genetic algorithm. Benha University (2006)
2. Li, Y., Tang, X., Hu, X.: Optimizing the reliability and efficiency for an assembly line that considers uncertain task time attribute, vol. 7. IEEE (2019)
3. Alavidoost, M., Babazadeh, H., Sayyari, S.: An interactive fuzzy programming approach for bi-objective straight and U-shaped assembly line balancing problem. Appl. Soft Comput. **40**, 221–235 (2016)
4. Özcan, U., Toklu, B.: A new hybrid improvement heuristic approach to simple straight and U-type assembly line balancing problems. J. Intell. Manuf. **20**(1), 123–136 (2009). https://doi.org/10.1007/s10845-008-0108-2
5. Jonnalagedda, V., Dabede, B.: Application of simple genetic algorithm to U-shaped assembly line balancing problem of type II. Int. Fed. Autom. S. Afr. 6118–6173 (2014)
6. Salehi, M., Maleki, H.R., Niroomand, S.: A multi-objective assembly line balancing problem with worker's skill and qualification considerations in fuzzy environment. Appl. Intell. **48**(8), 2137–2156 (2017). https://doi.org/10.1007/s10489-017-1065-2
7. Sarwar, F.: Heuristic optimization algorithm based line balancing in a fuzzy environment. Bangladesh University of Engineering and Technology (2007)
8. Zacharia, P.T., Nearchou, A.C.: A meta-heuristic algorithm for the fuzzy assembly line balancing type-E problem. Comput. Oper. Res. **40**, 3033–3044 (2013)
9. Unuigbe, A.I., Unuigbe, H.A., Aigboje, E., Ehizibue, P.A.: Assembly line balancing using fuzzy logic: a case study of a tricycle assembly line. J. Optim. **5**, 59–70 (2016)
10. Samah, A.A., AllaEldin, H.K.: A consecutive heuristic algorithm for balancing a mixed-model assembly line type II using a (W-TAWH) model developed for straight and U-shaped layouts. In: Material Science and Engineering, vol. 671 (2020). https://doi.org/10.1088/1757-899X/671/1/012147. https://iopscience.iop.org

11. Decline, Y., Aydoğan, E.K., Özcan, U.: Stochastic two sided U-type assembly line balancing: a genetic algorithm approach. Int. J. Prod. Res. **54**(11), 3429–3451 (2016)
12. Zhang, Z., Cheng, W.: Improved heuristic procedure for mixed-model U-line balancing problem with fuzzy times. In: Proceedings of China Modern Logistics Engineering. LNEE, vol. 286, pp. 395–406. Springer, Heidelberg (2015). https://doi.org/10.1007/978-3-662-44674-4_37
13. Zacharia, P., Nearchou, A.: Multi-objective fuzzy assembly line balancing using genetic algorithms. J. Intell. Manuf. **23**, 615–627 (2012)
14. Montgomery, D.C.: Design and Analysis of Experiments. Wiley, Hoboken (2008)
15. Alavidoost, M.H., Tarimoradi, M., Fazel, M.H.: Fuzzy adaptive genetic algorithm for multi-objective assembly line balancing problem. Appl. Soft Comput. **34**, 655–677 (2015)

An Approximation Approach for Fixed-Charge Transportation-p-Facility Location Problem

Soumen Kumar Das and Sankar Kumar Roy$^{(\boxtimes)}$

Department of Applied Mathematics with Oceanology and Computer Programming,
Vidyasagar University, Midnapore 721102, West Bengal, India
roysank@mail.vidyasagar.ac.in

Abstract. This chapter describes a single-objective, multi-facility, location model for a logistics network, whose aim is to support the economical aspect. In this work, a new variant of the facility location model is presented to ask the optimum positions of the new facilities with the target that the aggregate logistics cost from the endure facilities to the new facilities along with the fixed-charge cost will be reduced. A new approximation approach is incorporated for solving the proposed model for extracting results. An experimental design is consolidated to demonstrate the proficiency and viability of the proposed consideration in connection with reality. The novel contributions of this study have introduced a way to connect the facility location problem and fixed-charge transportation problem using a new approximation approach with minimizing the conveyance cost. The chapter ends with conclusions and perspectives on future studies.

Keywords: Approximation approach · Balinski's approximation · Locate-allocate heuristic · Facility location problem · Fixed-charge transportation problem

1 Introduction

Facility location problem (FLP) is one of the most broadly studied fields in the literature of Operations Research. It is mainly applied to locate the new facilities with respect to the other existing facilities. In consequence, it plays a major role in human-centric decision-making problems in both the government and non-government sectors. In FLP, warehouses, suppliers, and producers are treated as facilities; and purchasers, retailers and service users are concerned as customers. In a nutshell, FLP is concerned with choosing the best positions for the

This research work was financially supported by the "Department of Science & Technology (DST) of India under the [SRF-P (DST-INSPIRE Program)] scheme (No. DST/INSPIRE Fellowship/2015/IF 150209 dated 01/10/2015)", provided to the first author.

Z. Molamohamadi et al. (Eds.): LSCM 2020, CCIS 1458, pp. 219–237, 2021.
https://doi.org/10.1007/978-3-030-89743-7_12

new facilities to improve the customers' services, in line with the optimality criteria fulfilled. FLP, an NP-hard problem (Non-deterministic Polynomial-time), looks to the optimal locations for the potential facilities with respect to existing facilities. FLP can be characterized into different subcategories relying upon the properties. The location space might be continuous or discrete. In particular, FLP plays an important role in the distribution and transportation network industries, public facilities, private facilities, military environment, and business areas, etc.

In the last few decades, logistics modelling plays a vital role towards globalization. Fixed-charge transportation problem (FCTP) is a part of logistics network design. Substantially, FCTP, an augmentation of the classical transportation model is mainly described by source, demand and financial objective function. It is assumed that the cost of sending a non-zero amount of items from each source to each demand is equal to a variable non-negative cost proportional to the amount of commodity sent includes a fixed non-negative cost. Due to that cost structure, the estimation of the objective function likes a step function for every time open or close a route the objective function jumps a step. FCTP finds the minimum cost (overall fixed and variable cost) to ship out products from specific supply points to few destinations.

FLP and FCTP are correlated with the distribution framework. Locating the facilities in the best places and transportation with the fixed cost from the endure facilities to the best new facilities locations are the main objectives in *fixed-charge transportation-p-facility location* (FCT-p-FLP). In fact, FCT-p-FLP is a cost minimization problem got by coordinating FLP and FCTP that can be tackled in planner surfaces with a distance metric. It is an NP-hard problem with nonlinear objective function which is neither convex nor concave. So, it is generally difficult to solve since local optimal does not always imply global optimality by an exact method. Due to this fact, a new approximation approach is proposed to solve large-scale FCT-p-FLP.

The remainder of the study is presented in the following manner. In Sect. 2, an overview concept of FLP and FCTP with the relevant portions are depicted. Section 3 describes the mathematical model including problem delineation. The procedure of the approximation approach and its algorithm is incorporated in Sect. 4. Section 5 illustrates to test the performance of the proposed methodology with an application and computational results are provided. At last, conclusions and future scopes of the study are given.

2   Theoretical Framework

FLP frames an important class of integer programming problems, with application in the transportation and distribution organizations from the ancient days to till today. Despite the fact that it is numerous years old problem however nowadays it is getting more emphasis as a result of its part of practical applications. Intrigued perusers can take after these articles to discover more about FLP: Love et al. [31], Farahani et al. [23] and Farahani and Hekmatfar

[22]. Later, Bieniek [9] introduced a single source FLP with general stochastic identically distributed demands. Afterwards, Dias et al. [20] utilized a primal-dual heuristic to solve dynamic FLPs. Later, Karatas and Yakıcı [28] analyzed an FLP under a multi-criteria environment and solved it by an iterative method. Wanka and Wilfer [50] studied minimax location problems under a multi-composed solution technique. Atta et al. [5] employed two approaches to solve their uncapacitated location model under a multi-objective environment.

In recent years, the FCTP, augmentation of classical *transportation problem* (TP) which is attracted by the researchers. It was introduced by Hirsh and Dantzig [27]. Balinski [8] formulated a new approximation approach to solve the FCTP which is known as Balinski's approximation. Such a large number of researchers have analyzed on transportation network in several environments. A few of them are affixed here. Adlakha and Kowalski [1] gave a note on an FCTP. Then, they [2] introduced a simple heuristic to solve it. Klose [30] developed several approaches to solve a single-sink FCTP. Thereafter, Adlakha et al. [3] discussed a novel approximation to solve an FCTP by calculating an efficiency lower bound. Midya and Roy [37] examined a multi-objective FCTP in the light of the stochastic environment. Mingozzi and Roberti [40] solved an FCTP using an exact algorithm based on an unprecedented integer programming. Midya and Roy [38] analyzed an FCTP under different environments. Afterwards, Roy et al. [46] discussed an FCTP in the shadow of a multi-criteria environment and they considered the nature of parameters as a two-fold uncertainty. Recently, Mehlawat et al. [34] studied a multi-stage FCTP with three multiple conflicting objective functions and they applied it in a sustainable transportation example. Nowadays, product blending assumes an important part in setting up the refinery items for the market to fulfill the item determinations and ecological guidelines. Motivated by this idea, Roy and Midya [45] introduced it into a multi-objective FCTP with intuitionistic fuzzy parameters. Roy et al. [47] studied a multi-objective multi-item FCTP under fuzzy-rough environment.

In the present era, FLP and FCTP are a burning theme in production network management. Such a substantial number of researchers has implemented FLP in different network problem, for instance, supply chain management, transportation problem, and allocation problem, etc. A number of research works are annexed here. In fact, Cooper [13] first introduced a location model in distribution system. Then, he [12] also discussed a heuristic method for solving the problem and the heuristic is often called Locate-Allocate (Loc-Alloc) heuristic. Later on, he [14] developed the above-mentioned study in a stochastic environment. Sherali and Tuncbilek [49] utilized a Euclidean metric in a location-allocation problem and also gave a solution technique to solve large-scale instances. Dohse and Morrison [21] explored a connection between FLP and the transportation system and therefore, they solved it to obtain an optimal solution. Melo et al. [36] reviewed a paper on location model in supply chain. In the literature, researchers mainly concentrated on single criterion FLP in distribution planning such as [4,7,11,18,19,35,41,48]. There are few studies in this direction wherein multi-criteria decisions are included for instance [6,10,15–17,26,51].

A more detailed classification of some related research is presented in Table 1. To the best of knowledge, until now no one has considered the FLP in FCTP under this criterion. The significant contributions of the research are summarized as follows:

(i) A new variant of the nonlinear mixed-integer programming that integrates both FLP and FCTP is introduced.

(ii) The proposed problem yields joint decisions regarding the product flow and the optimum locations of new sites.

(iii) The total fixed and variable ship out cost is incorporated.

(iv) A new approximation approach is considered to get the optimal results.

(v) The existence of optimal solution is described analytically.

Table 1. Characteristics of some related research.

References	Year	FLP	TC	FCC	Existence of optimal solution analytically	Solution methods
Adlakha and Kowalski [2]	2003	x	✓	✓	✓	Simple heuristic
Adlakha et al. [3]	2014	x	✓	✓	x	Approximation approach
Balinski [8]	1961	x	✓	✓	✓	Balinski's approximation
Carlo et al. [11]	2017	✓	✓	x	x	Decomposition heuristic, Simulated annealing
Cooper [13]	1972	✓	✓	x	✓	Locate-allocate heuristic
Das and Roy [16]	2019	✓	✓	x	x	Proposed hybrid approach
Das et al. [18]	2020	✓	✓	x	✓	Exact algorithm, Heuristic approach
Das et al. [19]	2019	✓	✓	x	✓	Heuristic approaches
Klibi et al. [29]	2010	✓	✓	x	x	Hierarchical heuristic
Klose [30]	2008	x	✓	✓	x	Branch-and-bound method
Mehlawat et al. [34]	2019	x	✓	✓	x	ε-constrained method
Midya and Roy [37]	2014	x	✓	✓	x	Fuzzy programming
Midya and Roy [38]	2017	x	✓	✓	x	Interval programming, Fuzzy programming
Mingozzi and Roberti [40]	2017	x	✓	✓	✓	Exact algorithm
Mišković et al. [41]	2017	✓	✓	x	x	Proposed memetic algorithm
Roy et al. [44]	2017	x	✓	x	x	Conic scalarization
Roy and Midya [45]	2019	x	✓	✓	x	Intuitionistic fuzzy TOPSIS, Intuitionistic fuzzy programming
Roy et al. [46]	2018	x	✓	✓	x	Global criterion, ε-constrained method, Fuzzy programming
Roy et al. [47]	2019	x	✓	✓	x	TOPSIS, Weighted goal programming, Fuzzy programming
Saif and Elhedhli [48]	2016	✓	✓	x	x	Lagrangian heuristic
This study	2020	✓	✓	✓	✓	Proposed approximation approach

TC: Transportation cost; FCC: Fixed-charge cost

3   Mathematical Description

Here, the proposed problem is delineated. The mathematical model for FCT-p-FLP is formulated. Decision variables, parameters, and assumptions are provided. At last, the objective function and constraints are developed to describe the model.

3.1   Problem Definition

Here, a new geographical practical problem is inspected from the economical aspect. Figure 1 illustrates the FCT-p-FLP framework including several existing facilities, overall logistics cost, fixed-charge costs (represent setup costs underway frameworks, toll charges on an expressway, costs for building streets, or landing expenses at air terminals) and optimal new facilities. Products are sent from existing plants to potential plants with the point of minimizing the overall fixed cost and variable cost. We consider that there are 03 existing plants, for example, O_1, O_2 and O_3, four toll charges like F_1, F_2, F_3, and F_4 and four potential facility sites such as P_1, P_2, P_3, and P_4. The related availability and demand of the endure plants and the new facilities are known. Moreover, the positions of O_1, O_2 and O_3 are given. But, the locations of F_1, F_2, F_3 and F_4 are not known on the planner surface. The line denotes the conveyances cost function per unit item from O_1, O_2 and O_3 to P_1, P_2, P_3, and P_4 along with fixed costs for toll charges F_1, F_2, F_3, and F_4. In case that, the operations manager has to choose the best positions of the new facilities in such a way that the total fixed-charge logistics cost from endure plants to new plants is reduced. For that reason, a connection between FLP and FCTP is formulated.

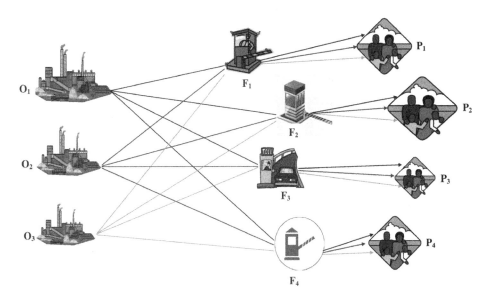

Fig. 1. Network for FCT-p-FLP.

3.2   Assumptions and Notations

The subsequent assumptions and notations are utilized to formulate the model:

- The feasible region is continuous;
- Parameters are in precise nature;
- The positions of new facilities are in coordinate plane;
- New facilities are considered as points;
- The metric is assumed as Euclidean metric $\big(\phi(a_i, b_i; x_j, y_j)$ $=$ $\sqrt{(a_i - x_j)^2 + (b_i - y_j)^2}\big)$;
- The new facilities have capacity and there is no connection between them;
- Setup costs are not considered to establish the new facilities;
- Fixed-charge cost is taken for opening each route;

$\quad m$: number of endure plants;
$\quad p$: number of new plants;
$\quad \alpha_i$: non-negative weights of endure plants $\forall i$;
$\quad r_i$: supply at the i-th endure plant $\forall i$;
$\quad s_j$: demand at the j-th new plant $\forall j$;
(a_i, b_i) : co-ordinate of i-th endure plant $\forall i$;
(x_j, y_j) : co-ordinate of j-th new plant $\forall j$;
$\quad w_{ij}$: decision variables (unknown quantities of commodity to be shipped from the i-th endure plant to the j-th new plant);
$\quad F$: feasible set;
$\quad f_{ij}$: fixed-charge cost (for opening i-th endure facility to j-th new facility route);
$\quad u_{ij}$: (0–1) variable that equals 1 iff $x_{ij} > 0$ (0 otherwise);
$\quad \phi$: logistics cost function per unit item from the i-th endure plant to the j-th new plant.

3.3   Model Formulation

Herein, a mathematical formulation is designed based on FLP and FCTP. In addition to minimize the transportation cost, we are mainly interested to find the optimal locations by choosing the potential facilities. The model of FCT-p-FLP is considered as follows:

Model 1

$$\text{minimize } Z = \sum_{i=1}^{m} \sum_{j=1}^{p} \alpha_i w_{ij} \phi(a_i, b_i; x_j, y_j) + \sum_{i=1}^{m} \sum_{j=1}^{p} f_{ij} u_{ij}, \qquad (3.1)$$

$$\text{subject to } \sum_{j=1}^{p} w_{ij} \le r_i \qquad (i = 1, 2, \ldots, m), \qquad (3.2)$$

$$\sum_{i=1}^{m} w_{ij} \ge s_j \qquad (j = 1, 2, \ldots, p), \qquad (3.3)$$

$$w_{ij} \ge 0 \quad \forall \quad i \text{ and } \quad j, \qquad (3.4)$$

$$u_{ij} = 0 \quad \text{if} \quad w_{ij} = 0, \qquad (3.5)$$

$$u_{ij} = 1 \quad \text{if} \quad w_{ij} > 0. \qquad (3.6)$$

The objective function (3.1) intends to minimize the overall logistics cost along with fixed cost from existing facilities to potential facilities. Constraints (3.2) authorize that the whole items of each existing plant which can't outperform its whole available. Constraints (3.3) force that the complete stream to every potential plant should satisfy its demand. Constraints (3.4) are the non-negativity criteria. Constraints (3.5) to (3.6) suggest that cost of sending no units along route (i, j) is zero; yet any positive shipment acquires a fixed cost.

3.4   Relationship Between FCT-p-FLP and FCTP

The objective of the stated problem relies upon the location of new facility sites. From Fig. 1, it can be seen that if the locations of potential facilities are riveted, then the cost functions become a constant cost function i.e., $\phi(a_i, b_i; x_j, y_j) = t_{ij}$ and if $\alpha_i t_{ij}$ is picked as c_{ij} (unit conveyance cost). Consequently, the objective function of Model 1 is lessened; and it progresses toward becoming as Model 2

Model 2

$$\text{minimize } Z = \sum_{i=1}^{m} \sum_{j=1}^{p} (c_{ij} w_{ij} + f_{ij} u_{ij}),$$

$$\text{subject to the constraints (3.2) to (3.6).} \qquad (3.7)$$

which is the traditional form of FCTP.

3.5   Characteristics of FCT-p-FLP

Herein, some key attributes are described to perceive the characteristics of FCT-p-FLP.

Proposition 1: A necessary and sufficient condition for the problem FCT-p-FLP is that $\sum_{i=1}^{m} r_i \ge \sum_{j=1}^{p} s_j$.

Proof: Straightforward

Proposition 2: The feasible solution of the FCT-p-FLP is never unbounded.

Proof: The constraints of the FCT-p-FLP are as follows:

$$\sum_{j=1}^{p} w_{ij} \leq r_i \qquad (i = 1, 2, \ldots, m),$$

$$\sum_{i=1}^{m} w_{ij} \geq s_j \qquad (j = 1, 2, \ldots, p),$$

$$w_{ij} \geq 0 \quad \forall \quad i \text{ and } j,$$

$$u_{ij} = 0 \quad \text{if} \quad w_{ij} = 0,$$

$$u_{ij} = 1 \quad \text{if} \quad w_{ij} > 0.$$

So, $s_j \leq w_{ij} \leq r_i \forall i$ and j, and furthermore $w_{ij} \geq 0 \forall i$ and j. It can be easily inferred that $\inf(0, s_j) \leq w_{ij} \leq r_i \forall i$ and j, presently since $s_j > 0 \forall j$ then $0 \leq w_{ij} \leq r_i \forall i$ and j.

Proposition 3: The number of basic variables in FCT-p-FLP is at most $(m + p - 1)$.

Proof: This property depends on the constraints. Here, the constraints of two problems are accepted to be same. Thus, this proposition is moreover same as FCTP.

Proposition 4: For the problem minimize Z $= \sum_{i=1}^{m} \sum_{j=1}^{p} \left[\alpha_i w_{ij} \phi(a_i, b_i; x_j, y_j) + f_{ij} u_{ij} \right]$, $w_{ij} \in$ F an optimal solution exists at an extreme point of the convex set F of feasible solutions to FCT-p-FLP.

Proof: Let $(x, y) \in \{(x_j, y_j), \ j = 1, 2, ..., p\}$, $w \in \{w_{ij}, \ i = 1, 2, \ldots, m, \ j = 1, 2, \ldots, p\}$ and $w_E \in \{w_{ij}^E$, set of all extreme points$\}$. If the destination such that $(x, y) = (x_j^*, y_j^*)$ is chosen by finding the optimal location then the objective function becomes minimize Z $= \sum_{i=1}^{m} \sum_{j=1}^{p} [\alpha_i w_{ij} \phi(a_i, b_i; x_j^*, y_j^*) + f_{ij} u_{ij}]$, $w_{ij} \in$ F, which is a traditional FCTP. Then it always has a solution at an extreme point $w_E \in$ F. Hence, we conclude that (x^*, y^*, w_E) is an optimal solution at an extreme point of FCT-p-FLP. This completes the proof of the proposition.

Proposition 5: The number of basic feasible solutions of FCT-p-FLP is at most $\binom{mp}{m+p-1}$

Proof: The FCT-p-FLP has mp variables and at most $m+p-1$ basic variables. So, the number of basic feasible solutions of a FCT-p-FLP is at most $\binom{mp}{m+p-1}$.

Theorem: The function minimize Z $= \sum_{i=1}^{m} \sum_{j=1}^{p} \left[\alpha_i w_{ij} \phi(a_i, b_i; x_j, y_j) + f_{ij} u_{ij} \right]$ is neither a convex function nor a concave function for all values of (x_j, y_j, w_{ij}).

Proof: Here, a particular case is considered where all $w_{ij} = 0$ except w_{22}. Further let, $b_2 = y_2$ and $a_2 > x_2$. Here, $Z = \alpha_2 w_{22}(a_2 - x_2) + K$, where K is the sum of all fixed charge costs. Now, it is known to all that a function will be neither convex nor concave if the Hessian matrix associated with Z is neither positive nor negative semi definite. The Hessian matrix for Z is

$$H = \begin{pmatrix} \frac{\partial^2 Z}{\partial x_2^2} & \frac{\partial^2 Z}{\partial w_{22} \partial x_2} \\ \frac{\partial^2 Z}{\partial w_{22} \partial x_2} & \frac{\partial^2 Z}{\partial w_{22}^2} \end{pmatrix} = \begin{pmatrix} 0 & -\alpha_2 \\ -\alpha_2 & 0 \end{pmatrix}$$

which indicates indefinite. Hence, Z is neither convex nor concave for particular value of (x_j, y_j, w_{ij}). Therefore, Z is neither convex nor concave for all values of (x_j, y_j, w_{ij}). This shows the proof of the theorem.

4   Methodology

In this section, an approximation approach with its algorithm is discussed for the proposed model.

4.1   Approximation Approach

Here, a new approximation approach is presented based on Balinski's approximation [8] and Loc-Alloc heuristic [12], which always provides an optimal solution. It is observed that, the feasible region of the proposed model is a bounded convex set with a neither convex nor concave objective. The optimal outcome occurs at an extreme point of the restriction set, each extreme point of the feasible set is a local minimum (from Theorem, Propositions 2 and 4 and Appendix A). The proposed approach comprises into two portions. In the first portion, the approach locates the primary positions and in the second division, it evolves steps for the optimum locations. The process of finding initial locations for the proposed approach depends on Loc-Alloc heuristic. Firstly, the initial positions are selected for p-new facilities from m-endure facilities. At that point, the distances between them are determined. When $p \leq m$, we can easily figure out such distances but, when $p \geq m$, then there is a problem to calculate such distances. Therefore, a large number is consigned for such distances which can't be determined. Presently, it is as of now expected that the distances are cost capacities per unit flow from i-th endure facility to the j-th new facility. We accept these distances as the cost coefficients then the problem turns into FCTP (i.e., Model 2). To find optimal basic feasible solution (OBFS), an approximation is utilized which is known as Balinski's approximation. Balinski [8] noticed that there always yields an optimum outcome to the lessened variant of FCTP, with the below criterion

$$u_{ij} = \frac{w_{ij}}{m_{ij}}, \text{where } m_{ij} = min(r_i, s_j). \tag{4.8}$$

In this way, the modified FCTP converts into a classical TP with unit logistics cost as $C_{ij} = c_{ij} + \frac{f_{ij}}{m_{ij}}$. The feasible outcome $\{w_{ij}^B\}$ can be effectively altered into an outcome $\{w_{ij}^B, u_{ij}^B\}$ of FCTP, which is as below:

$$u_{ij}^B = \begin{cases} 0, & w_{ij}^B = 0, \\ 1, & w_{ij}^B > 0. \end{cases}$$

The initial feasible solutions are designated as $\{w_{ij}^B, u_{ij}^B\}$, then using each such outcome we tackle the following problem.

$$\text{minimize } Z^B = \sum_{i=1}^m \sum_{j=1}^p \left[\alpha_i w_{ij}^B \sqrt{(a_i - x_j)^2 + (b_i - y_j)^2} + f_{ij} u_{ij}^B \right]. \quad (4.9)$$

Now, the problem can be written as

$$\text{minimize } Z^B = \sum_{j=1}^p \text{minimize } Z_j^B, \quad (4.10)$$

$$\text{where, minimize } Z_j^B = \sum_{i=1}^m \left[\alpha_i w_{ij}^B \sqrt{(a_i - x_j)^2 + (b_i - y_j)^2} + f_{ij} u_{ij}^B \right]. \quad (4.11)$$

Now we minimize $Z_j^B \; \forall j$ for minimize Z^B. Iterative formula (see Appendix B) are utilized to solve the problem (4.11). The initial estimates of (x_j, y_j) are chosen as follows:

$$x_j^0 = \frac{\sum_{i=1}^m \alpha_i w_{ij}^B a_i}{\sum_{i=1}^m \alpha_i w_{ij}^B} \qquad \forall j, \quad (4.12)$$

$$y_j^0 = \frac{\sum_{i=1}^m \alpha_i w_{ij}^B b_i}{\sum_{i=1}^m \alpha_i w_{ij}^B} \qquad \forall j. \quad (4.13)$$

and the general iterations are

$$x_j^{k+1} = \frac{\sum_{i=1}^m \frac{\alpha_i w_{ij}^B a_i}{\phi(a_i, b_i; x_j^k, y_j^k)}}{\sum_{i=1}^m \frac{\alpha_i w_{ij}^B}{\phi(a_i, b_i; x_j^k, y_j^k)}} \qquad \forall j; k \in \mathbb{N}, \quad (4.14)$$

$$y_j^{k+1} = \frac{\sum_{i=1}^m \frac{\alpha_i w_{ij}^B b_i}{\phi(a_i, b_i; x_j^k, y_j^k)}}{\sum_{i=1}^m \frac{\alpha_i w_{ij}^B}{\phi(a_i, b_i; x_j^k, y_j^k)}} \qquad \forall j; k \in \mathbb{N}, \quad (4.15)$$

$$\text{where} \quad \phi(a_i, b_i; x_j^k, y_j^k) = [(a_i - x_j^k)^2 + (b_i - y_j^k)^2]^{1/2}. \quad (4.16)$$

If we denote the optimum value of Z^B for n-th basic feasible solution as Z_n^*, then the optimum value of Z^* for the FCT-p-FLP will be

$$Z^* = \text{minimize } Z_n^* \quad \forall \, n \in \mathbb{N}. \quad (4.17)$$

If the optimum value is occurred at $n = l$, then the optimal outcomes are (x_{jl}^*, y_{jl}^*) and $\{w_{ijl}^*, u_{ijl}^*\}, \forall i, j$, where (x_{js}, y_{js}) and $\{w_{ijs}^B, u_{ijs}^B\}$ designate the s-th solution of (x_j, y_j) and $\{w_{ij}^B, u_{ij}^B\}$.

4.2   Algorithm

Here, an approximation approach is depicted for handle the FCT-p-FLP briefly. A schematic diagram of the proposed approach is displayed in Fig. 2. The steps are given for determination the best potential facilities in the FCT-p-FLP, which are as per the following:

Step 1: Firstly, the primary positions are picked for each of p-facilities from m-endure facilities.

Step 2: Accordingly, two cases emerge when $p \leq m$ then it can without much of a stretch discover the distances between them. Yet, when $p > m$ at that point, it can't discover all the distances. To tackle this issue, a large number is relegated for each distance, and stay away from to ascertain such distance.

Step 3: Now, it is assumed that the distances are proportional to the cost functions. Thus, these distances are taken as the cost functions. At that point, it is turned into a traditional FCTP.

Step 4: Employing Balinski's approximation formula (4.8), FCTP converts into a relaxed TP.

Step 5: Afterwards, we get the OBFSs to the relaxed TP by LINGO 16.0 iterative scheme. And these solutions can be improved into the feasible solutions of FCTP.

Step 6: Using the OBFS from Step 5 and the iteration formula from (4.12)–(4.16), FCT-p-FLP is solved to obtain the best potential locations.

Step 7: On the off chance the optimal solution has changed correct upto three decimal places, then repeat Step 6; otherwise Stop.

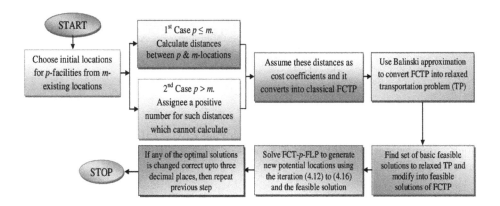

Fig. 2. Schematic diagram of proposed approach for solving FCT-p-FLP.

5   Empirical Design

Herein, a real-life experiment is taken to exhibit the proposed model; and the delineated procedure is more capable to set-up the new sites with a goal to reduce the overall logistics cost along with a fixed-charge cost. An organization desires to build up few new sites so that the cost of the overall conveyance from the existing sites is reduced. The organization has 04 plants S_1, S_2, S_3 and S_4; and the organization wants to establish 03 new wings D_1, D_2 and D_3. The capacities of supply at S_1, S_2, S_3 and S_4; and the requirement to the wings D_1, D_2, and D_3, the locations and weights of the plants S_1, S_2, S_3 and S_4 are also given. The input of all hypothetical parameters are listed in Tables 2–3.

Table 2. Cost parameter (f_{ij}, c_{ij}).

	D1	D2	D3	Supply (r_i)
S1	$(10, c_{11})$	$(30, c_{12})$	$(20, c_{13})$	10
S2	$(40, c_{21})$	$(90, c_{22})$	$(50, c_{23})$	70
S3	$(70, c_{31})$	$(150, c_{32})$	$(80, c_{33})$	60
S4	$(60, c_{41})$	$(160, c_{42})$	$(100, c_{43})$	30
Demand (s_j)	20	100	50	

Table 3. The parameters of the endure plants.

	Position (a_i, b_i)	Weight (α_i)
S1	(0, 1)	0.1
S2	(0, 0)	0.2
S3	(1, 0)	0.3
S4	(1, 1)	0.4

The empirical design is carried out employing LINGO 16.0 software and code block compiler on an Intel Core i5 processor with 4 GB RAM under Windows environment.

5.1   Performance of Our Approach

We focus on the accompanying steps for resolving the stated model by our approach:

– First, 03 primary locations are chosen for each of 03 potential wings from Table 3. In that case, 04 possible cases emerge and they are shown in Tables 4, 5, 6 and 7.

Table 4. 1st case.

	Position	Weight
D1	(0, 1)	0.1
D2	(0, 0)	0.2
D3	(1, 0)	0.3

Table 5. 2nd case.

	Position	Weight
D1	(0, 0)	0.2
D2	(1, 0)	0.3
D3	(1, 1)	0.4

Table 6. 3rd case.

	Position	Weight
D1	(1, 0)	0.3
D2	(1, 1)	0.4
D3	(0, 1)	0.1

Table 7. 4th case.

	Position	Weight
D1	(1, 1)	0.4
D2	(0, 1)	0.1
D3	(0, 0)	0.2

Table 8. Cost coefficient (c_{ij}) for Table 4. **Table 9.** Cost coefficient (c_{ij}) for Table 5.

	D1	D2	D3	r_i
S1	0	1	1.41	10
S2	1	0	1	70
S3	1.41	1	0	60
S4	1	1.41	1	30
s_j	20	100	50	

	D1	D2	D3	r_i
S1	1	1.41	1	10
S2	0	1	1.41	70
S3	1	0	1	60
S4	1.41	1	0	30
s_j	20	100	50	

Table 10. Cost coefficient (c_{ij}) for Table 6.

	D1	D2	D3	r_i
S1	1.41	1	0	10
S2	1	1.41	1	70
S3	0	1	1.41	60
S4	1	0	1	30
s_j	20	100	50	

Table 11. Cost coefficient (c_{ij}) for Table 7.

	D1	D2	D3	r_i
S1	1	0	1	10
S2	1.41	1	0	70
S3	1	1.41	1	60
S4	0	1	1.41	30
s_j	20	100	50	

- Now, the distances are calculated between them for cost coefficient by using Tables 4, 5, 6 and 7 and put them in Tables 8, 9, 10 and 11, respectively.
- Balinski's approximation is used in Tables 8, 9, 10 and 11 and the obtained results are shown in Tables 12, 13, 14 and 15 respectively.
- LINGO 16.0 iterative scheme is used for initial BFSs by utilising Tables 12, 13, 14 and 15. The acquired outcomes appear in Tables 16, 17, 18 and 19 separately.
- Using C++ programming language, the obtained experimental outcomes for Tables 16, 17, 18 and 19 are finally placed in Table 20.

Table 12. Balinski cost matrix $(C_{ij} = c_{ij} + f_{ij}/m_{ij})$ for Table 8.

	D1	D2	D3	r_i
S1	1	4	3.41	10
S2	3	1.29	2	70
S3	4.91	3.5	1.6	60
S4	4	6.74	4.33	30
s_j	20	100	50	

Table 13. Balinski cost matrix $(C_{ij} = c_{ij} + f_{ij}/m_{ij})$ for Table 9.

	D1	D2	D3	r_i
S1	2	4.41	3	10
S2	2	2.29	2.41	70
S3	4.5	2.5	2.6	60
S4	4.41	6.33	3.33	30
s_j	20	100	50	

Table 14. Balinski cost matrix $(C_{ij} = c_{ij} + f_{ij}/m_{ij})$ for Table 10.

	D1	D2	D3	r_i
S1	2.41	4	2	10
S2	3	2.70	2	70
S3	3.5	3.5	3.01	60
S4	4	5.33	4.33	30
s_j	20	100	50	

Table 15. Balinski cost matrix $(C_{ij} = c_{ij} + f_{ij}/m_{ij})$ for Table 11.

	D1	D2	D3	r_i
S1	2	3	3	10
S2	3.41	2.29	1	70
S3	4.5	3.91	2.6	60
S4	3	6.33	4.74	30
s_j	20	100	50	

Table 16. OBFS (w_{ij}^B, u_{ij}^B) for Table 12.

	D1	D2	D3	r_i
S1	(10, 1)	(0, 0)	(0, 0)	10
S2	(0, 0)	(70, 1)	(0, 0)	70
S3	(0, 0)	(30, 1)	(30, 1)	60
S4	(10, 1)	(0, 0)	(20, 1)	30
s_j	20	100	50	

Table 17. OBFS (w_{ij}^B, u_{ij}^B) for Table 13.

	D1	D2	D3	r_i
S1	(10, 1)	(0, 0)	(0, 0)	10
S2	(10, 1)	(60, 1)	(0, 0)	70
S3	(0, 0)	(40, 1)	(20, 1)	60
S4	(0, 0)	(0, 0)	(30, 1)	30
s_j	20	100	50	

Table 18. OBFS (w_{ij}^B, u_{ij}^B) for Table 14.

	D1	D2	D3	r_i
S1	(0, 0)	(0, 0)	(10, 1)	10
S2	(0, 0)	(40, 1)	(30, 1)	70
S3	(0, 0)	(60, 1)	(0, 0)	60
S4	(20, 1)	(0, 0)	(10, 1)	30
s_j	20	100	50	

Table 19. OBFS (w_{ij}^B, u_{ij}^B) for Table 15.

	D1	D2	D3	r_i
S1	(0, 0)	(10, 1)	(0, 0)	10
S2	(0, 0)	(70, 1)	(0, 0)	70
S3	(0, 0)	(20, 1)	(40, 1)	60
S4	(20, 1)	(0, 0)	(10, 1)	30
s_j	20	100	50	

5.2   Result and Discussion

Here, the optimal result of the experimental design is explored. The following solution is obtained by the approximation approach, utilizing Table 20, that is

Table 20. The experimental solutions for Tables 16, 17, 18 and 19.

OBFS	Location of D1	Location of D2	Location of D3	Value of Z
Table 15	(1.000, 1.000)	(0.015, 0.000)	(1.000, 0.267)	508.341
Table 16	(0.000, 0.001)	(0.500, 0.000)	(1.000, 0.999)	489.005
Table 17	(1.000, 1.000)	(1.000, 0.000)	(0.066, 0.089)	506.049
Table 18	(1.000, 1.000)	(0.000, 0.000)	(1.000, 0.000)	522.314

Table 21. The optimal outcome of FCT-p-FLP.

OBFS	Location of D1	Location of D2	Location of D3	Value of Z
Table 16	(0.000, 0.001)	(0.500, 0.000)	(1.000, 0.999)	489.005

listed in Table 21. The convergence of our approach and the solutions are shown in Figs. 3 and 4.

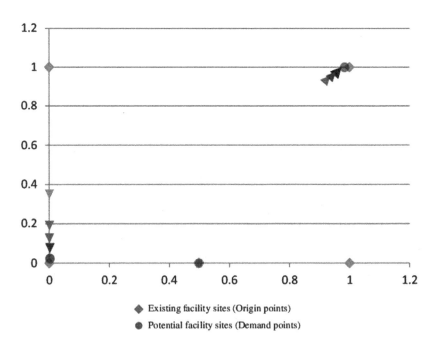

◆ Existing facility sites (Origin points)

● Potential facility sites (Demand points)

Fig. 3. Performance of the approximation approach.

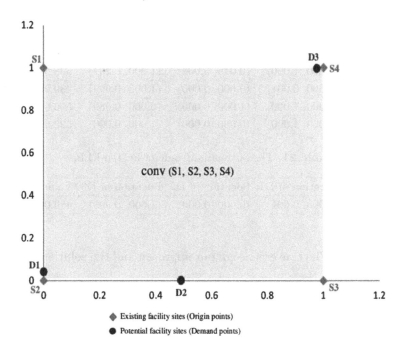

Fig. 4. The existing and potential sites in experimental design.

6   Conclusion and Future Work

This chapter has been depicted a newfangled strategic problem for distribution planning that goals to reduce the overall ship out cost along with fixed-charge cost and to select potential facilities for different sites. To the best of the information, for the first time in research, the proposed design has been rendered a strategy for investigating the association between the FLP and FCTP. Afterwards, few properties on FCT-p-FLP have been included to analyze the characteristic of FCT-p-FLP. Despite the over, a new approximation approach has been incorporated to tackle the stated formulation in a proficient way. The proposed model and approach have been tested by an application study. Ultimately, the obtained result from the approximation approach has been discussed with suggestions. The approximation approach has been used to find optimal solutions for FCT-p-FLP. However, the formulation has been introduced here can be applied in industrial problems such as the plant location problems, green transportation system and applications.

Such huge numbers of research bearings stay open, thus, researchers can investigate the problem by various distance metric such as block metric, rectangular metric, Hausdroff metric, signed distance etc. and they may employ alternative solution techniques such as Lagrangean relaxation heuristic, greedy algorithm, genetic algorithm, branching method and so on. In fact, one may develop the proposed model under several uncertain environments like stochastic [29], intuitionistic fuzzy [25,39], type-2 fuzzy [42], fuzzy-rough [24], and sustainable development [32] ones. Furthermore, the scientists can consider the multimodal transportation system [33] and multi-objective optimization [43] environment in our stated problem.

References

1. Adlakha, V., Kowalski, K.: On the fixed-charge transportation problem. Omega **27**(3), 381–388 (1999)
2. Adlakha, V., Kowalski, K.: A simple heuristic for solving small fixed-charge transportation problems. Omega **31**(3), 205–211 (2003)
3. Adlakha, V., Kowalski, K., Wang, S., Lev, B., Shen, W.: On approximation of the fixed charge transportation problem. Omega **43**, 64–70 (2014)
4. Amin, S.H., Baki, F.: A facility location model for global closed-loop supply chain network design. Appl. Math. Model. **41**, 316–330 (2017)
5. Atta, S., Mahapatra, P.R.S., Mukhopadhyay, A.: Multi-objective uncapacitated facility location problem with customers preferences: pareto-based and weighted sum GA-based approaches. Soft. Comput. **23**(23), 12347–12362 (2019). https://doi.org/10.1007/s00500-019-03774-1
6. Babaee Tirkolaee, E., Goli, A., Faridnia, A., Soltani, M., Weber, G.W.: Multi-objective optimization for the reliable pollution-routing problem with cross-dock selection using pareto-based algorithms. J. Clean. Prod. **276**, 122927 (2020)
7. Tirkolaee, E.B., Mahdavi, I., Esfahani, M.M.S., Weber, G.W.: A robust green location-allocation-inventory problem to design an urban waste management system under uncertainty. Waste Manag. **102**, 340–350 (2020)
8. Balinski, M.L.: Fixed-cost transportation problems. Naval Res. Logist. Q. **8**(1), 41–54 (1961)
9. Bieniek, M.: A note on the facility location problem with stochastic demands. Omega **55**, 53–60 (2015)
10. Brahami, M.A., Dahane, M., Souier, M., Sahnoun, M.: Sustainable capacitated facility location/network design problem: a Non-dominated Sorting Genetic Algorithm based multiobjective approach. Ann. Oper. Res. 1–32 (2020). https://doi.org/10.1007/s10479-020-03659-9
11. Carlo, H.J., David, V., Salvat-Dávila, G.S.: Transportation-location problem with unknown number of facilities. Comput. Ind. Eng. **112**, 212–220 (2017)
12. Cooper, L.: Heuristic methods for location-allocation problems. SIAM Rev. **6**(1), 37–53 (1964)
13. Cooper, L.: The transportation-location problem. Oper. Res. **20**(1), 94–108 (1972)
14. Cooper, L.: The stochastic transportation-location problem. Comput. Math. Appl. **4**(3), 265–275 (1978)
15. Das, S.K., Pervin, M., Roy, S.K., Weber, G.W.: Multi-objective solid transportation-location problem with variable carbon emission in inventory management: a hybrid approach. Ann. Oper. Res. 1–27 (2021). https://doi.org/10.1007/s10479-020-03809-z
16. Das, S.K., Roy, S.K.: Effect of variable carbon emission in a multi-objective transportation-p-facility location problem under neutrosophic environment. Comput. Ind. Eng. **132**, 311–324 (2019)

17. Das, S.K., Roy, S.K., Weber, G.W.: Application of type-2 fuzzy logic to a multiobjective green solid transportation-location problem with dwell time under carbon tax, cap, and offset policy: fuzzy versus nonfuzzy techniques. IEEE Trans. Fuzzy Syst. **28**(11), 2711–2725 (2020)

18. Das, S.K., Roy, S.K., Weber, G.W.: An exact and a heuristic approach for the transportation-p-facility location problem. CMS **17**(3), 389–407 (2020). https://doi.org/10.1007/s10287-020-00363-8

19. Das, S.K., Roy, S.K., Weber, G.W.: Heuristic approaches for solid transportation-p-facility location problem. CEJOR **28**(3), 939–961 (2020). https://doi.org/10.1007/s10100-019-00610-7

20. Dias, J., Captivo, M.E., Clíma, J.: Dynamic multi-level capacitated and uncapacitated location problems: an approach using primal-dual heuristics. Oper. Res. Int. J. **7**(3), 345–379 (2007). https://doi.org/10.1007/BF03024853

21. Dohse, E.D., Morrison, K.R.: Using transportation solutions for a facility location problem. Comput. Ind. Eng. **31**(1–2), 63–66 (1996)

22. Farahani, R.Z., Hekmatfar, M.: Facility Location: Concepts, Models, Algorithms and Case Studies. Springer, Heidelberg (2009). https://doi.org/10.1007/978-3-7908-2151-2

23. Farahani, R.Z., SteadieSeifi, M., Asgari, N.: Multiple criteria facility location problems: a survey. Appl. Math. Model. **34**(7), 1689–1709 (2010)

24. Ghosh, S., Roy, S.K.: Fuzzy-rough multi-objective product blending fixed-charge transportation problem with truck load constraints through transfer station. RAIRO: Recherche Opérationnelle **55**, S2923–S2952 (2021)

25. Ghosh, S., Roy, S.K., Ebrahimnejad, A., Verdegay, J.L.: Multi-objective fully intuitionistic fuzzy fixed-charge solid transportation problem. Complex Intell. Syst. **7**(2), 1009–1023 (2021). https://doi.org/10.1007/s40747-020-00251-3

26. Harris, I., Mumford, C.L., Naim, M.M.: A hybrid multi-objective approach to capacitated facility location with flexible store allocation for green logistics modeling. Transp. Res. Part E Logist. Transp. Rev. **66**, 1–22 (2014)

27. Hirsch, W.M., Dantzig, G.B.: The fixed charge problem. Naval Res. Logist. Q. **15**(3), 413–424 (1968)

28. Karatas, M., Yakıcı, E.: An iterative solution approach to a multi-objective facility location problem. Appl. Soft Comput. **62**, 272–287 (2018)

29. Klibi, W., Lasalle, F., Martel, A., Ichoua, S.: The stochastic multiperiod location transportation problem. Transp. Sci. **44**(2), 221–237 (2010)

30. Klose, A.: Algorithms for solving the single-sink fixed-charge transportation problem. Comput. Oper. Res. **35**(6), 2079–2092 (2008)

31. Love, R.F., Morris, J.G., Wesolowsky, G.O.: Facility Location: Models and Methods. Springer, Berlin (1995)

32. Maity, G., Roy, S.K., Verdegay, J.L.: Time variant multi-objective interval-valued transportation problem in sustainable development. Sustainability **11**(21), 6161 (2019)

33. Maity, G., Roy, S.K., Verdegay, J.L.: Analyzing multimodal transportation problem and its application to artificial intelligence. Neural Comput. Appl. **32**(7), 2243–2256 (2019). https://doi.org/10.1007/s00521-019-04393-5

34. Mehlawat, M.K., Kannan, D., Gupta, P., Aggarwal, U.: Sustainable transportation planning for a three-stage fixed charge multi-objective transportation problem. Ann. Oper. Res. 1–37 (2019). https://doi.org/10.1007/s10479-019-03451-4

35. Melkote, S., Daskin, M.S.: An integrated model of facility location and transportation network design. Transp. Res. Part A Policy Pract. **35**(6), 515–538 (2001)

36. Melo, M.T., Nickel, S., Saldanha-Da-Gama, F.: Facility location and supply chain management-a review. Eur. J. Oper. Res. **196**(2), 401–412 (2009)
37. Midya, S., Roy, S.K.: Solving single-sink, fixed-charge, multi-objective, multi-index stochastic transportation problem. Am. J. Math. Manag. Sci. **33**(4), 300–314 (2014)
38. Midya, S., Roy, S.K.: Analysis of interval programming in different environments and its application to fixed-charge transportation problem. Discrete Math. Algorithms Appl. **9**(03), 1750040 (2017)
39. Midya, S., Roy, S.K., Yu, V.F.: Intuitionistic fuzzy multi-stage multi-objective fixed-charge solid transportation problem in a green supply chain. Int. J. Mach. Learn. Cybern. **12**(3), 699–717 (2020). https://doi.org/10.1007/s13042-020-01197-1
40. Mingozzi, A., Roberti, R.: An exact algorithm for the fixed charge transportation problem based on matching source and sink patterns. Transp. Sci. **52**(2), 229–238 (2017)
41. Mišković, S., Stanimirović, Z., Grujičić, I.: Solving the robust two-stage capacitated facility location problem with uncertain transportation costs. Optim. Lett. **11**(6), 1169–1184 (2016). https://doi.org/10.1007/s11590-016-1036-2
42. Roy, S.K., Bhaumik, A.: Intelligent water management: a triangular type-2 intuitionistic fuzzy matrix games approach. Water Resour. Manag. **32**(3), 949–968 (2018). https://doi.org/10.1007/s11269-017-1848-6
43. Roy, S.K., Maity, G.: Minimizing cost and time through single objective function in multi-choice interval valued transportation problem. J. Intell. Fuzzy Syst. **32**(3), 1697–1709 (2017)
44. Roy, S.K., Maity, G., Weber, G.W., Gök, S.Z.A.: Conic scalarization approach to solve multi-choice multi-objective transportation problem with interval goal. Ann. Oper. Res. **253**(1), 599–620 (2016). https://doi.org/10.1007/s10479-016-2283-4
45. Roy, S.K., Midya, S.: Multi-objective fixed-charge solid transportation problem with product blending under intuitionistic fuzzy environment. Appl. Intell. **49**(10), 3524–3538 (2019). https://doi.org/10.1007/s10489-019-01466-9
46. Roy, S.K., Midya, S., Vincent, F.Y.: Multi-objective fixed-charge transportation problem with random rough variables. Int. J. Uncertainty Fuzziness Knowl.-Based Syst. **26**(6), 971–996 (2018)
47. Roy, S.K., Midya, S., Weber, G.-W.: Multi-objective multi-item fixed-charge solid transportation problem under twofold uncertainty. Neural Comput. Appl. **31**(12), 8593–8613 (2019). https://doi.org/10.1007/s00521-019-04431-2
48. Saif, A., Elhedhli, S.: A Lagrangian heuristic for concave cost facility location problems: the plant location and technology acquisition problem. Optim. Lett. **10**(5), 1087–1100 (2016). https://doi.org/10.1007/s11590-016-0998-4
49. Sherali, H.D., Tuncbilek, C.H.: A Squared-Euclidean distance location-allocation problem. Naval Res. Logist. (NRL) **39**(4), 447–469 (1992)
50. Wanka, G., Wilfer, O.: Duality results for nonlinear single minimax location problems via multi-composed optimization. Math. Methods Oper. Res. **86**(2), 401–439 (2017). https://doi.org/10.1007/s00186-017-0603-3
51. Xiao, Z., Sun, J., Shu, W., Wang, T.: Location-allocation problem of reverse logistics for end-of-life vehicles based on the measurement of carbon emissions. Comput. Ind. Eng. **127**, 169–181 (2019)

Unrelated Parallel Machine Scheduling Problem Subject to Inventory Limitations and Resource Constraints

Mohammad Arani[1](✉) ⓘ, Mousaalreza Dastmard[2] ⓘ, Mohsen Momenitabar[3] ⓘ, and Xian Liu[1] ⓘ

[1] Department of Systems Engineering, University of Arkansas at Little Rock, Little Rock, AR 72204, USA
mxarani@ualr.edu
[2] Department of Information Engineering, Electronics and Telecommunications, Sapienza University of Rome Lazio, Rome, Italy
[3] Department of Transportation, Logistics and Finance, North Dakota State University, Fargo, ND 58105-6050, USA

Abstract. In this paper, we studied a real-world example of an unrelated parallel machine scheduling problem in which the Just-in-Time (JIT) inventory system, including Work-in-Process and Finished Goods inventories, is the center of interest. The presented Simulated Annealing Algorithm tackling the combinatorial problem is a part of the expert and intelligent production system that assists a Decision Maker in perceiving his/her decision's outcome in advance. Precisely noted, the problem statement is novel since the integration of inventory systems (Work-in-Process and Finished Goods), and renewable and non-renewable resources engaged in manufacturing are taken into account. The undertaken manufacturing system could be attributed to the direct influence of the management strategy of the JIT inventory system not only to reduce the inventory level but also to minimize the earliness and tardiness cost responding to due dates. Moreover, the conventional parameters are reflected in the problem description and solution methodology. In conclusion, the results are remarkable in two terms, the first is to obtain better production planning in considerably less time, and secondly, a cost-benefit analysis tool to contrast the JIT approach with the earliest possible completion time criterion. Finally, possible research directions are suggested for future studies.

Keywords: Unrelated parallel machine scheduling problem · Kanban inventory planning · Makespan · Just-in-time production · Resource constraints · Simulated annealing

1 Introduction

In today's industrial manufacturing system, Parallel Machine Scheduling (PMS) Problem and its alike machine scheduling such as Unrelated Parallel Machine Scheduling Problem (UPMSP) have a wide range of usage in different types of the production process

© Springer Nature Switzerland AG 2021
Z. Molamohamadi et al. (Eds.): LSCM 2020, CCIS 1458, pp. 238–254, 2021.
https://doi.org/10.1007/978-3-030-89743-7_13

including project, job-shop, batch-production, mass production, and finally flow production process. Not to mention that any referred production process dedicates a certain level of production to be parallelly manufactured, amongst them Intermittent Production System (job-shop, and batch-production) is mostly founded on PMS [1]. While at least one part of the production planning in discrete part manufacturing employs PMS, the significance of properly addressing the problem absorbed attention.

The first and foremost decision that needs to be made in PMS is to determine the sequence of jobs on the assigned machine(s). In this study, we adopt a novel and real problem description in which it is described by a part manufacturing called Bronze Industrial Group. The coding package then becomes a segment of expert and intelligent production systems of the manufacturer. The novel and real-world problem address an unrelated parallel machine scheduling problem were applied in one production haul having several parallel machines with different capabilities and eligibilities to produce cars' door panels. Moreover, machines are sequence-dependent and their eligibilities to perform a task need to be verified, products have delivery due dates, therefore earliness and tardiness impose inventory cost and penalties respectively. Please bear in mind that we considered limited work-in-process and finished goods inventories to simulate the reality. Albeit the problem description is novel and intensely practical, the problem statement and the proposed simulated annealing algorithm are the substantial contributions of this paper. It is proved that the production plan offered by our package easily surpasses the production plan performed by expertise. The new approach to production planning leads the company to a more productive plan, and competitive strength amid their rivals, consequently, fewer backorders were encountered.

In the extensive reviewed literature, the adapted problem specification illustrates novelty and uniqueness. Founding on the literature and proposed tools to solve the related problems by scholars, it was decided to employ simulated annealing. It is noted that the meta-heuristic algorithm designed in a way to commence searching the solution area from a feasible point and enhance it in every iteration, by the end of the algorithm run time, it searches more than five thousand feasible points that realistically sounds impossible to be searched by a production planner. Noteworthy to mention that the run time for a conventional desktop computer is about four minutes. Therefore, the advantages of new optimization tools over the traditional ones are proved by experience.

The rest of the paper is organized as follows: Sect. 2 provides a thorough and recent literature review over the PMS problem and our contributions that distingue the current research. In Sect. 3, we propose simulated annealing details to adequately deal with the problem. Section 4 provides a real case study pertinent to one week of the PMS plan at Bronze Industrial Group. Finally, the conclusion and future research are provided in Sect. 5.

2 Literature Review

The first segment of the literature review is dedicated to the fundamental parallel machine scheduling problems solved with exact methods or developed methods in which the literature refers them to heuristics methods. Żurowski (2019) [2] proposed a fundamental PMS in which there are two identical machines and jobs could be preempted meaning that a job could be split and continued where it is left afterward. The authors stated that the processing time is sequence-dependent, moreover, the criterion of the two presented exact algorithms was to quest for the shortest completion time. Naderi (2019) [3] proposed a tractable decomposition method for the fundamental identical PMS called Branch-Relax-and-Check. Here a mixed-integer programming model presented alongside a novel assumption in the literature, which was order acceptance, for the proposed algorithm. Finally, the authors made a comparison between CPLEX embedded methods to solve the model and the proposed method, the first solves 13% of instances optimally whereas the exact method solves 50%.

Ding (2019) [4] represented an enhanced version of the Ejection Chain Algorithm (ECA) to solve PMS problems with sequence-dependent deteriorating effects. The objective function was to minimize the makespan. The authors not only did report solving 100% of the small-size problems of a benchmark but also achieved improvement on the best-known results for 388 large instances with a better computational time. Báez (2019) [5] tackled the identical PMS problem with a hybrid meta-heuristic algorithm. The algorithm was a combination of the Greedy Randomized Adaptive Search Procedure (GRASP) algorithm and the Variable Neighborhood Search (VNS) algorithm. The standard feature of the proposed identical PMS problem was to have dependent setup times and regular minimization objective function of makespan. Another variation of the problem was proposed by [6], a sequence-dependent setup time identical PMS problem. In this research, the effect of learning or tiredness was investigated. The so-called learning or deterioration effect influences the setup time of the current job taking into account the previous job. The detailed information could be found in [7].

Wang (2018) [8] presented a bi-objective identical PMS in which the two objective functions are (a) total energy consumption, (b) makespan. The authors expressed the paramount importance of energy consumption in production and manufacturing, specifically for energy-intensive industries. Two approaches were utilized to solve the problem regarding the size of the problem, for small size, augmented ε-constraint method, and for the medium- and large-size, non-dominated sorting genetic algorithm II (NSGA-II) are employed. Along with new approaches, Kılıç (2019) [9] proposed a thought-provoking algorithm called the antlion optimization algorithm to solve an identical PMS problem. The authors proposed a new tournament selection method for the algorithm and run several tests to measure the performance and accuracy of the improved meta-heuristics on multi-dimensions benchmark functions, as well. Liu (2019) [10] proposed a two-stage hybrid flow shop scheduling on an identical PMS problem. The studied problem had the properties of job-dependent deteriorating effect (learning to effect), and non-identical job sizes. The authors' approach to solving the problem was a combined algorithm constituting two eminent algorithms, the first one was the Estimation of Distribution Algorithm (EDA), and the second one was the Differential Evolution (DE).

On the other hand, in the literature, a new problem has arisen which is more practical in terms of manufacturing and production called an unrelated parallel machine scheduling problem (UPMSP). The UPMSP has received a lot of attention recently. Rauchecker (2019) [11] dealt with a similar problem in which the property of sequence-dependent setup times existed. The authors developed and computationally tested a parallel branch-and-price algorithm. The new approach to solve the problem was unique in sense of implementing a distributed-memory parallelization with a master/worker. Additionally, the result was interesting that with the new approach some large-size problems were solved in less than six minutes that were used to solve ninety-four hours formerly. Yepes-Borrero (2020) [12] tackled a variation of the problem in which additional renewable resources (e.g. workers) were considered. Similar to the standard version of PMS problems the criterion to solve the problem was makespan. Moreover, Greedy Randomized Adaptive Search Procedure (GRASP) algorithm was hired to solve the problem. Bektur (2019) [13] employed two meta-heuristic algorithms to solve unrelated PMS problems namely Simulated Annealing (SA) and Tabu Search (TS). In this model, a common server was taken into consideration that the concept was borrowed from [14] and it served as a limited resource that requires to perform the jobs. Soleimani (2020) [15] dealt with an unrelated PMS problem in which sequence-related setup time, position-dependent learning effect was considered, furthermore, the power consumption was considered to be minimized for the objective function. Although a new mathematical formulation was developed as the authors' contribution, a new meta-heuristic approach was employed called Cat Swarm Optimization (CSO) algorithm. Another interesting and two fresh algorithms were used to solve unrelated PMS problems exploited by [16] called Enhanced Symbiotic Organisms Search (ESOS) algorithm, and Hybrid Symbiotic Organisms Search with Simulated Annealing (HSOSSA).

Lu (2018) [17] proposed a novel problem statement that unrelated PMS problem was combined with maintenance planning of the machines. Alike the majority of reviewed papers, the minimization of makespan was emphasized. Since the problem was an NP-hard problem, a hybrid Artificial Bee Colony and Tabu Search (ABC-TS) was practiced for medium- and large-sized problems. Perez-Gonzalez (2019) [18] studied a real-world example unrelated PMS problem. The problem statement was adapted from a customized heating, ventilation, and air conditioning (HAVC) factory, metal folding section. A mixed-integer programming formulation established and a set of constructive heuristics search algorithms developed to solve the model efficiently. The other real-world instance studied at a television manufacturing called Vestel Electronics offered by [19]. The objective function was adapted from the Just-in-Time concept and make-to-order strategy, meaning minimizing earliness and tardiness of the completed jobs in the presence of sequence-dependent setup time, unequal release times, eligibility restrictions. Three algorithms were elected to crack the problem, that is to say, (I) a sequential algorithm, (II) a tabu search algorithm, and finally (III) a random set partitioning algorithm. In contrary to the heuristics and meta-heuristics methods developed for unrelated parallel machine scheduling, there are several exact algorithms proposed. Fanjul-Peyro (2019) [20] tackled the problem of considering setups and resources while minimizing the makespan. The authors developed a three-phase exact algorithm (TPEA) and it performed assignment, sequencing, and timing of jobs. Fanjul-Peyro (2020) [21] tackled the same problem with a new mathematical base algorithm to solve the problem. The new formulation was a particular heterogeneous m-TSP accompanied by two sets of valid inequalities. In Table 1 and Table 2, a systematic review of literature is presented regarding two main problem definitions of Identical PMS, and UPMSP. Finally, we would like to emphasize the contributions and novelty of the current paper that distinguish it from the literature review as follows:

Table 1. Research background (Identical PMS).

	Objective(s)	Solution method	Assumptions
[3]	Maximizing profit	Branch-Relax-and-Check exact method	Order acceptance, MIP model formulation, Orders have due dates
[6]	Minimizing makespan	Variable neighborhood search (VNS), Multi-agent simulation	Effect of learning and tiredness on sequence-dependent setup times
[2]	Minimizing makespan	Two exact solutions are presented for the case there are only two machines (B&B)	Preemptable jobs
[5]	Minimizing makespan	GRASP, and VNS algorithms	Dependent setup times
[9]	Minimizing makespan	Improved antlion optimization algorithm	Sequence-dependent setup time
[4]	Minimizing makespan	Ejection chain algorithm	Sequence-dependent deterioration
[10]	Minimizing makespan	Estimation of distribution algorithm (EDA), Differential evolution (DE)	Sequence-dependent deterioration,
[7]	Minimizing makespan	A new model	Integral-based learning effect
[22]	Total green cost, and Makespan	ε-constraint method, heuristic algorithm	Green working cost to address one of the objectives
[8]	Minimization of total energy consumption, and Makespan	Augmented ε-constraint method, NSGA II	Independent, Non-preemptive jobs
[23]	Makespan	Five novel mixed-integer programming	Family setup times
[24]	Minimization of total weighted completion time and total processing cost	Robust Mixed-Integer Linear Programming	Job due date, Availability of resources, and Uncertain processing time
[25]	Minimize the total cost of the production system and total energy consumption	Interactive fuzzy technique and a self-adaptive artificial fish swarm algorithm (SAAFSA)	Outsourcing option and just-in-time delivery

Table 2. Research background (UPMSP).

	Objective(s)	Solution method	Assumptions
[12]	Minimizing makespan	Greedy randomized adaptive search procedure (GRASP) algorithm	Setup time, renewable resources to do setup on machines
[13]	Minimizing Tardiness	Simulated annealing, Tabu Search	MILP, Sequence-dependent setup time, Scheduling with common server
[11]	Minimizing makespan	Parallel branch-and-price algorithm	Sequence-dependent setup time, High performance computing master/worker parallelization
[15]	Minimizing power consumption	Cat Swarm Optimization	Sequence-related setup time, Start time-dependent deterioration, Position-dependent learning effect
[16]	Minimizing makespan	ESOS, HSOSSA algorithms	Setup times, non-preemptive,
[17]	Minimizing makespan	ABS-TS algorithm	Deteriorating jobs, Maintenance activity, Batch production
[26]	Minimizing total electricity cost with bounded makespan	A heuristic fix and relax algorithm	Time-of-use (TOU) electricity tariffs, Dominance rules, and valid inequalities
[18]	Minimizing Tardiness	Constructive heuristics	Machine eligibility, Sequence-dependent setup times
[19]	Minimizing earliness and tardiness	Sequential algorithm, Tabu search algorithm, Random set partitioning algorithm	Sequence-dependent setup time, unequal release times, eligibility restrictions
This research	Minimizing tardiness and makespan	Mixed-integer linear programming model, and Simulated annealing	Working schedule and different working shifts, Sequence-dependent setup time and preparation time, Preemptive jobs under condition (batch preemption), Renewable resource scheduling, Integrated inventory

- Integration of machine scheduling and two types of conventional inventory in the manufacturing system (work-in-process and finished goods inventories).
- Unrelated parallel machine scheduling with consideration of different capacities, eligibilities, speed, product-sequence-related setup time, product-sequence-related preparation time.
- Consideration of working schedules and different working shifts, and renewable resource scheduling namely, crane and lift trucks (fork-lift carrier).
- Accounting due dates for delivery of products and makespan criterion to help the decision-making process and cost-benefit analysis.
- The practicality of the problem description was maintained since the project was conducted by a part manufacturer and the results were verified on-site by experts.

3   Problem Statement

Consider a UPMSP in the context of production installed on the shop floor of a part-manufacturing. There are n distinguishable parts regarding the varied molding frames, and resources (renewable such as worker, crane to replace the frames, and non-renewable such as materials for extrusion operations) to produce. Each job is tied together as a batch, in other words, each job consists of several identical parts that need to be produced while the machine settings remain unchanged. Preemption of batches is allowed provided that each time one molding frame is installed at least a minimum number of quality parts are produced, additionally, the change of extruding materials is considered. There exist m different machines that are varied in the sense of (a) capabilities of production whether one can produce a specific type of parts (eligibility restriction), (b) speed of production, (c) required material concerning the productivity of the machine (percentage of damaged products), (d) the required number of operators to run the machine. As mentioned before, the manufacturer requires to produce each job while a set of frames is installed on a machine, called economic production to be financially beneficial to go through the changes of frames and extruding material. Bearing in mind that the preparation time, setup time, and warm-up time are modified knowing the sequence of jobs (sequence-related times, or learning effect). In reality, there are different shifts and working hours which are also embedded in the model. Up to this point, the stated problem is well-known in the literature and categorized as $R/s_{ijk}, M_j, \sum w_j C_j$ or $R/s_{ijk}, M_j, \sum w_j * max\{0, DD_j - C_j\}$ [27] in the well-founded $\alpha/\beta/\gamma$-notation [28].

In this problem statement, the machine scheduling problem is combined with an inventory problem, since the minimization of lateness (minimization of earliness and tardiness) criterion is measured in place of one of the objective functions along with the conventional makespan objective function. In this inventory system, a minimum level of products required to be always kept in the main depot. Moreover, the temporary inventory system is observed in the production line, and the management strategy towards inventory is to keep as many parts as required to meet the customers' demands for one day of customers' consumptions. By that approach, the inventory cost is mitigated to holding costs of end-products for only one day. The policy held significant in part manufacturing is to manufacture as many parts as required in a certain time (Just-in-Time production policy). Therefore, taking into account the production plan, and inventory system either the work-in-process or finished goods inventories are the genuine model boundaries, the due dates for the required products need to be met otherwise penalty charges apply to the manufacturer.

4 Solution Methodology: Simulated Annealing Algorithm

To meet the challenge of resolving the problem in small, medium, and large-size, we take an orthodox and standard method, namely: Simulated Annealing (SA) algorithm. The simulated annealing (SA) algorithm is extensively used to solve combinatorial optimization since it provides numerically robust solutions. SA was first presented by [29] and intensely considered as a meta-heuristic algorithm to solve a wide variety of parallel machine scheduling [13, 16, 30–34]. SA executes the search not only deliberately by accepting a better solution throughout inspecting neighborhoods but also stochastically by accepting a worse one. In some solution spaces always investigating the vicinity of the current solution may not lead us to the best answer yet local optima would be found. To have chances to leave a local optimum and locate the global optimum, with a probability correlated to the algorithm's temperature, SA acquires the worse solution and attempts to detect a better one by iterative operations until the termination conditions are met or it converges to a solution. Table 3 presents the fundamental Simulated Annealing algorithm pseudocode, however, in this study, the multi-start-point version is benefited from. Simply instead of one initial solution point, several initial solution points are exercised in each iteration, finally, the best-received solution is offered to the Decision Maker (DM).

The solution space, X, for the combinatorial problem is defined as a set of all possible permutation of jobs that are assigned to m machines. The algorithm commences with an initial permutation ($perm_{initial}$) considering the eligibility of machines to perform their assigned tasks, and randomly generate a new one ($perm_{new}$) in the vicinity of the initial one with consideration of eligibility at each iteration. If the new one accepted, it would be substituted as the solution of the current iteration ($perm_{it} \leftarrow perm_{new}$) and algorithm continues until the stopping criteria are met. There are several specifically designed operators, which are pertinent to unrelated PMS problems, embedded in the fundamental structure of the SA algorithm. These operators can either be inter-machine or intra-machine. The inter-machine operators comprise the rescheduling of jobs between two distinct eligible machines and the intra-machine operators function only on jobs that are scheduled on the same machine. The standard operators were employed and the feasibility of assigned permutation is assessed automatically.

Due to attaining a good quality schedule, as the algorithm advances a temperature scaling is hired. SA starts with an initial temperature T_{init}, and every iteration it modifies with a constant $\alpha \in (0, 1)$ calling cooling coefficient $T_{it} = \alpha \times T_{it-1}$. For the SA algorithm to converge to a solution, temperature T_{it} plays an essential role. In each iteration, if the utility value or objective function of new permutation is better than the current one, the new swaps with the current one ($F(perm_{new}) \leq F(perm_{it})$). On the contrary, if the new schedule is worse, with a probability associated with the current temperature the worse is going to be accepted. An interesting observation here is that the chance

Table 3. Solution methodology.

Modified Simulated Annealing Algorithm (for the current problem statement)
Initialization:
temperature, T
cooling coefficient, α
randomly generate a feasible solution, $perm_{it}$
iteration, $it = 1$
temperature of the first iteration, $T_{it} \leftarrow T$
While termination conditions are not met:
search the neighborhood of $perm_{it}$, $perm_{new}$
If $F(perm_{new}) \leq F(perm_{it})$
$perm_{it} \leftarrow perm_{new}$
Else
$\delta = F(perm_{new}) - F(perm_{it})$
create a random number, β
If $\beta \leq exp(-\delta/T)$
$perm_{it} \leftarrow perm_{new}$
$it \mathrel{+}= 1$
$T_{it+1} = \alpha \times T_{it}$
Report:
the best solution found

reduces while the algorithm progresses, consequently, towards the end, the algorithm converges to a solution. The algorithm's termination conditions are obtained from the literature, namely: (1) after a predefined number of iterations exhausted appropriate to the size of the problem, (2) while several function evaluation is calculated, in other words, there is a cap on calling objective function, (3) terminating the algorithm while the best-discovered solution remains intact for several iterations. Table 4 provide deep insight into how a solution is coded to be comprehensible by the algorithm, and decoded to be apprehensible by a user, and convert to a real scheduling plan on the shop floor by operators.

Although multi-initial points to commence the SA algorithm augmented the robustness of the solution, the Taguchi experimental design method embedded in the MiniTab commercial package was utilized to tune the parameters of the algorithm. Please refer to [35] for more information on the parameter tune of Meta-Heuristics algorithms.

Table 4. Objective function.

Objective Function Evaluation Algorithm (Total Weighted Earliness and Tardiness, *TWET*)

Input:
- each preemptable batch is split into economic sub batches (the last batch might be less than the economic number)
- assign a segment of memory to record renewable resource working hours (expert labors, and crane working hours), $SM_{ren.res.}$
- a solution point is presented by sequences of sub batches for each machine, given that some of all sub batches regardless of the sequences add up to the demand of the week
- assign a segment of memory to record the completion time of each batch, $SM_{CT,batch}$
- $TWET \leftarrow 0$

For each machine do:
- consider the sequence of feasible assigned batches, the required renewable resources for each batch

For each batch do:
- detect the earliest time machine is available

If required renewable resources are accessible at the earliest time machine is available, $SM_{ren.res.}$
- assign the time window that renewable resources are engaged to $SM_{ren.res.}$
- calculate the setup, and preparation time
- begin producing the batch, and update $SM_{CT,batch}$

Else find the earliest time that all renewable resources are accessible
- assign the time window that renewable resources are engaged to $SM_{ren.res.}$
- calculate the setup, and preparation time
- begin producing the batch, and update $SM_{CT,batch}$

Output:
- update the inventory level according to the completion time of each batch, $SM_{CT,batch}$
- calculate the earliness and tardiness of the required number of products based on inventory level over the time horizon (every time an order is placed, the inventory level is updated as well)

return the utility value (total weighted earliness and tardiness, *TWET*)

5   Case Study

In this section, the computational experiments and performance of the SA algorithms are analyzed. The algorithms are performed on MATLAB 2019 and ran on Intel® CoreTM i5-4200 U CPU @ 1.60 GHz with 12.0 GB memory running under Windows 7, 64bits operating system. Under the reasonably required experimentations by the Bronze Co., we have run five weeks' data, and indicate one example herein, which is a real representative example and the parameters are provided to have a comparison among the production plan presents by an expert with the one offered by SA algorithm.

There are four unrelated machines, and 32 distinct parts are desired to produce. If a machine is qualified to produce a part, the followings are available: (1) setup time – sequence-dependent, (2) preparation time – sequence-dependent, (3) the number of parts can be produced in one hour – job-, and machine-dependent, (4) stop time – job-, and machine-dependent, (5) percentage of low-quality production – job-, and machine-dependent, and (6) minimum economic production – job-, and machine-dependent. Additionally, each part has individual properties, for instance, (1) sharing a frameset or expert labors, so that, they cannot be produced at the same time, (2) demand which is scattered during the period and known at the beginning of the planning horizon (one week), (3) lost profit due to late production or fail to produce quality parts, and (4) inventory level either temporary or permanent. There are three types of working hours mentioned in Table 5, therefore, the production cost under each working hour could be taken into consideration. Renewable resources have their operating schedule, for example, a crane cannot operate during night shifts. Finally, the inventory level of the main inventory depot and temporary inventory is at hand at the beginning of the planning horizon.

Table 5. Working hours.

Type 1	Type 2	Type 3
- 08:00 to 16:00 - 16:00 to 00:00 - 00:00 to 08:00	- 08:00 to 16:00 - 16:00 to 00:00	- 08:00 to 19:00

To illuminate the detailed insights, two schematic planning views are presented in Fig. 1, and Fig. 2. First, the figures are divided into two parts, the upper part is based on Kanban policy, which put weight on meeting the demand scattered throughout the week according to due dates, and the second one, however, is based on the makespan policy, that put weight on the completion time of all demand regardless of their due dates for each part and each segment. It is convenient to see that the first row presents weekdays, and the above four rows depicting four unrelated parallel machines. As mentioned earlier, there are 32 jobs, here, however, one could see fewer numbers representing jobs. Two probable reasons inquire no need to produce specific parts, either is simply no ordered placed or the inventory level mandates not to produce anymore.

A critical look at Fig. 2, which is the best-found solutions, grants the following points: (a) in Kanban production planning system, the best completion time is 102.69 h,

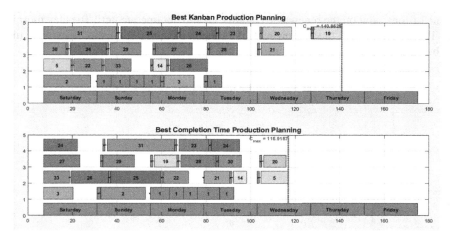

Fig. 1. A feasible schedule is provided in the first iteration.

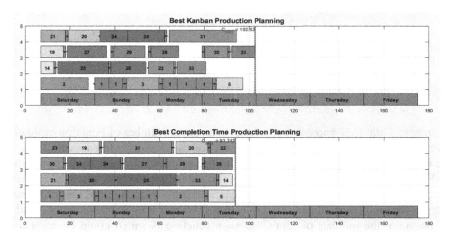

Fig. 2. A feasible schedule is provided in the last iteration.

in contrast, when total completion time matters, 93.74 h is the best. This time and cost difference allow a DM to decide whether one working shift is worth paying penalties for not obeying due dates. Moreover, it is noteworthy to mention that any additional information could leverage the decision process for higher-ranked managers and extremely effective to successfully negotiate with the customers. (b) The horizontally presented number one on Gantt Chart shows while one of the renewable resources is scheduled which is a crane to install a new set of frames before one batch could be produced. As it is designed, the time window is different from one machine and job to another one. Moreover, another clear reflection is that when a batch continues producing, there is no need to occupy the crane. In the Kanban system, the sequence on machine number three is {19, 27, 29, 28, 30, 23}, that is a time gap between {28, 30} revealing the fact that the crane (a constraint on the renewable resource) could not operate until Tuesday morning. (c) The length of each block represents the time one batch requires to be manufactured embracing the type of shift or working hours.

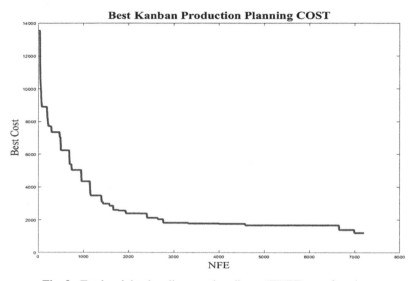

Fig. 3. Total weighted earliness and tardiness (*TWET*) cost function.

In Fig. 3, the well-known trend of cost against the Number of Function evaluations (NFE) is presented. The graph is significant for two logical inferences to draw. The one is how the algorithm attempts to find a good solution, in other words, how tedious or sharp it reduces the objective function. The second is after how many NFE, the SA algorithm convergence to a solution that it cannot be improved anymore. One outstanding purpose served by the delineated graph is that, no matter how fast a computer could run the program, the number of times the main function is called could be an accurate measurement of the efficiency.

5.1 Sensitivity Analysis and Insights

As stated earlier, there are two objectives at the forefront. First is Kanban production planning, where one-day inventories are reserved to meet customer needs while minimizing the earliness and tardiness costs associated with delivery due dates. Considering production limitations, the measured objective function begins at 138.84 h (Fig. 4) and is reduced to 108.31 h using the proposed method (Fig. 5). By virtue of this, the index is boosted by nearly 30 h of production time. While the completion time is highlighted by a DM, the time substantially decreases from 115.24 h to 94.24 h by using the modified SA algorithm, which is the equivalent of one day in production time. Additionally to the improvement of production hours, DM is provided with a powerful tool to make tradeoffs when necessary. As shown in Fig. 5, the two best production planning schemes differ by one day. DM decides whether to accept penalties imposed by customers due to late or early delivery or to take advantage of lower direct and indirect production costs.

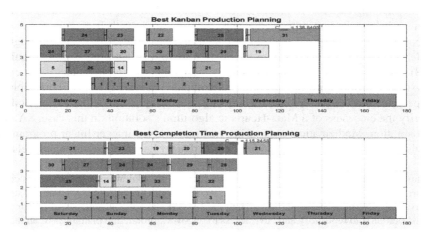

Fig. 4. A feasible schedule is provided in the first iteration.

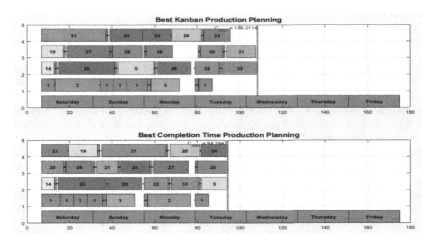

Fig. 5. A feasible schedule is provided in the last iteration.

6   Conclusion and Future Research Direction

In this research, an unrelated parallel machine scheduling with sequence-dependent setup and preparation time is investigated where inventory and resource constraints play vital roles in minimization of earliness and tardiness from due dates. Once a practical and astute problem description is articulated, following that a simulated annealing algorithm is specifically designed and utilized to solve the NP-hard UPMSP. The conventional and predominant scheme of coding and decoding is adapted to have a one-to-one mapping between solution points and actual machine scheduling plans. The number of solution points and the quality of the proposed algorithm is verified by the validation of production planning experts. To that extent, we would like to suggest that, industries could enormously benefit from cutting-edge approaches to improve their planning operations. Moreover, it is rewarding that leading scholars further probe industries for pragmatic problems. For future research, we would like to suggest Discrete Event Simulation (DES) models [36], the integration of family setup time, a closer look at cost and energy analysis, and different inventory policies accompanying varied types of perishable and non-perishable products. We also recommend the conventional mathematical modeling to verify the closeness of a Mata-Heuristic algorithm's solution, in this case, SA, to an exact solution. Mathematical approaches including mixed-integer linear programming models, stochastic optimization models [37], and fuzzy linear programming models [38].

Acknowledgment. In the end, it is a matter of courtesy to acknowledge the generous consult support received from Bronze Industrial Group. We also would like to cordially invite enthusiastic readers, who determine to further investigate the UPMSP, take the advantages of our provided code on our GitHub repository and have access to one-week production planning - as an instance - of the studied extrusion haul.

References

1. Spinellis, D., Vidalis, M.J., O'Kelly, M.E.J., Papadopoulos, C.T.: Analysis and Design of Discrete Part Production Lines, vol. 31. Springer, New York (2009)
2. Żurowski, M., Gawiejnowicz, S.: Scheduling preemptable position-dependent jobs on two parallel identical machines. Comput. Ind. Eng. **132**, 373–384 (2019). https://doi.org/10.1016/j.cie.2019.03.043
3. Naderi, B., Roshanaei, V.: Branch-relax-and-check: a tractable decomposition method for order acceptance and identical parallel machine scheduling. Eur. J. Oper. Res. (2019). https://doi.org/10.1016/j.ejor.2019.10.014
4. Ding, J., Shen, L., Lü, Z., Peng, B.: Parallel machine scheduling with completion-time-based criteria and sequence-dependent deterioration. Comput. Oper. Res. **103**, 35–45 (2019). https://doi.org/10.1016/j.cor.2018.10.016
5. Báez, S., Angel-Bello, F., Alvarez, A., Melián-Batista, B.: A hybrid metaheuristic algorithm for a parallel machine scheduling problem with dependent setup times. Comput. Ind. Eng. **131**, 295–305 (May 2019). https://doi.org/10.1016/j.cie.2019.03.051
6. Expósito-Izquierdo, C., Angel-Bello, F., Melián-Batista, B., Alvarez, A., Báez, S.: A meta-heuristic algorithm and simulation to study the effect of learning or tiredness on sequence-dependent setup times in a parallel machine scheduling problem. Exp. Syst. Appl. **117**, 62–74 (2019). https://doi.org/10.1016/j.eswa.2018.09.041
7. Przybylski, B.: A new model of parallel-machine scheduling with integral-based learning effect. Comput. Ind. Eng. **121**, 189–194 (2018). https://doi.org/10.1016/j.cie.2018.05.035
8. Wang, S., Wang, X., Yu, J., Ma, S., Liu, M.: Bi-objective identical parallel machine scheduling to minimize total energy consumption and makespan. J. Clean. Prod. **193**, 424–440 (2018). https://doi.org/10.1016/j.jclepro.2018.05.056
9. Kılıç, H., Yüzgeç, U.: Improved antlion optimization algorithm via tournament selection and its application to parallel machine scheduling. Comput. Ind. Eng. **132**, 166–186 (2019). https://doi.org/10.1016/j.cie.2019.04.029

10. Liu, S., Pei, J., Cheng, H., Liu, X., Pardalos, M.: Two-stage hybrid flow shop scheduling on parallel batching machines considering a job-dependent deteriorating effect and non-identical job sizes. Appl. Soft Comput. **84**, 105701 (2019). https://doi.org/10.1016/j.asoc.2019.105701
11. Rauchecker, G., Schryen, G.: Using high performance computing for unrelated parallel machine scheduling with sequence-dependent setup times: Development and computational evaluation of a parallel branch-and-price algorithm. Comput. Oper. Res. **104**, 338–357 (2019). https://doi.org/10.1016/j.cor.2018.12.020
12. Yepes-Borrero, J.C., Villa, F., Perea, F., Caballero-Villalobos, P.: GRASP algorithm for the unrelated parallel machine scheduling problem with setup times and additional resources. Expert Syst. Appl. **141**, 112959 (2020). https://doi.org/10.1016/j.eswa.2019.112959
13. Bektur, G., Saraç, T.: A mathematical model and heuristic algorithms for an unrelated parallel machine scheduling problem with sequence-dependent setup times, machine eligibility restrictions and a common server. Comput. Oper. Res. **103**, 46–63 (2019). https://doi.org/10.1016/j.cor.2018.10.010
14. Hamzadayi, A., Yildiz, G.: Modeling and solving static m identical parallel machines scheduling problem with a common server and sequence dependent setup times. Comput. Ind. Eng. **106**, 287–298 (2017). https://doi.org/10.1016/j.cie.2017.02.013
15. Soleimani, H., Ghaderi, H., Tsai, P.W., Zarbakhshnia, N., Maleki, M.: Scheduling of unrelated parallel machines considering sequence-related setup time, start time-dependent deterioration, position-dependent learning and power consumption minimization. J. Clean. Prod. **249**, 119428 (2020). https://doi.org/10.1016/j.jclepro.2019.119428
16. Ezugwu, A.E.: Enhanced symbiotic organisms search algorithm for unrelated parallel machines manufacturing scheduling with setup times. Knowl.-Based Syst. **172**, 15–32 (2019). https://doi.org/10.1016/j.knosys.2019.02.005
17. Lu, S., Liu, X., Pei, J., Thai, M.T., Pardalos, P.M.: A hybrid ABC-TS algorithm for the unrelated parallel-batching machines scheduling problem with deteriorating jobs and maintenance activity. Appl. Soft Comput. **66**, 168–182 (2018). https://doi.org/10.1016/j.asoc.2018.02.018
18. Perez-Gonzalez, P., Fernandez-Viagas, V., Zamora García, M., Framinan, J.M.: Constructive heuristics for the unrelated parallel machines scheduling problem with machine eligibility and setup times. Comput. Ind. Eng. **131**, 131–145 (2019). https://doi.org/10.1016/j.cie.2019.03.034
19. Ekici, A., Elyasi, M., Özener, O.Ö., Sarıkaya, M.B.: An application of unrelated parallel machine scheduling with sequence-dependent setups at Vestel Electronics. Comput. Oper. Res. **111**, 130–140 (2019). https://doi.org/10.1016/j.cor.2019.06.007
20. Fanjul-Peyro, L., Ruiz, R., Perea, F.: Reformulations and an exact algorithm for unrelated parallel machine scheduling problems with setup times. Comput. Oper. Res. **101**, 173–182 (2019). https://doi.org/10.1016/j.cor.2018.07.007
21. Fanjul-Peyro, L.: Models and an exact method for the unrelated parallel machine scheduling problem with setups and resources. Expert Syst. with Appl. X **5**, 100022 (2020). https://doi.org/10.1016/j.eswax.2020.100022
22. Safarzadeh, H., Niaki, S.T.A.: Bi-objective green scheduling in uniform parallel machine environments. J. Clean. Prod. **217**, 559–572 (2019). https://doi.org/10.1016/j.jclepro.2019.01.166
23. Kramer, A., Iori, M., Lacomme, P.: Mathematical formulations for scheduling jobs on identical parallel machines with family setup times and total weighted completion time minimization. Eur. J. Oper. Res. (2019). https://doi.org/10.1016/j.ejor.2019.07.006
24. Goli, A., Babaee Tirkolaee, E., Soltani, M.: A robust just-in-time flow shop scheduling problem with outsourcing option on subcontractors. Prod. Manuf. Res. **7**(1), 294–315 (2019). https://doi.org/10.1080/21693277.2019.1620651

25. Babaee Tirkolaee, E., Goli, A., Weber, G.W.: Fuzzy mathematical programming and self-adaptive artificial fish swarm algorithm for just-in-time energy-aware flow shop scheduling problem with outsourcing option. IEEE Trans. Fuzzy Syst. **28**(11), 2772–2783 (2020). https://doi.org/10.1109/TFUZZ.2020.2998174

26. Saberi-Aliabad, H., Reisi-Nafchi, M., Moslehi, G.: Energy-efficient scheduling in an unrelated parallel-machine environment under time-of-use electricity tariffs. J. Clean. Prod. **249**, 119393 (2020). https://doi.org/10.1016/j.jclepro.2019.119393

27. Pinedo, M.L.: Scheduling. Springer, Boston (2012)

28. Graham, R.L., Lawler, E.L., Lenstra, J.K., Kan, A.R.: Optimization and approximation in deterministic sequencing and scheduling: a survey. Ann. Discrete Math. **5**, 287–326 (1979)

29. Metropolis, N., Rosenbluth, A.W., Rosenbluth, M.N., Teller, A.H., Teller, E.: Equation of state calculations by fast computing machines. J. Chem. Phys. **21**(6), 1087–1092 (1953). https://doi.org/10.1063/1.1699114

30. Sekkal, N., Belkaid, F.: A multi-objective simulated annealing to solve an identical parallel machine scheduling problem with deterioration effect and resources consumption constraints. J. Comb. Optim. **40**(3), 660–696 (2020). https://doi.org/10.1007/s10878-020-00607-y

31. Nattaf, M., Dauzère-Pérès, S., Yugma, C., Wu, C.-H.: Parallel machine scheduling with time constraints on machine qualifications. Comput. Oper. Res. **107**, 61–76 (2019). https://doi.org/10.1016/j.cor.2019.03.004

32. Woo, Y.-B., Kim, B.S.: Matheuristic approaches for parallel machine scheduling problem with time-dependent deterioration and multiple rate-modifying activities. Comput. Oper. Res. **95**, 97–112 (2018). https://doi.org/10.1016/j.cor.2018.02.017

33. Li, K., Xiao, W., Yang, S.: Minimizing total tardiness on two uniform parallel machines considering a cost constraint. Expert Syst. Appl. **123**, 143–153 (2019). https://doi.org/10.1016/j.eswa.2019.01.002

34. Al-harkan, I.M., Qamhan, A.A.: Optimize unrelated parallel machines scheduling problems with multiple limited additional resources, sequence-dependent setup times and release date constraints. IEEE Access **7**, 171533–171547 (2019). https://doi.org/10.1109/ACCESS.2019.2955975

35. Afshar-Nadjafi, B., Arani, M.: Multimode preemptive resource investment problem subject to due dates for activities: formulation and solution procedure. Adv. Oper. Res. **2014**, 1 (2014). https://doi.org/10.1155/2014/740670

36. Arani, M., Abdolmaleki, S., Liu, X.: Scenario-Based Simulation Approach for An Integrated Inventory Blood Supply Chain System. In: 2020 Winter Simulation Conference (WSC), pp. 1348–1359 (2020). https://doi.org/10.1109/WSC48552.2020.9384018

37. Arani, M., Chan, Y., Liu, X., Momenitabar, M.: A lateral resupply blood supply chain network design under uncertainties. Appl. Math. Model. **93**, 165–187 (2021). https://doi.org/10.1016/j.apm.2020.12.010

38. Momenitabar, M., et al.: Fuzzy mathematical modeling of distribution network through location allocation model in a three-level supply chain design. J. Math. Comput. Sci. **9**(3), 165–174 (2014). https://doi.org/10.22436/jmcs.09.03.02

Water Truck Routing Optimization in Open Pit Mines Using the General Algebraic Modelling System Approach

Mohammad Hossein Sadat Hosseini Khajouei[1] , Maryam Lotfi[2(✉)],
Ahmad Ebrahimi[1], and Soheil Jafari[3]

[1] Department of Industrial Management, IAU, Science and Research Branch, Tehran, Iran
[2] Cardiff Business School, Cardiff University, Cardiff, UK
LotfiM@cardiff.ac.uk
[3] School of Aerospace, Transport and Manufacturing (SATM),
Cranfield University, Cranfield, UK

Abstract. This paper presents a methodological approach for routing optimization in open pit mines which is a trending topic for dust emission reduction in mining process. In this context, the aim of the research and its contribution to the knowledge is firstly described based on a comprehensive literature survey in the field. Then, as an arc routing problem, the mathematical model for the process is generated including the objective function, minimizing the total distance traveled by the water truck fleets, practical constraints that should be met and the used assumptions. Finally, the formulated optimization problem solved employing General Algebraic Modelling System (GAMS) approach respect to the nature of the mathematical equations. The tested results by simulations discussed to confirm the effectiveness of the proposed method in dealing with the in-hand problem. This methodological approach could be used in optimization of other similar engineering problem as well.

Keywords: Routing optimization · Open pit mines · GAMS · Arc routing problem

1 Introduction

Mining as a process of extracting minerals from the ground with the aim of profitability includes five main steps: prospection, exploration, development, exploitation and reclamation, in which making the strategic decisions is indispensable to enhance the performance of the whole process.

Increasing the mineral resources extraction and subsequently, surface mining leads to dust emission containing Silica in mines and their suburbs. Inhaling of dust containing Silica as a mineral with the most plenty in the earth causes respiratory diseases such as silicosis and lung cancer. As a potential solution, utilizing an appropriate water trucks fleet in the mines' roads and ramps in loading, hauling, vehicle travel and blasting processes would result in dust control. The other advantages of water fleet implementation

© Springer Nature Switzerland AG 2021
Z. Molamohamadi et al. (Eds.): LSCM 2020, CCIS 1458, pp. 255–270, 2021.
https://doi.org/10.1007/978-3-030-89743-7_14

in mines as a sustainable approach in mining engineering, particularly in open pit mines could be: to provide the healthier working environment, increased visibility of roads, tires' lifetime and machinery maintenance cost reduction.

There are numerous mines with a high diversification of minerals in which their exploitation, volume and dimension of production are dramatically increasing. Hence, application of an optimized procedure of fleet of water truck that loads water from the set of predetermined stations and prevents the formation of dust by traversing the haul roads and spraying them is a vital step in mining strategic decision process. The objective of present paper is to apply an Equation-Based Modeling (EBM) approach in the field of mining engineering and consequently to propose an optimized model to specify optimal routes for each homogeneous water trucks in open pit mines.

The remainder of this paper is as follows. Section 2 explains the applications of the operation research techniques in mining engineering problems to describe the gap in the literature that is aimed in this paper. Section 3 presents a review of the related literature. Section 4 is devoted to implementation of the mathematical modeling. Section 5 discusses sensitivity analysis and findings and finally, Sect. 6 presents the conclusion. This paper aims to respond to the issue of how to provide a model for water truck routing problem using a system approach tool.

2 Application of Operation Research Techniques in Mining Engineering

Use of operational research (OR) methodologies in planning related to extraction and exploration operations of surface and underground mines, including pit design, production planning in short and long-term horizons or fleet planning of transport and loading machines dates back to the 1960s Newman (2010). It is noteworthy that rare studies have been conducted to dispatch a fleet of water trucks as one of the pillars to control the phenomenon of dust emission. This area can be studied as a sustainable approach in mining engineering. Many labors and employees in open pit mines are constantly suffering from diseases caused by the spread of silica dust.

Based on a literature review, considering the type of process, optimization of mining and haulage is implemented commonly by queue theory and simulation. Due to the complexity and large dimensions of pit design and production scheduling, heuristic and meta-heuristic methods employed. Linear programming is a usual method for truck allocating-dispatching in the shovel-truck systems. Dynamic programming for productivity improvement of shovel-truck systems and integer programming, goal programming and non-linear programming applied for extraction scheduling in underground mines, coal mines especially. Newly, application of mathematical modeling with the approach of vehicle routing problem in determination of final limits of pit with the Lerch and Grossman method is implemented. In the realm of mining engineering, application of arc routing problem for water truck fleet with the approach of dust emission decreasing is less observed. Table 1 categorizes different OR techniques used in different aspects of the mining problems. Newman et al. (2016) proposed an integer programing for optimizing the open pit to underground mining transition and solved their problem by an ad-hoc branch-and-bound approach.

Gupta et al. (2018) formulated an integrated multi objective model in which estimation of weights of different types of vehicles available for transportation in coal mine industry is calculated by the Analytic Hierarchy Process (AHP) and Data Envelopment Analysis (DEA) technique is applied to calculate efficiency privileges of vehicles on different routes of a transportation system.

Table 1. Application of OR techniques in mining problems

	Mining engineering sections	OR methodology
1	Optimization of hauling in open pit mining	Queue theory Simulation
2	Open pit mine design as a complex problem	Heuristic and meta heuristic algorithm
3	Dump truck allocation considering shovel idling	Linear programming
4	To determine the final limit of pit	Vehicle routing problem
5	Time scheduling for mineral production in underground and open pit mining	Meta heuristic algorithm
6	Production efficiency improvement in truck-shovel systems	Dynamic programming
7	Time scheduling for underground mines production	Mix integer programming
8	Time scheduling for underground coal mines extraction	Non-Linear programming Goal programming
9	Dump truck dispatching system	Linear programming

According to Table 1, there is a gap in the literature for consideration of the dust emission control in open pit mining. On the other side, propounding Arc Routing Problem for services that are related to roads of mines, for instance water truck services, can be considered as an innovation in the realm of OR method deployment in mining engineering. This paper will focus on these gaps to propose a new solution for gaining safe atmosphere in mines for labors' health.

3 Literature Review and Background of the Problem

In the present study, the problem of wetting down service to a set of roads in the open pit mine network using a fleet of water trucks with bounded capacity is considered as an Arc Routing Problem (ARP). Refilling station is located at a central depot and the objective of the problem is to minimize the total traveling distance. Therefore, in this section, the literature related to the subject is briefly studied.

Golden and Wong (1981) defined a capacitated ARP, to prepare mathematical programming formulations, to accomplish a computational complexity analysis, and to present an approximate solution strategy for this kind of problems. Eglese (1994) presented the formulation of a practical routing problem which has been implemented as a computer program to run on a microcomputer and used the formulation to explore the cost consequences of different scenarios for the gritting operation in a County.

Eiselt et al. (1995) implemented a comprehensive review on ARP including the Chinese Postman Problem (CPP) and the algorithms for its directed and undirected types, the windy postman problem (WPP), the mixed CPP and the hierarchical CPP, and also the Rural Postman Problem (RPP), the directed and undirected types of RPP, the stacker crane and capacitated ARP, considering their applications. For a more detailed study, readers are highly recommended to refer to Sachs et al. (1988) for Euler's Konigberg letters, Kwan (1962) and Edmonds and Johnson (1973) for Chinese Postman Problem (CPP), Eiselt et al. (1995) for Rural Postman Problem (RPP) and Golden and Wong (1981) for capacitated ARP as the milestones in ARP.

Generally, the ARPs could be categorized in un-capacitated and capacitated forms: Un-capacitated ARP includes determining minimum cost tours which in the most appropriate arcs are traversed at least once. Un-capacitated ARP has two major categories: CPP and RPP. If a central depot, consisting of a fleet of constrained-capacity vehicles, provides service to a subset of the streets of a network with the goal of minimizing the total cost of routing, problem capacitated ARP arises Wohlk (2008). Table 2 shows the milestones in un-capacitated Arc Routing Problem formulation.

Table 2. Milestones in un-capacitated arc routing problem formulation

Un-capacitated ARPs (Chinese postman problem)	Reference
Mixed CPP (MCPP)	Guan (1962)
Windy PP (WPP)	Guan (1962)
Hierarchical CPP (HCPP)	Dror et al. (1987)
Generalized CPP (GCPP)	Dror et al. (2000)
Maximum Benefit CP (MBCPP)	Pearn et al. (2003)
CPP with time windows (CPPTW)	Eglese (2006)
Un-capacitated ARPs (Rural Postman Problem)	**Reference**
Undirected RPP	Christofieds et al. (1986)
RPP with deadline classes	Letchford et al. (1998)
DRPP with turn penalties and forbidden turns (DRPPTP)	Benavent et al. (1999)
Periodic RPP	Ghiani et al. (2001)
Profitable Arc Tour Problem (PATP)	Feillet et al. (2005)
Prize-collecting RPP (PRPP)	Araoz et al. (2006)
Windy RPP	Benavent et al. (2003)

Table 3 shows the milestones in capacitated ARP formulation. In the realm of ARP application in the road maintenance operations in cold regions and in winter, spreading of chemicals and abrasives, snow-plowing, loading snow into trucks, and hauling snow to disposal sites, including vehicle routing and depot location are investigated. Langevin et al. (2004) presented a multiple survey in which optimization models and solution algorithms for the design of winter road maintenance systems reviewed. Moreover, optimizing the total distance travelled and gritters (In terms of capacity and number), Eglese (1994) proposed routing winter gritting vehicles. The introduced model dealt with multiple depot locations. He assumed that vehicles were bounded-capacities, and roads were different in terms of priority.

Table 3. Milestones in capacitated Arc Routing Problem formulation

Capacitated ARPs	Reference
The Undirected Capacitated Arc Routing Problem	Golden and Wong (1981)
CARP With Temporal Constraints or Split Deliveries	Mullaseril (1996) Dror et al. (1998)
Capacitated Arc Routing With Several Facilities	Ghiani and Laporte (2001)
The Mixed Capacitated ARP	Mourao and Amado (2005)
Capacitated ARP with refill points	Amaya et al. (2007)

In addition, for the road maintenance operations and road network surveillance, the road network monitoring is carried out periodically. Each category of roads visited, which is already classified according to the road hierarchy, requires different monitoring during the sub-periods throughout the time horizon. (Monroy et al. 2013). With this explanation, Monroy et al. (2013) introduced the periodic CARP with irregular services. They classified PCARP into two categories: period independent demand and cost and period dependent demand and cost. Each of these classes could be consisted of irregular or regular services.

Soler et al. (2011) presented the capacitated general windy routing problem with turn penalties which is vital for real-life problems and includes many important and well-known arc and node routing problems. (Beraldi et al. 2015) studied the general routing problem under uncertainty (demand is stochastic), in which a chance-constrained integer programming formulation was taken into account. The problem is defined based on a mixed graph theory, in which vehicles' capacity is assumed limited.

As a general case of the capacitated LRP, Karaoglan et al. (2013) considered a Capacitated LRP with Mixed Backhauls (CLRPMB). CLRPMB is defined as locating depots and finding routes of vehicles with overall cost minimization, whereas the same vehicle meets customer pickup demands and performs their delivery services. Identifying vehicles routes, Constantino et al. (2015) considered AR with a constraint to limit number of common nodes, so they introduced the MCARP with non-overlapping routes that enforces an upper bound on the number of nodes that are shared for different routes.

Armas et al. (2018) proposed the non-smooth ARP with soft constraints in order to capture in more perceptive way realistic constraints violations arising in transportation and logistics. Gonzalez et al. (2016) introduced a simheuristics approach for solving an ARP variant in which, customers' demands are modeled as random variables. Aiming to minimize total cost, Babaee Tirkolaee et al. (2018a) proposed a mixed-integer linear programming (MILP) model was developed for the multi-trip Capacitated Arc Routing Problem (CARP) with different located depots and disposal facilities. To check the validity of the proposed model they general algebraic modelling system approach (GAMS) and CPLEX solver.

Babaee Tirkolaei et al. (2019) have addressed a novel robust bi-objective multi-trip periodic capacitated arc routing problem under demand uncertainty to treat the urban waste collection problem and IWO is developed to solve the proposed model. They have mentioned that Multi Objective Invasive Weed Optimization is regarded as the best method to solve the large-sized multi-trip periodic capacitated arc routing problem. Willemse and Joubert (2019) developed local Search heuristics for Mixed Capacitated Arc Routing Problem under Time Restrictions with Intermediate Facilities (MCARPTIF), basic local search versions showed poor performance on realistically sized problem instances and three local search acceleration techniques were combined to improve its performance. They came to the conclusion that the accelerated local search can be applicable within meta-heuristics for similar problems.

Studying the uncertain nature of demand parameter and minimizing total traversed distance and total usage cost of vehicles, Babaee Tirkolaee (2018b) proposed a novel mathematical model for robust periodic capacitated arc routing problem (PCARP) considering multiple trips and drivers and crew's working time. Aiming to minimize total cost including the cost of generation and emission of greenhouse gases, the cost of vehicle usage and routing cost, Babaee Tirkolaee (2018c) addressed a multi-trip Green Capacitated Arc Routing Problem (G-CARP) and a Hybrid Genetic Algorithm (HGA) to solve the proposed model. Khajepoura et al. (2020) modeled agricultural operations as a CARP. By separating the paths into two classes of required and non-required, CARP demonstrates a field as a graph. Using CARP, no extra or invalid edge on the graph would be required. They presented an adaptive large neighborhood search (ALNS) to solve similar large scales models.

As a contribution, in the realm of mining engineering, application of arc routing problem for water truck fleet with the approach of dust emission decreasing is less observed. Therefore, this research aims to optimize the water truck routing in open pit mines to find the minimum length of arcs, the minimum required amount of water for each path and the minimum service time for each truck in each path. Hence, in the next chapter, this problem will be formulated and solved as a mathematical optimization problem.

4 Mathematical Modeling

Attending to machinery efficiency, environmental issues and employee healthcare vitalism, this research focused on finding optimal tours by proposing an appropriate mathematical model for water truck fleet in open pit mines with minimum traveled distance.

Depending on the type of water truck fleet operation, instead of nodes, the roads should be serviced. So, arc routing problem is propounded, and the part of network is serviced, hence the model can be in rural postman problem category. After loading the tanks, water trucks start traveling from refilling station which located in depot (Fig. 1) and with completion of servicing or reserved water finishing, they turn back to the depot.

Let $G = (N, A = R_1 \cup R_2)$ be an undirected graph in which N is the set of nodes and A is the set of arcs including subset R_1 for unidirectional roads and R_2 for two ways ones. Note here that refilling station is located in node O which is the depot too. The objective function includes minimization of total distance traveled by water truck fleet. Model assumptions are:

- Water trucks are homogeneous, and their capacities are known.
- The arcs can be used in unidirectional, two-way and combinational.
- All of arcs can be traveled several times but can be serviced just once and by one truck.
- Just a part of graph could be serviced.

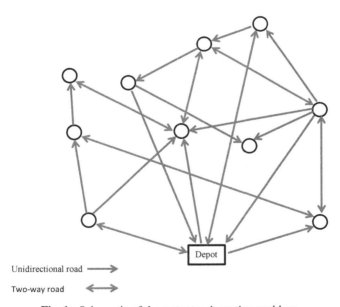

Fig. 1. Schematic of the water truck routing problem.

Figure 2 illustrates a typical format of the problem in order to define index sets, variables, and parameters as follow:

Index sets

I   Set of nodes
i,j   Index of nodes
K   Set of water trucks
K   Index of water trucks

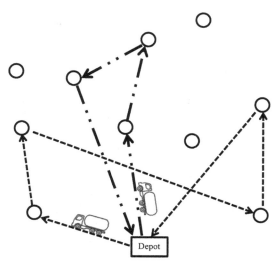

Fig. 2. Problem schematic to define the parameters.

S   Set of sub-tours
O   Index of refilling station

Parameters

$d(i,j)$ Length of arc (n_i, n_j)
$w(i,j)$ Amount of water needed for arc (n_i, n_j)
$TT(k)$ Total time of availability of truck k
$t(i,j,k)$ Time needed truck k traverses arc (n_i, n_j)
$s(i,j,k)$ Time needed truck k services arc (n_i, n_j)
Cap   Capacity of trucks
M   Big number

Variables

$x(i,j,k)$ A binary decision variable representing a dispatch decision 1 if and only if
            truck k traverses arc (n_i, n_j), and 0 otherwise
$y(i,j,k)$ A binary decision variable representing a service decision 1 if and only if
            truck k services arc (n_i, n_j), 0 otherwise
$AT(i,j,k)$ Truck k's arrival time to arc (n_i, n_j)
$B(i,j)$ Time to start servicing arc (n_i, n_j)
z_1 Variable related to objective function
U   Binary variable related to impermissible sub-tours
L   Binary variable related to impermissible sub-tours.

The mathematical equations associated with the problem could be summarized as follow:

$$\min \sum_{k=1}^{m} \sum_{(n_i, n_j) \in A} d_{ij} X_{ijk} \tag{1}$$

$$\sum_{n_j:(n_j,n_i) \in A} X_{jik} = \sum_{n_j:(n_i,n_j) \in A} X_{ijk} \forall\, k \in K,\ n_i \in N \tag{2}$$

$$\sum_{k=1}^{m} \left(Y_{ijk} + Y_{jik}\right) = 1 \forall \left(n_i, n_j\right), \left(n_j, n_i\right) \in R_1 \tag{3}$$

$$\sum_{k=1}^{m} Y_{ijk} = 1 \forall \left(n_i, n_j\right) \in R_2 \tag{4}$$

$$X_{ijk} \geq Y_{ijk} \forall \left(n_i, n_j\right) \in A \tag{5}$$

$$\sum_{(n_j,n_i) \in A} w_{ij} Y_{ijk} \leq c\ \forall\, k \in K \tag{6}$$

$$At_{ijk} + S_{ijk} Y_{ijk} + t_{ijk} X_{ijk} \leq M\left(1 - X_{ijk}\right) + At_{jik} \qquad \forall\, k \in K,\ \text{all} \left(n_i, n_j\right) \tag{7}$$

$$At_{jik} \leq M\left(1 - Y_{ijk}\right) + B_{ij} \qquad \forall\, k \in K \tag{8}$$

$$\sum_{(n_i,n_j) \in A} \left(S_{ijk} Y_{ijk} + t_{ijk} X_{ijk}\right) < T_k \forall\, k \in K \tag{9}$$

$$\sum_{n_j:(n_o,n_j) \in A} \left(X_{0jk}\right) = 1 \forall\, k \in K \tag{10}$$

$$\sum_{n_i:(n_i,n_o) \in A} \left(X_{i\,o\,k}\right) = 1 \forall\, k \in K \tag{11}$$

$$\sum_{n_i,n_j \in S} x_{ijk} = |S| - 1 + |N|^2 u_k^S,\ \forall S \subseteq N \backslash \{o\};\ S \neq \emptyset;\ \forall k \in K \tag{12}$$

$$\sum_{n_i \in S} \sum_{n_j \notin S} x_{ijk} \geq 1 - l_k^S,\ \forall S \subseteq N \backslash \{o\};\ S \neq \emptyset;\ \forall k \in K \tag{13}$$

$$u_k^S + l_k^S \leq 1,\ \forall S \subseteq N \backslash \{o\};\ S \neq \emptyset;\ \forall k \in K \tag{14}$$

$$u_k^S, \, l_k^S \text{ Binary}, \, \forall S \subseteq V\backslash\{o\}; \, S \neq \emptyset; \, \forall k \in K \qquad (15)$$

$$X_{ijk}, Y_{ijk} \text{ Binary} \qquad (16)$$

Where:

- Equation (1) represents the objective function which calculates the total distance traveled by the homogeneous water truck fleet.
- Equation (2) is a flow conservation constraint.
- Constraint (3) ensures that water truck k services to unidirectional road just once.
- Constraint (4) guarantees that water truck k services to each two-way road back and forth.
- Constraint (5) ensures that the number of services is equal or less than the number of traverses. Constraint (6) indicates that total water consumption in each tour is less than water truck's capacity. Terms (7), (8) and (9) are related to time window constraints.
- Constraint (10) requires that tour begins from depot.
- Constraint (11) indicates that the model is not in open category ARP.
- Constraint (12), (13), (14) and (15) are constraints for sub-tour elimination.
- Constraint (16) defines the decision variables.

With respect to the nature of the problem (limited number of vehicles, nodes and arcs) and type of problem (mix integer linear programming model), the General Algebraic Modelling System (GAMS) is selected to deal with this optimization problem. In addition, since there is at least one binary variable in the model, solving is done by MIP and appropriate solver is CPLEX. GAMS version used in this research is 24.1.2 and the configuration of the system consists of Intel Core i3-2370M CPU @ 2.4 GHz and 2 GB DDR3 SD-RAM. Among different solvers, CPLEX with the best time (0.26 s) showed the highest speed.

The GAMS as high-level modelling software for mathematical programming and optimization, includes a language compiler and integrated high-performance solvers. The GAMS is tailored for complex, large scale modelling applications, and allows designers to build large maintainable models that can be adapted quickly to new situations. The GAMS is specifically designed for modelling linear, nonlinear and mixed integer optimization problems. Different solvers and portfolio of GAMS are summarized in Table 4.

Table 4. Solvers and portfolio of the GAMS

Solver	Portfolio
Linear Programing/Mixed Integer Programing/Quadratically constrained program (QCP)/Mixed Integer QCP	CPLEX/GUROBI/MOSEK/XPRESS
Non Linear Programing (NLP)	CONOPT/IPOPTH/KNITRO/MINOS/SNOP
Mixed Integer NLP (MINLP)	ALPHAECP/ANTIGONE/BARON/DICOPT/OQNLP/SBB
Mixed complementarity problem (MCP)/Mathematical program with equilibrium constraints (MPEC)/Constrained Nonlinear System (CNS)	----

5   Results Analysis

In order to confirm the ability of the GAMS in dealing with the water truck routing in open pit mines, a graph with 6 nodes is assumed as the case study. Node O indicates depot and refilling point that located in depot. Following matrix shows the existing roads including unidirectional, two ways and combinational roads. A fleet with 3 water trucks and 300 units in capacity is considered (Fig. 3).

So, the graph network would be as follows which has 5 arcs with double sided demand and 3 arcs with one way demand (Fig. 4).

The matrices, graphs and Eqs. 1–16 are dealt with the GAMS to find the optimized solution for the problem. After solving the problem, the optimal value of objective function was equal to 229.274 distance unit and the optimized tours for each water truck are shown in Fig. 5. For instance, this figure shows that truck 2 would not pass the point 4 in the optimized design. In Fig. 5, black lines show the two-way paths and blue-lines show unidirectional paths.

Also, the optimized parameters are shown in Table 6. As shown in this table, the obtained results for the optimal tour of Fig. 5 states that:

- The minimum and maximum length of arc will be passed by truck 1 between nodes 2 and 3 (10.134) and between nodes 5 and 2 (11.662) respectively.
- The minimum required amount of water is for truck 1 in the route between nodes 5 and 2 (12.218) while the maximum required amount of water is for truck 2 between nodes 1 and 3 (28.124).
- Finally, the minimum service time is 5.085 from nodes 1 to node 3 by truck 2 and the maximum service time will be taken by truck 3 when servicing path between nodes 4 and 6.

Other details about the optimized results could be found in Table 5 as well.

0	1	1	1	1	1
1	0	1	1	1	1
1	1	0	1	1	1
0	1	0	0	1	1
1	1	1	1	0	1
0	0	0	1	1	0

Existing arcs matrix

0	0	1	0	0	0
0	0	1	0	1	0
1	1	0	0	1	0
0	0	0	0	0	1
0	1	1	0	0	0
0	0	0	1	0	0

Existing two ways arcs matrix

0	0	0	0	0	0
0	0	0	0	0	1
0	0	0	1	0	0
0	1	0	0	0	0
0	0	0	0	0	0
0	0	0	0	0	0

Existing unidirectional arcs matrix

0	0	1	0	0	0
0	0	1	0	1	1
1	1	0	1	1	0
0	1	0	0	0	1
0	1	1	0	0	0
0	0	0	1	0	0

Combinational matrix

Fig. 3. Existing arcs for the defined case study.

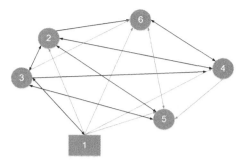

Fig. 4. The graph network for the defined case study.

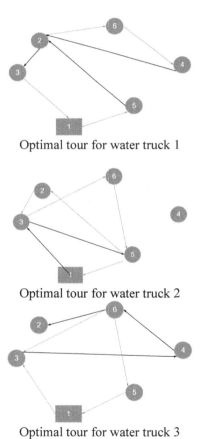

Optimal tour for water truck 1

Optimal tour for water truck 2

Optimal tour for water truck 3

Fig. 5. The optimized tours for trucks.

Table 5. The optimized parameters for the water truck routing problem

Serves arc	d(i, j)	W(i, j)	S(i, j)	AT(i, j)	Truck
2-5	11.662	12.218	5.593	2.269	1
3-2	10.134	17.106	5.734	2.145	1
2-4	10.871	19.121	5.348	3.543	1
5-3	10.319	16.226	5.155	2.348	2
3-1	11.101	28.124	5.085	2.0931	2
4-3	11.279	20.241	5.259	2.999	3
2-6	11.008	18.232	5.701	3.214	3
6-4	10.300	24.158	5.901	2.552	3

6 Conclusion

This paper presents a methodological approach for optimization of the water truck routing problem using a system approach tool. Expansion of mineral extraction and subsequently, surface mining leads to dust emission containing Silica in mines and their suburbs. Inhaling of dust containing Silica as a mineral with the most plenty in the earth causes respiratory diseases such as silicosis and lung cancer. Therefore, the main motive of this research is to pay special attention to the issue of sustainability in mines with an environmental approach and protecting human health. A comprehensive literature survey presented to conclude that this piece of research will fill a systematic gap in the literature from transport and logistics point of view. A mathematical model of nodes, paths, trucks, depot and refilling is developed using a matrix approach and the objective function and constraint equations are generated in detail. The General Algebraic Modelling System Approach (GAMS) is used to deal with the optimization problem and the obtained results show the effectiveness of the used method in finding the optimal tours for each truck, the minimum length of the arcs, the minimum amount of water for servicing the routes, and the minimum service time required for each route. The study confirms the capability of the proposed method for the future research studies in the field (e.g. effect of environmental parameters, soil characteristics, and human behavior in optimization of water truck routing problem in open pit mines).

References

Archetti, C., Feillet, D., Hertz, A., Speranza, M.G.: The undirected capacitated arc routing problem with profits. Comput. Oper. Res. **37**, 1860–186 (2010)

Armas, J.D., Ferrer, A., Juan, A.A., Lalla-Ruiz, E.: Modeling and solving the non-smooth arc routing problem with realistic soft constraints. Expert Syst. Appl. **98**, 205–220 (2018)

Assad, A.A., Golden, B.L., Pearn, W.L.: Transforming arc routing into node routing problems. Comput. Oper. Res. **14**(4), 285–288 (1987)

Babaee Tirkolaee, E., Alinaghian, M., Rahmani Hosseinabadi, A.A., Bakhshi Sasi, M., Sangaiah, A.K.: An improved ant colony optimization for the multi-trip Capacitated Arc Routing Problem. Comput. Electr. Eng. **77**, 1–15 (2018)

Babaee Tirkolaee, E., Mahdavi, I., Seyyed Esfahani, M.M.: A robust periodic capacitated arc routing problem for urban waste collection considering drivers and crew's working time. Waste Manage. **76**, 138 (2018b).

Babaee Tirkolaee, E., Rahmani Hosseinabadi, A.A., Soltani, M., Sangaiah, A.K., Wang, J.: A hybrid genetic algorithm for multi-trip green capacitated arc routing problem in the scope of urban services. Sustainability **10**, 1366 (2018c)

Babaee Tirkolaei, E., Goli, A., Malekalipour Kordestanizadeh, R.: A robust bi-objective multi-trip periodic capacitated arc routing problem for urban waste collection using a multi-objective invasive weed optimization. Waste Manage. Res. **37**, 1–13 (2019)

Belenguer, J.M., Benavent, E.: A cutting plane algorithm for the capacitated arc routing problem. Comput. Oper. Res. **30**, 705–728 (2003)

Benavent, E., Soler, D.: The directed rural postman problem with turn penalties. Transp. Sci. **33**, 408–418 (1999)

Beraldi, P., Bruni, M.E., Laganà, D., Musmanno, R.: The mixed capacitated general routing problem under uncertainty. Eur. J. Oper. Res. **240**, 382–392 (2015)

Christofidrordres, N.: The optimum traversal of a graph. Omega **1**, 719–732 (1973)

Constantino, M., Gouveia, L., Mourão, M.C., Nunes, A.C.: The mixed capacitated arc routing problem with non-overlapping routes. Eur. J. Oper. Res. **24**, 445–456 (2015)

Dror, M., Leung, J.M.Y.: Combinatorial optimization in a cattle yard: feed distribution, vehicle scheduling, lot sizing and dynamic pen assignment. In: Yu, G. (ed.) Industrial Applications of Combinatorial Optimization, vol. 16, pp. 142–171, Kluwer (1998)

Dror, M., Leung, J.M.Y.: Combinatorial optimization in a cattle yard: Feed distribution, vehicle scheduling, lot sizing and dynamic pen assignment. Ind. Appl. Comb. Optim. **16**, 142–171 (1998)

Dror, M., Stern, H., Trudeau, P.: Postman tour on a graph with precedence relation on arcs. Networks **17**, 283–294 (1978)

Edmonds, J., Johnson, E.L.: Matching, euler tours and the chinese postman. Math. Program. **5**(1), 88–124 (1973)

Eglese, R.W.: Routing winter gritting vehicle. Discret. Appl. Math. **48**, 231–244 (1994)

Eiselt, H.A., Gendreau, M., Laporte, G.: Arc routing-problems, part I: the Chinese postman problem. Oper. Res. **43**(2), 231–242 (1995)

Eiselt, H.A., Gendreau, M., Laporte, G.: Arc routing-problems, part II: the rural postman problem. Oper. Res. **43**(3), 399–414 (1995)

Ghiani, G., Improta, G., Laporte, G.: The capacitated arc routing problem with intermediate facilities. Networks **37**, 134–143 (2001)

Golden, B.L., Wong, R.T.: Capacitated arc routing problems. Networks **11**, 305–315 (1981)

Golden, B.L., Bruce, L., Raghavan, S., Wasil Edward, A.: The Vehicle Routing Problem: Latest Advances and New Challenges. Springer, Boston (2008). https://doi.org/10.1007/978-0-387-77778-8

Gonzalez, S., Juan, A., Riera, D., Elizondo, M., Ramos, J.: A simheuristic algorithm for solving the arc routing problem with stochastic demands. J. Simul. **12**(1), 53–66 (2016)

Gupta, P., Mehlawat, M. K., Aggarval, U., Charles, V.: An integrated AHP-DEA multi-objective optimization model for sustainable transportation in mining industry. In: Resource Policy in Press (2018)

Hedetniemi, S.: On minimum walks in graphs. Naval Res. Logist. Wiley New York **15**(3), 453–458 (1968)

Karaoglan, I., Altiparmak, E.: A memetic algorithm for the capacitated location-routing problem with mixed backhauls. Comput. Oper. Res. **55**, 200–216 (2013)

Khajepoura, A., Sheikhmohammady, M., Nikbakhsh, E.: Field path planning using capacitated arc routing problem. Comput. Electron. Agric. **173**, 105401 (2020)

King, B., Goycoolea, M., Newman, A.: Optimizing the open pit-to-underground mining transition. Eur. J. Oper. Res. **257**, 297–309 (2016)

Krushinsky, D., Woensel, T.V.: Discrete optimization an approach to the asymmetric multi-depot capacitated arc routing problem. Eur. J. Oper. Res. **244**, 100–110 (2015)

Kwan, M.K.: Graphic programming using odd or even points. Chin. Math. **1**, 273–277 (1962)

Letchford, A.N., Eglese, R.W.: The rural postman problem with deadline classes. Eur. J. Oper. Res. **105**, 390–400 (1998)

Mico, J.C., Soler, D.: The capacitated general windy routing problem with turn penalties. Oper. Res. Lett. **39**, 265–271 (2011)

Monroy, I.M., Amaya, C.A., Langevin, A.: The periodic capacitated arc routing problem with irregular services. Discret. Appl. Math. **161**, 691–701 (2013)

Mullaseril, P.A.: Capacitated rural postman problem with time windows and split delivery. Ph.D. thesis, MIS Department, University of Arizona (1996)

Newman, A.M., Rubio, E., Caro, R., Weintraub, A., Eurek, K.: A Review of Operations Research in Mine Planning. Interfaces **40**, 222–224 (2010)

Sachs, H., Stiebitz, M., Wilson, R.J.: An historical note: Euler's Königsberg letters. J. Graph Theory **12**(1), 133–139 (1988)

Willemse, E.J., Joubert, J.W.: Efficient local search strategies for the Mixed Capacitated Arc Routing Problems under Time restrictions with Intermediate Facilities. Comput. Oper. Res. **105**, 203 (2019)

Wohlk, S.: A decade of capacitated arc routing. Oper. Res. Comput. Sci. Interfaces **43**, 29–48 (2008)

Offering a New Bus Route Between Campus and Bus Terminal Using Shortest Path Algorithm

Eren Özceylan[✉] [iD]

Department of Industrial Engineering, Gaziantep University, 27027 Gaziantep, Turkey

Abstract. More than 50 thousand students are studying at Gaziantep University. Most of these students reside around the campus of Gaziantep University, while their hometown is outside Gaziantep. Therefore, they use the intercity bus station in the southeast of Gaziantep very often. In this paper, it is aimed to detect the shortest route between Gaziantep University campus and Gaziantep intercity bus station. Within the aim of the study, the shortest path (SPP) problem was considered. The purpose of the shortest path algorithm is to find the shortest distance between the start and end points in a road network. In this problem, 11 different municipal bus routes operating on the route starting from Gaziantep University campus and ending point Gaziantep bus station were examined and a road network consisting of 46 points was obtained. Then, the distance between each point was calculated with the help of Google Earth and the data was entered to mathematical model developed for the shortest path algorithm. The model was solved by the GAMS 23.1 optimization package and the shortest distance between the two points (10.4 km) was found to be optimal. As a result, a new bus route including 18 destination stops with shorter distances than the existing 11 bus routes has been proposed.

Keywords: Shortest path algorithm · Mathematical modeling · Bus route

1 Introduction

In this paper, a real case of SPP which belongs to Gaziantep city (Turkey) is considered. The starting point is considered as the main gate of Gaziantep University. The finishing point is determined as the bus terminal of Gaziantep. The main reason considered this route is that it is one of the most frequently used lines in the city. There are almost 50 thousands students in Gaziantep University and most of them are using bus terminal while they are living near the university campus. Although there are currently 11 different bus routes between Gaziantep University and Gaziantep bus terminal, most of the students dissatisfy with the long and circulating routes. To overcome aforementioned dissatisfaction, a new route which is shorter and has fewer stops is investigated in this paper. Therefore, the problem is considered as a SPP.

The SPP is a typical combinatorial problem that contains key elements of network flow problems; it has been extensively studied, and a large number of algorithms exist

© Springer Nature Switzerland AG 2021
Z. Molamohamadi et al. (Eds.): LSCM 2020, CCIS 1458, pp. 271–276, 2021.
https://doi.org/10.1007/978-3-030-89743-7_15

to solve it [1]. The SPP is to search a path with a minimal overall length from a starting node to an end node. The arc lengths can alternatively be thought of as travel times or expenditures for traveling the respective arcs [2].

In literature, there are mainly four types of SPPs. They are single pair, single source, single destination and all pairs SPPs [3]. The SPP variants with their explanations are given in Table 1.

Table 1. SPP variants

SPP types	Definition	Solution	Ref.
Single pair	Find a shortest path from u to v for given vertices u and v	Binary integer formulation	Taccari [4]
Single source	Find the shortest paths from u to each vertex	Djikstra and Bellman Ford algorithms	Orlin et al. [5]
Single destination	Find the shortest paths to a given destination vertex v from each vertex		
All pairs	Find the shortest paths from u to v for every pair of vertices u and v	Floyd-Warshall algorithm	Hougardy [6]

Our problem belongs to single pair SPP type. Despite the fact that shortest path methods have been used in numerous areas, most of them focuses complexity of the problems [7]. Application of single pair shortest path algorithms to the transportation problems is not a new issue [8]. For instance, Wu and Hartley [9] used shortest paths algorithms for calculating a sufficient number of ranked shortest routes for the public transportation network of Nottingham City. Then, Huang et al. [10] created an algorithm that calculates the shortest cost path between a moving object and its destination by constantly adapting to changing traffic conditions while using prior search results.

For multimodal transportation networks, Idri et al. [11] devised a new time-dependent shortest path algorithm. The proposed method was a single-source single-destination goal-oriented algorithm. As a case study of Pokhara city, Thapa and Shrestha [12] focused on finding the shortest path possible in terms of minimum time and cost to reach a specific place for a person. Finally, Tu et al. [13] investigated the constrained reliable shortest path (CRSP) problem for electric vehicles in the urban transportation network. For recent applications of SPP on public transportation, the reader is referred to Susilowati and Fitriani [14]; Win et al. [15] and Liu and Liu [16].

The remainder of the paper is organized as follows after the introduction section: Sect. 2 presents the applied binary integer programming formulation. In Sect. 3, the case is introduced and the results of computational experiments are given. Main conclusions are offered in Sect. 4.

2   The Mathematical Model

As mentioned in the previous section, single pair SPP is considered in this study. For the solution approach, a binary integer programming formulation [17] is used. Given a directed graph $G = (V, A)$ and arc costs (distance) d_{ij} for each $(i, j) \in A$, he SPP entails determining the shortest path between two nodes, s and t. The following is a standard binary integer programming formulation for determining the shortest path from node s to node t [17]:

$$minimize \sum_{i,j \in A} d_{ij} x_{ij} \tag{1}$$

Subject To

$$\sum_{i,j \in A} x_{ij} - \sum_{i,j \in A} x_{ij} = \begin{cases} 1 \ if \ i = s \\ -1 \ if \ i = t \\ 0 \ else \end{cases} \quad \forall i \in V \tag{2}$$

$$\sum_{i,j \in A} x_{ij} \leq 1 \quad \forall i \in V \tag{3}$$

$$x_{ij} \in \{0, 1\} \quad \forall i, j \in A \tag{4}$$

where x_{ij} is binary link variables that get value 1 if the arc (i, j) belongs to the path. Objective function (1) finds the shortest path between two nodes s and t. Equation (2) are flow conservation constraints, while Eq. (3) assures that each node's outgoing degree is limited to one. Finally, Eq. (4) is the binary restriction constraint.

3   The Case Study

In this paper, we tried to find the shortest public transportation from Gaziantep University to Gaziantep bus terminal based on the road distance. For this, we used the Gaziantepkart application, which was put into practice by Gaziantep Municipality and can be downloaded for free from providers such as Google Play. We have identified the names of all municipal buses on the route from Gaziantepkart application to Gaziantep University and Gaziantep bus terminal and at which station they stop. Many buses followed the same route. The stops were very close to each other or they used more than one same stop at the same time. Therefore, after determining all the stops and the distance between them, we eliminated some of the stations very close to each other and included only one of the stops in which they stopped again in a loop.

Our aim here was to create the shortest route by using the same stops rather than the routes already used and to shorten the journey time. We thought that if we reduced the number of stops and determined the appropriate intervals between stops, we would gain both distance and time. By analyzing the current 11 bus routes, 46 different and common stops are determined. The locations of stops are shown in Fig. 1.

We marked the stops obtained from Gaziantepkart application on the map between the Gaziantep University and Gaziantep bus terminal in the most appropriate way for real

Fig. 1. Considered 46 stop locations.

measurements using Google Earth. Then, using the distance measurement application in Google Earth, we determined the approximate distance between stops. The network consists of 46 stops are shown in Fig. 2. While the first node shows starting point (Gaziantep University), the last point indicates the ending point (Gaziantep bus terminal).

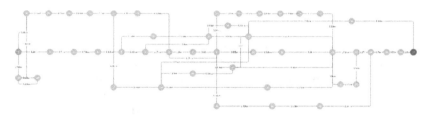

Fig. 2. Network of bus stops.

The model was solved by the GAMS 23.1 optimization package program (less than 5 CPU seconds) and the shortest distance between the two points (10.4 km) was found to be optimal. As a result, a new bus route including 18 destination stops with shorter distances than the existing 11 bus routes has been proposed. The new route is shown as pink in Fig. 3. It must be noted the proposed new route (10.4 km) is 17% less than the current route.

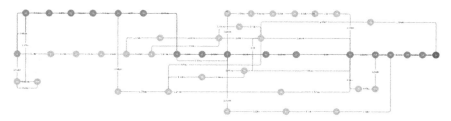

Fig. 3. Proposed route in the network.

4 Conclusion

Single pair SPP for the public transportation case is studied in this paper. Considering the current bus routes between starting and ending points, a network with 46 stops is created. A binary integer programming formulation is applied to find the shortest route between Gaziantep University (starting point) and Gaziantep bus terminal (ending point). As a result, obtained optimal route consists of 18 bus stops and provides 10.4 km distance which is less distance (17% less) than the existing routes. For the future, a geographic information system-based solution approach should be developed. In addition, traveling time instead of traveling distance should be used as a performance indicator.

1. References

1. Ferone, D., Festa, P., Guerriero, F., Laganà, D.: The constrained shortest path tour problem. Comput. Oper. Res. **74**, 64–77 (2016)
2. Yu, G., Yang, J.: On the robust shortest path problem. Comput. Oper. Res. **25**(6), 457–468 (1998)
3. Dreyfus, S.E.: An appraisal of some shortest-path algorithms. Oper. Res. **17**(3), 373–556 (1969)
4. Taccari, L.: Integer programming formulations for the elementary shortest path problem. Eur. J. Oper. Res. **252**(1), 122–130 (2016)
5. Orlin, J.B., Madduri, K., Subramani, K., Williamson, M.: A faster algorithm for the single source shortest path problem with few distinct positive lengths. J. Discrete Algorithms **8**(2), 189–198 (2010)
6. Hougardy, S.: The Floyd-Warshall algorithm on graphs with negative cycles. Inf. Process. Lett. **110**(8–9), 279–281 (2010)
7. Rosen, J.B., Sun, Z.S., Xue, G.L.: Algorithms for the quickest path problem and the enumeration of quickest paths. Comput. Oper. Res. **18**(6), 579–584 (1991)
8. Kumari, S.M., Geethanjali, N.: A survey on shortest path routing algorithms for public transport travel. Global J. Comp. Sci. Technol. **9**(5), 73–76 (2010)
9. Wu, Q., Hartley, J.: Using K-shortest paths algorithms to accommodate user preferences in the optimization of public transport travel. In: 8th International Conference on Applications of Advanced Technologies in Transportation Engineering Proceedings, pp. 181–186, Beijing (2004)
10. Huang, B., Wu, Q., Zhan, F.B.: A shortest path algorithm with novel heuristics for dynamic transportation networks. Int. J. Geogr. Inf. Sci. **21**(6), 625–644 (2007)

11. Idri, A., Oukarfi, M., Boulmakoul, A., Zeitouni, K., Masri, A.: A new time-dependent shortest path algorithm for multimodal transportation network. Proc. Comput. Sci. **109**, 692–697 (2017)
12. Thapa, R., Shrestha, J.: Optimal route computation for public transport with minimum travelling time and travel cost: a case study of Pokhara city. Tech. J. **1**(1), 79–86 (2019)
13. Tu, Q., Cheng, L., Yuan, T., Cheng, Y., Li, M.: The constrained reliable shortest path problem for electric vehicles in the urban transportation network. J. Cleaner Prod. **261**, 121130 (2020)
14. Susilowati, E., Fitriani, F.: Determining the shortest path between terminal and airport in Yogyakarta using trans Jogja with min plus algorithm. J. Math. Educ., Sci. Technol. **4**(2), 123–134 (2019)
15. Win, Y.Y., Hlaing, H.S., Thein, T.T.: Shortest path analysis based on Dijkstra's algorithm in Myanmar road network. Int. J. Res. **6**(10), 132–139 (2019)
16. Liu, Y., Liu, X.: Application of improved A* algorithm in customized bus path planning. Comput. Sci. Appl. **10**(1), 21–28 (2020)
17. Raith, A., Ehrgott, M.: A comparison of solution strategies for biobjective shortest path problems. Comput. Oper. Res. **36**(4), 1299–1331 (2009)

Sustainable and Resilient Supply Chain Management

A Multi-objective Robust Optimization Model for Green Supply Chain Design Under Uncertainty

Mazyar Sasanian, Nikbakhsh Javadian, and Masoud Alinezhad$^{(\boxtimes)}$

Department of Industrial Engineering, Mazandaran University of Science and Technology,
Babol, Iran
{sasanian,nijavadian,m.alinezhad}@ustmb.ac.ir

Abstract. Nowadays, because of the growing competition and industrialization globally, the design of supply chain network has been noticed by many researchers. Additionally, because of environmental issues, social responsibility, governmental regulation, and lack of resources, many companies have focused on the green supply chain (GSC). A GSC network includes a forward supply chain combining environmental issues. In this regard, we propose a multi-echelon, multi-product, multi-objective, mixed-integer linear programming (MILP) model, considering multiple suppliers, multiple transportation modes, and multiple environmental protection levels for manufacturing plants that are developed for a GSC network problem. The model has two goals: the first one is to decrease the economic cost, and the second is to reduce environmental influences. Also, we consider demand and cost parameters under uncertainty. A robust optimization method is employed to handle uncertain parameters, and the suggested problem is resolved as a one objective MILP model using the LP-metric approach. A statistical example is offered to show the applicability of the given model.

Keywords: Robust optimization · Multi-objective problem · Uncertainty · GSCN · Network design

1 Introduction

Designing an effective supply chain network (SCN) has always been a challenge for researchers. The importance of SCN design was recognized once in the early 1970s [7]. The classical goal of almost all supply chain (SC) problems is to decrease the entire cost of SC. SCN can be categorized into three types: 1. strategic, 2. tactical, and 3. operational. Operational and tactical duties are identified by short-term and mid-term planning matters, respectively. In contrast, strategic duties involve long-term choices which generally cannot be altered within a short time [1], and these decisions have a significant influence on environmental efficiency [2, 8]. One of these strategic decisions is the protection level selection for manufacturing plants. In addition to the tactical decisions (such as the amount of flow transportation among the SC's facilities) are included in this paper. Because supply chain design (SCD) has long-lasting planning

© Springer Nature Switzerland AG 2021
Z. Molamohamadi et al. (Eds.): LSCM 2020, CCIS 1458, pp. 279–302, 2021.
https://doi.org/10.1007/978-3-030-89743-7_16

about an environmental issue and complicacy, a decision-maker (DM) faces top-level of uncertainty.

Furthermore, industries are subject to societal and government pressures to enhance the environmental performance of their activities. The developing consciousness of SC environmental prospects is currently enormously perceived by academic and industrial communities [7]. In the SC context, concentrating on environmental issues often alluded to environmental or green supply chain (GSC) management. Recently, focusing on GSCM research has been increasing rapidly over the past decade because of the necessity by the company to observe and shedding light on environmental matters [20, 21]. Nowadays, politics, global industries, and companies have realized the need for air pollution emissions reduction. Therefore, it is vital to examine the optimization of total SC costs and environmental influences together. Based on those mentioned above, a GSC can be defined as a network that ensures that good transportation from production plant to client is environmentally friendly [25]. To acquire this objective, companies should focus on designing and organizing to optimize their logistics chain while representing the trade-off between cost and environmental impacts [25]. Also, environmental SC decisions interact with other enterprise parts such as buying, logistics, and managing all things that happen in SC sections. SC planning and optimization on its own is a generally complicated operation with a significant number of variables and limitations, and the incorporation of environmental dimensions rolls it up to complicacy [6]. Various decisions that a company makes regarding its facility location, supplier and transportation mode selection, etc., can remarkably affect its pollution traces.

In the recent decade, uncertainty has been paid attention by many researchers. Different types of uncertainty approaches have been developed to fuzzy programming method, stochastic programming, and robust optimization (RO) method. Rather than the other specified methods, RO is a way that searches for a method that fewer sensitive to diversifying inputting data (solution robustness) and feasible in the probable set of uncertain scenarios (model robustness) [19]. RO approach is appropriate for a model with a top-level uncertainty where a DM cannot alter a decision when it is made. Hence, the utilization of the RO concept is able to be beneficial. The study's idea is to know which manufacturing plants should be opened between a set of potential manufacturing plants (the number of manufacturing plants that should be opened is predetermined) and which environmental protection level (EPL) should be assigned. Tactical decisions are also taken into consideration. We take into account both economic costs and environmental influences in the model. Moreover, we have developed the problem for multiple suppliers, manufacturing plants, demand zones with different transportation modes, and various protection levels for manufacturing plants.

Moreover, most studies in SCD have been limited to the deterministic method, where all model parameters are considered to be acknowledged with certainty. While in the real world, we are dealing with uncertainties. Therefore, we consider demand and cost as uncertainty parameters to have an applicable model. On the other hand, except for the tactical decisions, we analyze the EPL for manufacturing plants as a strategic decision in SCD, making it more complicated. Our study is the first to discuss a bi-objective RO model for GSCN considering the EPL in manufacturing plants under uncertainty. It means that the first venture in environmental protection facilities should be defined in the design phase, which high-grade equipment is more expensive but leads to a lower economic and environmental cost in the operations phase.

The rest of the article is categorized as follows: In Sect. 2, some related research on the SCND problem under uncertainty considering environmental issues is given. In Sect. 3, the background of RO formulation is described. Problem description and formulation are discussed in Sect. 4. In Sect. 5.1, the resolution procedure is proposed. In Sect. 5.2, a numerical investigation is shown for testing the model. Results and computational experiments are explained in this section, and the paper outcomes are provided in Sect. 6. Subsequent paragraphs, however, are indented.

2   Literature Review

In the past years, the number of research on SCM problems has been increasing among researchers. Generally, the focus of most studies is on deterministic approaches, while in practical SC problems, because of the effect of different aspects, comprehensive information about parameters like demand and costs is usually unavailable. Moreover, consideration of environmental concerns meanwhile designing and planning SCs has been the focus of some studies. Our study literature deals with two fundamental aspects: (1) research related to SCN under uncertainty and (2) research associated with environmental aspects of the SC.

2.1   Supply Chain Network

Goodarzian et al. [9] proposed a new multi-objective (MO), multi-echelon (ME), multi-product (MP), multi-period pharmaceutical SCN. They used novel robust fuzzy programming to cope with uncertainty parameters. Pishvaee et al. [22] proposed a MO fuzzy mathematical programming model for creating an environmental SC under the intrinsic uncertainty of input data in such a problem, considering the minimization of various environmental influences besides the traditional cost minimization goal to adjust equality among them.

Alinezhad et al. [3] presented an MP closed-loop supply chain (CLSC) network design. They considered the demand and the return rate as certain parameters. The model objective is maximizing the profit in the whole SC. Guillen and Grossman [11] indicated the model and planning of SCs. They introduced a MO stochastic mixed-integer non-linear program that concurrently decreases cost and environmental effect for an addressed possibility level for a liquid material SC.

Paksoy et al. [20] modeled an SC to reduce the entire cost, avoid more CO_2 gas emissions, and persuade clients to utilize recyclable goods. They suggested various shipping modes between echelons according to CO_2 emissions. Jamshidi [14] proposed modeling and resolving an SCD for annual cost minimization while considering environmental impacts. Alinezhad [2] proposed an MP and multi-period mixed-integer linear programming (MILP) problem maximizing the profit in the CLSC network. The paper is applied to a representative case from the dairy industry. Goodarzian et al. [10] proposed a novel MO problem for the production distribution problem by considering different suppliers, productions, distributors, and customers. They assumed costs, demands, and capacity as uncertain parameters. Rahemi et al. [23] formulated a bi-objective MILP for a bioethanol SCN. The model has two objectives: minimizing the cost of the whole SC and the second is maximizing the suitability of crops.

2.2 Environmental Aspects

Ramudhin et al. [24] developed an SCD problem under carbon dealing considerations by integrating the environmental impacts of strategic SC decisions in carbon dioxide equivalents. Tsai and hung [29] presented a fuzzy goal programming method for supplier selection and flow allocation. They pointed out that the activity-based way is appropriate for cost and efficiency evaluation. Mele et al. [18] established a MO MILP problem considering environmental SCD choices in the sugarcane manufacturing of Argentina, where the environmental effect is specified based on the life cycle assessment strategy.

Harris et al. [12] utilized a simulation problem to assess the connection between the whole logistics costs and environmental effect. They have attended to carbon emissions, examining various scenarios regarding foundation and divers load utilization ratios. Diabat and simchi-Levi [5] constructed carbon emissions in manufacturing, storage, and warehouse and analyzed the effect of varying emission caps on SC's economic performance. Mallidis et al. [17] investigated carbon and particulate matter emissions in a network design problem. Emissions are combined with diverse transportation modes as well as the assigned or shared utilization of distributions. Validi et al. [30] pointed out a robust MO problem for modeling a capacitated network, which distributes dairy goods in Ireland.

Lotfi et al. [16] presented a sustainable model related to the medical waste management problem for COVID-19 pandemics. They have used CPLEX for solving the model. Wang et al. [31] suggested a MILP for a GSC network model with two objectives: minimizing the whole costs and the depreciation of overall environmental effects. They considered environmental investment decisions in the design phase. A statistical case is applied in order to investigate the effectiveness of the objectives function. Yang et al. [32] pointed out that shipping costs and customer demands were all stochastic variables in their SCND problem. CLSC is one of the concepts that are related to our study. There are many types of research about closed-loop SCND in the literature. The most relevant ones published in the past five years are those proposed by Ruimin et al. [25] and Altmann and Bogaschewsky [1]. They consider both economic and environmental dimensions in their models. Hasani et al. [13] formulated a model that optimizes and maximizes the after-tax profit of a global CLSC under uncertainty. Chaabane et al. [4] assessed that gathering used products and re-manufacturing may enhance SC's whole costs and lead to a rise in carbon dioxide emissions. Tirkolaee et al. [28] presented a novel MILP. The model has three objective functions: minimizing travel time, destruction of time windows/service preferences, and infection/environmental risk. They applied a fuzzy chance-constrained programming method in order to tackle sustainability. Therefore, this paper is focused on the forward green SCND. Gholizadeh and Fazlollahtabar [8] presented a green CLSC with considering environmental impact. The model objective function is maximizing profit with attention to environmental-friendly policies. They used demand as an uncertainty parameter.

Table 1. A brief overview of relevant work

Reference(s)	Problem	Objective	Solution methodology
Rahemi et al. [23]	Bioethanol SC	Minimizing cost and maximizing suitability of crops	The best-worst method and PROMEETHEE
Alinezhad [2]	CLSC	Maximizing the profit	CPLEX
Goodarzian et al. [10]	SCN	Minimizing cost	Meta-heuristic algorithms
Pishvaee et al. [22]	CLSC	Minimizing cost	CPLEX
Ramudhin et al. [24]	GSCN	Minimizing cost and CO_2 emission	CPLEX
Wang et al. [25]	GSCN	Minimizing cost and overall environmental impact	CPLEX
Gholizadeh and Fazlollahtabar [8]	Closed-loop green Supply chain	Maximizing the profit	Heuristic algorithms
This study	GSCN	Minimizing the total cost and environmental cost	LINGO

3 Robust Optimization

Mulvey et al. [19] build up a structure for the RO that can confront circumstances that have an uncertain environment. The framework consists of two robustness concepts: solution robustness and model robustness. The first concept determines that the answer for all scenarios must be near to the optimum solution if entry data changed. In contrast, the second one refers to the feasibility of the solution to small modifies in the input data for all scenarios. However, it is implausible to accomplish an answer that is both possible and optimal for whole scenarios. Therefore, multi-criteria decision making (MCDM) theory can be applied to create a trade-off between solution robustness and model robustness.

Pan and Rakesh [21] developed the following LP model that covers random parameters:

$$Min \ \ cTx + dTy \tag{1}$$

Subject to

$$Ax = b, \tag{2}$$

$$Bx + Cy = e, \tag{3}$$

$$x, y \geq 0. \tag{4}$$

Where x is the decision variable vector and should be defined under the uncertainty of model parameters. Y is the control variable vector, and B, C, and e are the random values. Assuming that $S = 1, 2, \ldots, s$ is a set of scenarios for uncertain parameters, in which every scenario has a probability value ρs ($\sum_{s \in S} \rho$ s $= 1$). Equation (2) is the fundamental constraint whose coefficients are determined and without noise, while Eq. (3) denotes the handle constraint whose coefficients are subject to noise. Equation (4) enforces non-negative restrictions on vectors.

Notice that the model is possibly infeasible because of parameter uncertainty for some scenarios. Therefore, δs is defined to shows the infeasibility of the problem under scenario s. If the problem is feasible, then δs is equal to zero. In another way, δs will be assigned a value that is positive.

The RO is formulated as follows:

$$Min \ \sigma(x, y1, y2, \ldots, ys) + \omega \rho(\delta 1, \delta 2, \ldots, \delta s) \tag{5}$$

Subject to

$$Ax = b, \tag{6}$$

$$B_s x + C_s y + \delta_s = e_s, \qquad \forall s \in S, \tag{7}$$

$$x \geq 0, y_s \geq 0, \delta_s \geq 0 \qquad \forall s \in S. \tag{8}$$

The first part of the above objective function represents the solution robustness, and the second part deals with model robustness.

Mulvey et al. [19] defined the RO model for the first term to represent solution robustness as:

$$\sigma(x, y1, y2, \ldots, ys) = \sum_{s \in S} \rho s \psi s + \lambda \sum_{s \in S} \rho s \left(\psi s - \sum_{s' \in S} \rho s' \psi s' \right) 2\sigma \tag{9}$$

Where λ explains the weight value allocated for the variance solution, the solution is concisely sensitive to alterations in the data under all scenarios as λ enhances [19]. As it is obvious, the above equation has a quadratic term. Yu and Li [33] changed over the mentioned equation into an absolute value and proposed a formulation by a few alterations which have the following structure:

$$\sigma(x, y1, y2, \ldots, ys) = \sum_{s \in S} \rho s \psi s + \lambda \sum_{s \in S} \rho s \left| \psi s - \sum_{s' \in S} \rho s' \psi s' \right| \tag{10}$$

Though Eq. (10) is a non-linear function, it can be optimized by transforming the model into a linear programming model which has a linear objective function with linear constraints by proposing two non-negative deviational variables. Instead of minimizing the sum of the sheer deviations in Eq. (10), two deviational variables are minimized subject to original constraints and additional soft constraints, which give positive values of the difference inside the sheer functions. However, Yu and Li [33] expressed that this straightforward linearization method is widely confined because numerous non-negative deviational variables and constraints are presented. The framework model of Yu and Li is designed to decrease the objective function as follows:

$$Min \sum_{s \in S} \rho_s \psi_s + \lambda \sum_{s \in S} \rho_s \left[\left(\psi_s + \sum_{s' \in S} \rho_{s'} \psi_{s'} \right) + 2\theta_s \right] \tag{11}$$

Subject to

$$\psi_s - \sum_{s \in S} \rho_s \psi_s + \theta_s \geq 0, \qquad \forall s \in S \tag{12}$$

$$\theta_s \geq 0 \qquad \forall s \in S \tag{13}$$

The second section of the objective function is associated with the model robustness, which composes the penalties applied in the control constraints. Using the coefficient ω as the weight to balance two parts of the objective function, the final formulation can be presented as follows:

$$Min \sum_{s \in S} \rho_s \psi_s + \lambda \sum_{s \in S} \rho_s [(\psi_s - \sum_{s' \in S} \rho_{s'} \psi_{s'}) + 2\theta_s] + \omega \sum_{s \in S} \rho_s \delta_s \tag{14}$$

4 Problem Description and Formulation

In this study, we formulate a ME, M, MO SC network, including suppliers, manufacturing plants, and customer zones (see Fig. 1). Goods are produced by manufacturing plants, manufactured by raw materials supplied through suppliers regarding consumption rates. We consider different transportation alternatives, which every transportation mode has a different cost and environmental pollution. Moreover, we consider the EPL for manufacturing plants.

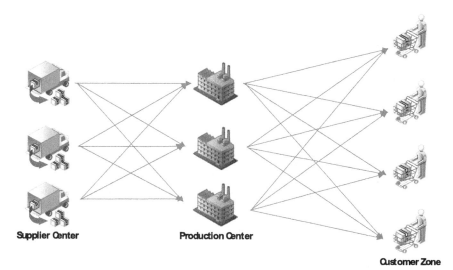

Supplier Center   **Production Center**

                                                              Customer Zone

Fig. 1. SCN under study

The assumptions of the model are as follows:

1. The model is discrete, and there is a limited amount of suppliers, manufacturing plants, and customers.
2. The model is planned for a single period.
3. The number of manufacturing plants that should be opened is known and fixed.
4. The uncertainty associated with the customer demands and cost parameters is considered using discrete scenarios-sample Heading (Third Level). Only two levels of headings should be counted. Lower level headings reside unnumbered; they are formatted as run-in headings.

4.1 Sets and Indices

$P = \{1, \ldots, p\}$ set of manufacturing plants,
$A = \{1, \ldots, a\}$ set of suppliers,
$C = \{1, \ldots, c\}$ set of customer zones,
$R = \{1, \ldots, r\}$ set or raw materials,
$F = \{1, \ldots, f\}$ set of products,
$N = \{1, \ldots, n\}$ set of transportation modes,
$L = \{1, \ldots, l\}$ set of EPL for production plants,
$S = \{1, \ldots, s\}$ set of scenarios.

4.2 Parameters

$tcap^s_{aprn}$	The shipping cost of raw material r per unit from the supplier a to production plant p through transportation mode n under scenario s,
$tcpc^s_{pcfn}$	The shipping cost of product f per unit from production plant p to customer position c through transportation mode n under scenario s,
$tcpc^s_{pcfn}$	The shipping cost of product f per unit from production plant p to customer position c through transportation mode n under scenario s,
cp^s_{ra}	Purchasing cost of raw material r per unit from supplier a under scenario s,
cm^s_{fpl}	The cost of producing goods f per unit by production plant p with EPL l under scenario s,
ce^s_p	Establishment cost of production plant p under scenario s,
dem^s_{fc}	Demand for product f by customer zone c under scenario s,
$rawnum_{rf}$	Amount of units of raw material r needed for per unit product f,
$rawmax_{ar}$	Maximum amount of raw material r supplier a could provide,
ct^s_{pl}	The cost of selecting EPL l for production plant p under scenario s,
$trea^s_{aprn}$	The environmental cost of transporting raw material r per unit from the supplier a to production plant p through transportation mode n under scenario s,
$trep^s_{pcfn}$	The environmental cost of transporting goods f per unit from production plant p to customer zone c through transportation mode n under scenario s,

(continued)

(continued)

em_{fpl}^s	The environmental cost of producing product f per unit by production plant p with EPL l under scenario s,
K	The number of production plants that should be established among a set of potential production plants,
γ_1^{max}	maximum transportation capacity of raw material r from the supplier a to production plant p through shipping mode n,
γ_2^{max}	Maximum transportation capacity of product f from production plant p to customer zone c through transportation mode n,
ρ_s	The probability of scenario s,
λ_1	Weighting factor for solution robustness part in the first objective function,
λ_2	Weighting factor for solution robustness part in the second objective function,
ω	Weighting factor for model robustness,
ν	Weighting factor for objective functions

4.3 Decision Variables

$c\ xap_{aprn}^s$	The quantity of raw material r sent from the supplier a to production plant p through transportation mode n under scenario s,
xpc_{pcfn}^s	The amount of product f sent from production plant p to customer zone c by transportation mode n under scenario s,
xm_{fp}^s	The amount of product f produced in production plant p under scenario s,
x_p	Binary variable for opening production plant p,
z_{pl}	Binary variable for selecting EPL l for production plant p,
δ_{fc}^s	The amount of not meeting the demand of customer zone c for product f under scenario s,
θ_1^s	Deviation for violations of the mean of total economical costs under scenario s,
θ_2^s	Deviation for violations of the mean of total environmental costs under scenario s

4.4 Objective Functions

The objective function consists of two parties: economical cost (Z_1) and ***environmental*** cost (Z_2). The first objective function (Z_1) consists of purchasing cost (PRC) Eq. (15), production cost (PDC) Eq. (16), transportation cost (TRC) Eq. (17), establishment cost (ESC) Eq. (18) and EPL selecting cost for manufacturing plants (ENC) Eq. (19).

$$PRC_s = \sum_{p=1}^{P} \sum_{a=1}^{A} \sum_{r=1}^{R} \sum_{n=1}^{N} cp_{ra}^s . xap_{aprn}^s \qquad (15)$$

$$PDC_s = \sum_{f=1}^{F} \sum_{p=1}^{P} \sum_{l=1}^{L} cm_{fpl}^{s} . xm_{fp}^{s} \tag{16}$$

$$TRC_s = \sum_{p=1}^{P} \sum_{a=1}^{A} \sum_{r=1}^{R} \sum_{n=1}^{N} tcap_{aprn}^{s} . xap_{aprn}^{s}$$
$$+ \sum_{p=1}^{P} \sum_{c=1}^{C} \sum_{n=1}^{N} \sum_{f=1}^{F} tcpc_{pcfn}^{s} . xpc_{pcfn}^{s} \tag{17}$$

$$ESC_s = \sum_{p=1}^{P} ce_p^{s} . x_p \tag{18}$$

$$ENC_s = \sum_{p=1}^{P} \sum_{l=1}^{P} ct_{pl}^{s} . z_{pl} \tag{19}$$

Hence, the whole economic cost can be presented as follows:

$$\text{Min} z_1 = PRC_s + PDC_s + TRC_s + ESC_s + ENC_s = \sum_{p=1}^{P} \sum_{a=1}^{A} \sum_{r=1}^{R} \sum_{n=1}^{N} cp_{ra}^{s} . xap_{aprn}^{s}$$
$$+ \sum_{f=1}^{F} \sum_{p=1}^{P} \sum_{l=1}^{L} cm_{fpl}^{s} . xm_{fp}^{s} + \sum_{p=1}^{P} \sum_{a=1}^{A} \sum_{r=1}^{R} \sum_{n=1}^{N} tcap_{aprn}^{s} . xap_{aprn}^{s}$$
$$+ \sum_{p=1}^{P} \sum_{c=1}^{C} \sum_{n=1}^{N} \sum_{f=1}^{F} tcpc_{pcfn}^{s} . xpc_{pcfn}^{s} + \sum_{p=1}^{P} ce_p^{s} . x_p + \sum_{p=1}^{P} \sum_{l=1}^{L} ct_{pl}^{s} . z_{pl} \tag{20}$$

The first objective function (Z1) reduces the economic cost of the SC. Similarity, the second objective function (Z2), which decreases the total environmental cost, consists of two parties: environmental cost caused by transportation Eq. (21) and environmental cost caused by production regarding EPL Eq. (22).

$$ECTR_s = \sum_{a=1}^{A} \sum_{p=1}^{P} \sum_{r=1}^{R} \sum_{n=1}^{N} xap_{aprn}^{s} . trea_{aprn}^{s} + \sum_{p=1}^{P} \sum_{c=1}^{C} \sum_{f=1}^{F} \sum_{n=1}^{N} xpc_{pcfn}^{s} . trep_{pcfn}^{s} \tag{21}$$

$$EM_s = \sum_{f=1}^{F} \sum_{p=1}^{P} \sum_{l=1}^{L} em_{fpl}^{s} . xm_{fp}^{s} \tag{22}$$

So, the total environmental cost can be shown as follows:

$$\text{Min} z_2 = ECTR_s + EM_s = \sum_{a=1}^{A} \sum_{p=1}^{P} \sum_{r=1}^{R} \sum_{n=1}^{N} xap_{aprn}^{s} . trea_{aprn}^{s}$$
$$+ \sum_{p=1}^{P} \sum_{c=1}^{C} \sum_{f=1}^{F} \sum_{n=1}^{N} xpc_{pcfn}^{s} . trep_{pcfn}^{s} + \sum_{f=1}^{F} \sum_{p=1}^{P} \sum_{l=1}^{L} em_{fpl}^{s} . xm_{fp}^{s} \tag{23}$$

4.5 Constraints

$$\sum_{a=1}^{A} \sum_{n=1}^{N} xap_{aprn}^{s} = \sum_{f=1}^{F} xm_{fp}^{s} . rawnum_{rf}, \qquad \forall p, r, l \tag{24}$$

$$xm_{fp}^{s} = \sum_{c=1}^{C} \sum_{n=1}^{N} xpc_{pcfn}^{s}, \qquad \forall p, f \tag{25}$$

$$\sum_{p=1}^{P} \sum_{n=1}^{N} xap_{aprn}^{s} \leq rawmax_{ar}, \qquad \forall a, r \tag{26}$$

$$\sum_{f=1}^{F} tm. xm_{fp}^{s} \leq ca_p . x_p, \qquad \forall p \tag{27}$$

$$\sum_{p=1}^{P} x_p = k, \tag{28}$$

$$\sum_{l=1}^{L} z_{pl} \leq 1, \qquad \forall p \tag{29}$$

$$\sum_{l=1}^{L} z_{pl} = x_p, \qquad \forall p \tag{30}$$

$$xap_{aprn}^{s} \leq \gamma_1^{max}, \qquad \forall a, p, r, n \tag{31}$$

$$xpc_{pcfn}^{s} \leq \gamma_2^{max}, \qquad \forall p, f, c, n \tag{32}$$

$$x_p, z_{pl} \in \{0, 1\}, \qquad \forall p, l \tag{33}$$

$$xap_{aprn}^{s}, xpc_{pcfn}^{s}, xm_{fp}^{s} \geq 0. \qquad \forall a, c, f, p, n, r \tag{34}$$

Constraints (24), (25) are balance constraints. Constraint (24) guarantees that the shipping number of raw materials from suppliers to manufacturing plants should be equal to raw material requirements for products. Constraint (25) indicates that the quantity of goods is produced in every plant is equivalent to the shipping amount of commodities to all customers. Constraint (26) limits the supplier's capabilities to supply raw materials for production centers. Constraint (27) is a capacity constraint of production plants. Constraint (28) specifies the number of manufacturing plants that should be established. Constraint (29) limits that only one EPL can be set for any opening production centers. Constraint (30) ensures that any opening manufacturing plant has an EPL. Constraints (31), (32) are capacity limits for transporting raw materials from suppliers to manufacturing plants (Eq. (31)) and goods from manufacturing centers to customer zones (Eq. (32)). Constraint (33) indicates the binary variables, and eventually, constraint (34) performs the non-negative limitation on the decision variables.

Based on the model presented by Mulvey [19], which discussed in Sect. 3, we can obtain our robust bi-objective formulation as follows:

$$Minz_1 = \sum_{s \in S} \rho_s (PRC_s + PDC_s + TRC_s + ESC_s + ENC_s)$$

$$+ \lambda_1 \sum_{s \in S} \rho_s [(PRC_s + PDC_s + TRC_s + ESC_s + ENC_s) - \sum_{s' \in S} \rho_{s'} (PRC_{s'} + PDC_{s'}$$

$$+ TRC_{s'} + ESC_{s'} + ENC_{s'}) + 2\theta_1^{s}] + \omega \sum_{f=1}^{F} \sum_{c=1}^{C} \sum_{s=1}^{S} \rho_s \delta_{fc}^{s} \tag{35}$$

$$Minz_2 = \sum_{s \in S} \rho_s (ECTR_s + EM_s) + \lambda_2 \sum_{s \in S} \rho_s [(ECTR_s + EM_s)$$

$$-\sum_{s' \in S} \rho_{s'} (ECTR_s + EM_s) + 2\theta_2^s] + \omega \sum_{f=1}^{F} \sum_{c=1}^{C} \sum_{s=1}^{S} \rho_s \delta_{fc}^s \qquad (36)$$

Subject to constraints (24–34) and the following constraints:

$$\sum_{p=1}^{P} \sum_{n=1}^{N} xpc_{pcfn} + \delta_{fc}^s = dem_{fc}^s, \qquad \forall f, c, s \qquad (37)$$

$$(PRC_s + PDC_s + TRC_s + ESC_s + ENC_s) - \sum_{s \in S} \rho_s (PRC_s \qquad \forall s \in S$$
$$+PDC_s + TRC_s + ESC_s + ENC_s) + \theta_1^s \geq 0, \qquad (38)$$

$$(ECTR_s + EM_s) - \sum_{s \in S} \rho_s (ECTR_s + EM_s) + \theta_2^s \geq 0, \qquad \forall s \in S \qquad (39)$$

$$\theta_1^s, \theta_2^s, \delta_{fc}^s \geq 0. \qquad \forall f, c, s \qquad (40)$$

Constraint (37) is a control constraint and assesses the goods flow from manufacturing plants to customer zones that should be identical to the market demand. δ_{fc}^s would be a positive value when the scenario obtains an infeasible solution. Otherwise, $\delta_{fc}^s = 0$. Constraints (38), (39) are auxiliary constraints defined in Eq. (12) for linearization.

5 Solution Approach and Computational Results

5.1 Solution Process

The RO model, presented in the earlier segment, is a MO MILP. Additionally, the objective functions in the proposed model are entirely inconsistent. Therefore, using the weighting approach, also recognized as the LP-metrics method, a popular approach to solve MO models, we can integrate two goals into one goal. Since the economic and environmental objective functions do not have the same scale, we first normalize them as follows:

$$Z_i^{norm} = \frac{z_i - z_i^*}{z_i^*} \qquad (41)$$

Where, Z_i^* is the ideal value for each objective function. So, we provide the novel scale-less single-objective function (Z_3) as follows:

$$Minz_3 = [v_1 z_1^{norm} + v_2 z_2^{norm}] \qquad (42)$$

Where $0 \leq v_i \leq 1$ and $\sum_i v_i = 1$ are the weight coefficients for elements of the objective function (42) presented by the DM(s) and discuss the relevance of economic cost compared to environmental concern.

5.2 Computational Results

In this segment, a numerical example is presented and analyzed to represent the applicability of the recommended model to realistic problems.

5.2.1 Case Description

Consider a three-echelon supply chain similar to the one seen in Fig. 1. There is a set of three suppliers that supply raw materials for three manufacturing plant candidates. In this study, the number of manufacturing centers that should be opened is equal to two. The manufacturing plants use four raw material types to produce two products. Those products are then shipped to customer zones, which are five zones in our numerical study. Every product has a specific rate of raw material consumption. There are three levels of technology for environmental protection that can be selected by manufacturing plants. The environmental cost of the production process in manufacturing plants is related to its EPL.

Furthermore, the three modes of transportation are different in their costs (economic and environmental) and capacities. To express the offered model, all data are provided in Tables 1, 2, 3, 4, 5, 6, 7, 8, 9, 10, 11, 12, and 13. The demand for customer zones is presented in Table 1. Purchasing costs of raw material from each supplier are provided in Table 2. The establishment costs of manufacturing plants are displayed in Table 3. Table 4 presents the production cost of products in every manufacturing plant. Table 5 presents the EPL costs for manufacturing plants as a strategic decision. Transportation costs of raw materials and goods are presented in Tables 6 and 7. As we mentioned before, each supplier can provide a certain amount of raw materials, shown in Table 8. The environmental costs of transporting raw materials and products through transportation alternatives are presented in Tables 9 and 10. Also, the environmental cost of producing products regarding the selected protection level is indicated in Table 11. Tables 12 and 13 display the shipping capacity of raw materials and products through transportation alternatives. Finally, Table 14 presents the consumption rate of raw materials for producing each product.

Table 2. Market demands for scenario 1.

Customer zones					Products
5	4	3	2	1	
2100	1600	1800	1200	500	1
2500	1900	1500	1000	700	2

Note: For scenarios 2 and 3, the evaluations are increased by 0.8 and 1.2, sequentially.

According to the data mentioned above and considering three plots: optimistic, realistic, and pessimistic with related possibilities of 0.3, 0.4, and 0.3, sequentially, the proposed problem is optimally resolved three periods, every period with one of the objective functions z_1, z_2, and z_3. The first objective function aim (z_1) is to diminish the expected value and the weighted variance, and the infeasibility penalty of the whole costs of the SCN. The second one (z_2) is to reduce the predicted value plus the weighted variance of total environmental costs of the SCN. The third one (z_3) is the LP-metrics objective function which is the best value of the above-mentioned objective functions (z_1^*, z_2^*).

Table 3. Purchasing cost for scenario 1. (€)

Suppliers	Raw material cost			
	1	2	3	4
1	70000	110000	30000	90000
2	75000	90000	36000	85000
3	65000	130000	44000	95000

Note: For scenarios 2 and 3, the evaluations are increased by 0.8 and 1.2, sequentially.

Table 4. Establishment cost of manufacturing plants for scenario 1. (€)

Scenarios	Manufacturing plants		
	1	2	3
1	17000000000	19000000000	18000000000
2	18000000000	17500000000	18500000000
3	16500000000	18500000000	19500000000

Table 5. The cost of producing products for scenario 1. ($\times 1000$) (€)

Products	Manufacturing plants								
	1			2			3		
	L1	L2	L3	L1	L2	L3	L1	L2	L3
1	140	110	90	145	115	85	150	120	90
2	145	115	85	150	120	90	155	125	95

Note: for scenarios 2 and 3, the evaluations are increased by 0.75 and 1.25, sequentially. It is necessary to mention that L1, L2, and L3 in the above table refer to EPL 1, EPL level 2, and EPL 3, respectiv.ely.

Table 6. EPL selecting cost for scenario 1. (€)

EPL	Manufacturing plants		
	1	2	3
1	700000000	750000000	650000000
2	1800000000	1900000000	1700000000
3	2600000000	2800000000	2400000000

Note: for scenarios 2 and 3, the evaluations are increased by 0.8 and 1.2, sequentially.

Table 7. Transportation cost of raw materials for scenario 1. ($\times 1000$)

Suppliers	Raw materials	Manufacturing plants								
		1			2			3		
		M1	M2	M3	M1	M2	M3	M1	M2	M3
1	1	25	40	60	62	82	102	110	140	160
	2	35	50	75	65	80	105	125	150	175
	3	30	45	65	55	70	90	105	130	155
	4	40	65	85	80	95	110	140	160	180
2	1	55	75	80	70	95	120	105	125	150
	2	30	45	50	80	100	115	110	125	140
	3	45	55	60	90	105	125	140	150	155
	4	60	70	75	85	100	115	125	150	160
3	1	60	75	90	80	105	115	110	130	150
	2	50	70	85	75	95	110	115	125	140
	3	45	55	75	80	100	115	135	150	170
	4	50	70	95	90	110	125	105	130	155

Note: for scenarios 2 and 3, the evaluations are increased by 0.8 and 1.2 sequentially. It is necessary to mention that M1, M2, and M3 in the above table refer to shipping mode 1, shipping mode 2, and shipping mode 3, respectively.

Table 8. Transportation cost of products for scenario 1. ($\times 1000$) ($€$)

Manufacturing plants	Products	Customer zones														
		1			2			3			4			5		
		M1	M2	M3	M1	M2	M3	M1	M2	M3	M1	M2	M3	M1	M2	M3
1	1	50	65	85	35	50	65	55	70	85	45	65	80	65	80	95
	2	45	60	80	30	45	60	50	65	80	40	60	75	60	75	90
2	1	55	70	90	40	55	70	60	75	90	55	70	80	70	80	90
	2	35	50	70	20	35	50	40	55	70	30	50	65	50	65	80
3	1	40	55	70	35	50	65	45	55	70	45	65	80	65	80	95
	2	30	45	60	25	40	55	35	45	60	35	55	70	55	70	85

Note: for scenarios 2 and 3,, the evaluations are increased by 0.8 and 1.2, sequentially.

Table 9. The capacity of suppliers for raw materials

Suppliers	Raw materials			
	1	2	3	4
1	9000	12000	10000	15000
2	11000	10000	13000	8000
3	14000	11000	12000	13000

Table 10. The environmental cost of transporting raw materials for scenario 1. ($\times 1000$) (€)

Suppliers	Raw materials	Manufacturing plants								
		1			2			3		
		M1	M2	M3	M1	M2	M3	M1	M2	M3
1	1	15	12	10	20	18	15	25	22	19
	2	25	22	19	30	27	24	35	32	29
	3	18	15	12	28	25	22	33	30	27
	4	35	32	29	40	37	34	45	42	39
2	1	13	10	8	18	16	13	23	20	17
	2	23	20	17	28	25	22	33	30	27
	3	16	13	10	26	23	20	31	28	25
	4	33	30	27	38	35	32	43	40	37
3	1	18	15	13	23	21	18	28	25	22
	2	28	25	22	33	30	27	38	35	32
	3	21	18	15	31	28	25	36	33	30
	4	38	35	32	43	40	37	48	45	42

Note: for scenarios 2 and 3, the evaluations are increased by 0.8 and 1.2 sequentially. It is necessary to mention that M1, M2, and M3 in the above table refer shipping mode 1, shipping mode 2, and shipping mode 3, respectively.

Table 11. The environmental cost of transporting products for scenario 1. ($\times 1000$) (€)

Manufacturing plants	Products	Customer zones														
		1			2			3			4			5		
		M1	M2	M3	M1	M2	M3	M1	M2	M3	M1	M2	M3	M1	M2	M3
1	1	25	22	19	29	26	23	36	33	30	28	25	22	42	39	36
	2	27	24	21	31	28	25	38	35	32	30	27	24	44	41	38
2	1	42	39	36	46	43	40	51	48	45	37	34	31	44	41	38
	2	44	41	38	48	45	42	53	50	47	39	36	33	46	43	40
3	1	22	19	16	26	23	20	32	29	26	25	22	19	44	41	38
	2	24	21	18	28	25	22	34	31	28	27	24	21	46	43	40

Note: for scenarios 2 and 3, the evaluations are increased by 0.8 and 1.2, sequentially.

5.2.2 Results

The modeling and solution processes of the above problem were performed utilizing the branch and bound algorithm obtained via LINGO 16. On a PC core i5–2.30 GHz and 4 GB RAM DDR3 under Windows 7 SP1. Tables 14, 15 and 16 present an perceive the output of data properties by establishing the related weight of every objective function element (v) to 0.5, $\lambda_1 = \lambda_2 = 1$ and the model robustness (ω) to 2500000.

Table 12. The environmental cost of producing products for scenario 1. ($\times 1000$) (€)

Products	Manufacturing plants								
	1			2			3		
	L1	L2	L3	L1	L2	L3	L1	L2	L3
1	80	50	30	85	55	25	90	60	30
2	85	55	25	90	60	30	95	65	35

Note: for scenarios 2 and 3, the evaluations are increased by 0.75 and 1.25, sequentially. It is necessary to mention that L1, L2, and L3 in the above table refer to EPL 1, EPL 2, and EPL 3, respectively.

Table 13. Transportation capacity of raw materials ($\times 10$).

Suppliers	Raw materials	Manufacturing plants								
		1			2			3		
		M1	M2	M3	M1	M2	M3	M1	M2	M3
1	1	800	1200	1600	800	1200	1600	800	1200	1600
	2	1000	1400	1800	1000	1400	1800	1000	1400	1800
	3	1500	1900	2300	1500	1900	2300	1500	1900	2300
	4	500	900	1300	500	900	1200	500	900	1300
2	1	820	1220	1620	820	1220	1620	820	1220	1620
	2	1020	1420	1820	1020	1420	1820	1020	1420	1820
	3	1520	1920	2320	1520	1920	2320	1520	1920	2320
	4	520	920	1320	520	920	1320	520	920	1320
3	1	810	1210	1610	810	1210	1610	810	1210	1610
	2	1010	1410	1810	1010	1410	1810	1010	1410	1810
	3	1510	1910	2310	1510	1910	2310	1510	1910	2310
	4	510	910	1310	510	910	1310	510	910	1310

Table 14. Transportation capacity of products ($\times 10$

Manufacturing plants	Products	Customer zones														
		1			2			3			4			5		
		M1	M2	M3	M1	M2	M3	M1	M2	M3	M1	M2	M3	M1	M2	M3
1	1	50	65	85	50	65	85	50	65	85	50	65	85	50	65	85
	2	55	70	90	55	70	90	55	70	90	55	70	90	55	70	90
2	1	52	67	87	52	67	87	52	67	87	52	67	87	52	67	87
	2	57	72	92	57	72	92	57	72	92	57	72	92	57	72	92
3	1	51	66	86	51	66	86	51	66	86	51	66	86	51	66	86
	2	56	71	91	56	71	91	56	71	91	56	71	91	56	71	91

Table 15. Raw material requirements for products.

Products	Raw materials			
	1	2	3	4
1	2	5	8	5
2	4	9	2	8

Table 16. Production amount of products in the selected production centers under each scenario

Scenario	Production center 1		Production center 2		Production center 3	
	Product 1	Product 2	Product 1	Product 2	Product 1	Product 2
1	7100	5800	100	1800		
2	5760	6080				
3	7960	5262	680			

The result shows that manufacturing plants 1 and 2 can be established. For both of them, EPL 1 is distinguished as a suitable choice regarding economic and environmental features. The production quantity of each product in the selected manufacturing plants under each scenario is shown in Table 15. The amount of products produced in manufacturing plant 1 is way much more than that provided in manufacturing plant 2 in all scenarios. Table 16 specifies how to supply raw materials from suppliers in all scenarios. Obviously, in each scenario, the values can vary. According to the results, in all scenarios, most of the required raw material for manufacturing plants is provided by supplier 2.

Also, Table 17 specifies the market share for each manufacturing plant. As can be seen in the Table 15, manufacturing plant 1 has the highest amount of production. Also, according to Tables 16 and 17, it can be concluded that transportation mode 3 is a better choice for the shipping of raw materials and goods than others. In Tables 15, 16 and 17, the values of blank cells are equal to zero.

As stated before, the model is composed of two goals, which we assigned equal weights to each one. Now, if based on the DM's decision, one of the goals preferred over another, we should allocate more weight to it. Hence, a set of resolutions for the problem can be achieved by various v_1 and v_2. Figure 2 graphically shows the trade-off between Z_1 and Z_2.

5.2.3 The Trade-off Between Solution Robustness and Model Robustness

In the objective functions (33) and (34), ω has the role of getting a trade-off between solution robustness (near to an optimal solution) and model robustness (near to a feasible solution). Infeasibilities in control constraints are allowed using penalties. For example, when $\omega = 0$, δ_{fc}^s in the control restrictions equal to dem_{fc}^s because of the depreciation of goals. In this instance, the amount of production is equal to zero. Obviously, this

Table 17. The movement of raw materials from suppliers to production centers under each scenario.

Scenario 1	Suppliers	Raw materials	Manufacturing plants 1			2			3		
			M1	M2	M3	M1	M2	M3	M1	M2	M3
	1	1	8000	12000	16000						
		2	10000	14000	18000						
		3			10000						
		4	5000	9000	13000						
	2	1			1400	7400					
		2	10200	14200	18200			16700			
		3	15200	19200	23200			4400			
		4	5200	9200	13200		1700	13200			
	3	1									
		2			3100						
		3			800						
		4	5100	9100	13100						
Scenario 2	1	1	7840	12000	16000						
		2	8920	14000	18000						
		3			10000						
		4	5000	9000	13000						
	2	1									
		2	10200	14200	18200						
		3	5840	19200	23200						
		4	5200	9200	13200						
	3	1									
		2									
		3									
		4	5100	9100	8640						

(*continued*)

Table 17. (*continued*)

Scenario 3	1	1	8000	12000	16000				
		2	10000	14000	18000				
		3			10000				
		4	5000	9000	13000				
	2	1		968	1360				
		2	10200	14200	18200		3400		
		3	15200	19200	23200		5440		
		4	5200	9200	13200		3400		
	3	1							
		2			2558				
		3			6604				
		4	5100	9100	13096				

Table 18. The flow of products from manufacturing plants to customer zones under each scenario

	Production centers	Products	Customer zones														
			1			2			3			4			5		
			M1	M2	M3	M1	M2	M3	M1	M2	M3	M1	M2	M3	M1	M2	M3
Scenario 1	1	1			500		350	850	300	650	850	100	650	850	500	650	850
		2			700		100	900		600	900	300	700	900			700
	2	1															100
		2													160	720	920
	3	1															
		2															
Scenario 2	1	1			400		110	850		590	850		430	850	180	650	850
		2			560			800		300	900		620	900	400	700	900
	2	1															
		2															
	3	1															
		2															
Scenario 3	1	1			600		590	850	500	650	850	420	650	850	500	650	850
		2			840		300	900		172	900	550	700	900			
	2	1								160							520
		2															
	3	1															
		2															

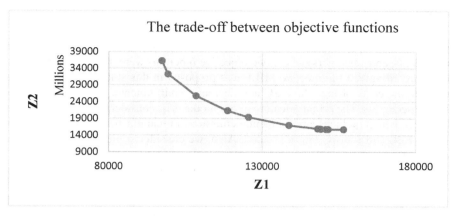

Fig. 2. The trade-off between economic cost and environmental effect

isn't an appropriate solution to the problem. So, it is essential to consider the RO problem with different ω values. Increasing the amount of ω reduces the expected unmet demands (model robustness). However, the predicted whole cost (solution robustness) raises noticeably. Figure 3 shows the trade-off between solution robustness and model robustness. Depending upon the DM's aims, the amount of ω can be set and results in a suitable solution for their perspective.

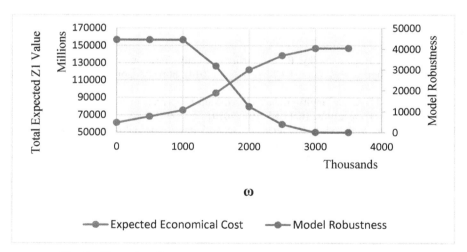

Fig. 3. The trade-off between solution robustness and model robustness

6 Conclusion

This article developed a RO model for a GSCN in three layers (i.e., suppliers, manufacturing plants, and customers) where two objective functions were minimized simultaneously. Accordingly, a bi-objective MILP was proposed in this study, aiming to reduce the economic costs of the whole SC and decrease environmental influences based on different scenarios. The primary feature of our model is its consideration of environmental aspects, which includes environmental emissions in manufacturing plants regarding selected EPL and transportation process. Also, we considered multiple transportation modes in our study.

The robust MO model was changed to a single-objective problem to solve the model, and the LP-metric method was applied. After that, we assess the applicability of the problem by a numerical example. In our study, we considered cost parameters and demands as sources of uncertainty. A real-world case study was analyzed to estimate the applicability of the suggested model in the green supply chain design. Also, the computational results were analyzed in diverse scenarios. The results showed that the second supplier supplied all raw materials for production. Also, we present the trade-off between solution robustness and model robustness, which helps DM make a suitable decision.

For future research in this field, we propose applying heuristic and meta-heuristic algorithms for the solution methods of large-scale problems. Furthermore, the model can be extended to consider multiple periods, and one can added inventory cost, backorder cost, and so on.

References

1. Altmann, M., Bogaschewsky, R.: An environmentally conscious robust closed-loop supply chain design. J. Bus. Econ. **84**(5), 613–637 (2014). https://doi.org/10.1007/s11573-014-0726-4
2. Alinezahd, M.: A new multi-product closed-loop supply chain network design in dairy industry. Iran. J. Oper. Res. **10**(1), 85–101 (2019)
3. Alinezhad, M., Mahdavi, I., Rezaeian, J.: Network design for multi-product, multi-echelon closed-loop supply chain in dairy industry (2017)
4. Chaabane, A., Ramudhin, A., Paquet, M.: Design of sustainable supply chains under the emission trading scheme. Int. J. Prod. Econ. **135**(1), 37–49 (2012)
5. Diabat, A., Simchi-Levi, D.: A carbon-capped supply chain network problem. In: 2009 IEEE International Conference on Industrial Engineering and Engineering Management. IEEE (2009)
6. Fahimnia, B., et al.: Policy insights from a green supply chain optimization model. Int. J. Prod. Res. **53**(21), 6522–6533 (2015)
7. Geoffrion, A.M., Graves, G.W.: Multicommodity distribution system design by Benders decomposition. Manage. Sci. **20**(5), 822–844 (1974)
8. Gholizadeh, H., Fazlollahtabar, H.: Robust optimization and modified genetic algorithm for a closed loop green supply chain under uncertainty: case study in melting industry. Comput. Ind. Eng. **147**, 106653 (2020)
9. Goodarzian, F., Hosseini-Nasab, H., Muñuzuri, J., Fakhrzad, M.B.: A multi-objective pharmaceutical supply chain network based on a robust fuzzy model: a comparison of meta-heuristics. Appl. Soft Comput. **92**, 106331 (2020)

10. Goodarzian, F., Shishebori, D., Nasseri, H., Dadvar, F.: A bi-objective production-distribution problem in a supply chain network under grey flexible conditions. RAIRO-Oper. Res. **55**, S1287–S1316 (2021)

11. Guillén-Gosálbez, G., Grossmann, I.E.: Optimal design and planning of sustainable chemical supply chains under uncertainty. AIChE J. **55**(1), 99–121 (2009)

12. Harris, I., et al.: Assessing the impact of cost optimization based on infrastructure modeling on CO_2 emissions. Int. J. Prod. Econ. **131**(1), 313–321 (2011)

13. Hasani, A., Zegordi, S.H., Nikbakhsh, E.: Robust closed-loop global supply chain network design under uncertainty: the case of the medical device industry, Int. J. Prod. Res. **53**(5), 1596–1624 (2015).

14. Jamshidi, R., FatemiGhomi, S.M.T., Karimi, B.: Multi-objective green supply chain optimization with a new hybrid memetic algorithm using the Taguchi method. ScientiaIranica **19**(6), 1876–1886 (2012)

15. Leung, S.C.H., et al.: A robust optimization model for multi-site production planning problem in an uncertain environment. Eur. J. Oper. Res. **181**(1), 224–238 (2007)

16. Lotfi, R., Yadegari, Z., Hosseini, S.H., Khameneh, A.H., Tirkolaee, E.B., Weber, G.W.: A robust time-cost-quality-energy-environment trade-off with resource-constrained in project management: a case study for a bridge construction project. J. Ind. Manag. Optim. (2020). https://doi.org/10.3934/jimo.2020158

17. Mallidis, I., Dekker, R., Vlachos, D.: The impact of greening on supply chain design and cost: a case for a developing region. J. Transp. Geogr. **22**, 118–128 (2012)

18. Mele, F.D., Guille´n-Gosa´lbez, G., Jime´nez, L., Bandoni, A.: Optimal planning of sustainable supply chain sugar and bioethanol production. In: 10th International Symposium on Process Systems Engineering (PSE 2009), pp. 597–602 (2009)

19. Mulvey, J.M., Ruszczyński, A.: A new scenario decomposition method for large-scale stochastic optimization. Oper. Res. **43**(3), 477–490 (1995)

20. Paksoy, T., et al.: A multi-objective model for optimization of a green supply chain network. In: AIP Conference Proceedings, vol. 1239. No. 1 (2010)

21. Pan, F., Nagi, R.: Robust supply chain design under uncertain demand in agile manufacturing. Comput. Oper. Res. **37**(4), 668–683 (2010)

22. Pishvaee, M.S., Rabbani, M., Torabi, S.A.: A robust optimization approach to closed-loop supply chain network design under uncertainty. Appl. Math. Model. **35**(2), 637–649 (2011)

23. Rahemi, H., Torabi, S.A., Avami, A., Jolai, F.: Bioethanol supply chain network design considering land characteristics. Renew. Sustain. Energy Rev. **119**, 109517 (2020)

24. Ramudhin, A., et al.: Carbon market sensitive green supply chain network design. In: IEEE International Conference on Industrial Engineering and Engineering Management (2008)

25. Ruimin, M.A., et al.: Robust environmental closed-loop supply chain design under uncertainty. Chaos Solitons Fractals **89**, 195–202 (2016)

26. Sarkis, J., Zhu, Q., Lai, K.-H.: An organizational theoretic review of green supply chain management literature. Int. J. Prod. Econ. **130**(1), 1–15 (2011)

27. Seuring, S., Müller, M.: From a literature review to a conceptual framework for the sustainable supply chain management. J. Clean. Prod. **16**(15), 1699–1710 (2008)

28. Tirkolaee, E.B., Abbasian, P., Weber, G.W.: Sustainable fuzzy multi-trip location-routing problem for medical waste management during the COVID-19 outbreak. Sci. Total Environ. **756**, 143607 (2021)

29. Tsai, W.-H., Hung, S.-J.: A fuzzy goal programming approach for green supply chain optimization under activity-based costing and performance evaluation with a value-chain structure. Int. J. Prod. Res. **47**(18), 4991–5017 (2009)

30. Validi, S., Bhattacharya, A., Byrne, P.J.: A case analysis of a sustainable food supply chain distribution system—A multi-objective approach. Int. J. Prod. Econ. **152**, 71–87 (2014)

31. Wang, F., Lai, X., Shi, N.: A multi-objective optimization for green supply chain network design. Decis. Support Syst. **51**(2), 262–269 (2011)
32. Yang, G., Liu, Y.: Designing fuzzy supply chain network problem by mean-risk optimization method. J. Intell. Manuf. **26**(3), 447–458 (2013). https://doi.org/10.1007/s10845-013-0801-7
33. Yu, C.-S., Li, H.-L.: A robust optimization model for stochastic logistic problems. Int. J. Prod. Econ. **64**(1), 385–397 (2000)

The Effects of Individual and Organizational Factors on Creativity in Sustainable Supply Chains

Aidin Delgoshaei[1]([✉]), Aisa Khoshniat Aram[2], and Amir Hossein Nasiri[3]

[1] Department of Mechanical and Manufacturing Engineering, Ryerson University, Toronto, ON M5B 2K3, Canada
[2] Department of Business Administration, Social Science University of Ankara, Ankara, Turkey
[3] Department of Economic and Management, Science and Research Branch, Islamic Azad University, 1477893855 Tehran, Iran

Abstract. Human Resource management is among the most critical issues of supply chain management. However, in today's rivalry world, the supply chains should be flexible enough to adapt to market changes. In this regard, designers and executive engineers play a crucial role in responding to market needs, affecting the supply chain system reconfiguring and developing sustainable systems accordingly. This study evaluates the impact of individual and social harmful factors on creativity inaction period in supply chains. The subjects consisted of 362 experts in supply chains using clustering sampling. For this purpose, two questionnaires are designed based on Mood and Feeling Questionnaire (MFQ). The reliability showed that Cronbach's Alphas were 0.953 and 0.797 for individual and social questionnaires, respectively. Results show that the harmful individual and social factors impose adverse effects on individual employees that cause different inaction periods named short-term, long-term, and organizational death of individual creativity inertia.

Keywords: Supply chain management · Human resource · Quantitative research · Organizational behavior; creativity

1 Introduction

During the last decade, significant efforts have been made for the sustainable development of manufacturing systems. It is proved that sustainable development can enhance supply chain performance in economic, environmental, and social aspects. However, the results of reviewing 220 research studies in sustainable supply chain management carried out by Barbosa-Póvoa et al. (2018) indicated that only economic and environmental factors were considered in most of the reviewed cases, whereas social aspects were ignored. Therefore, this research focuses on the social aspect of sustainability in supply chains. Each year, a noticeable budget is assigned for system development in the U.S (Tompkins et al. 2003). However, some gaps existed to be filled by scientists.

© Springer Nature Switzerland AG 2021
Z. Molamohamadi et al. (Eds.): LSCM 2020, CCIS 1458, pp. 303–318, 2021.
https://doi.org/10.1007/978-3-030-89743-7_17

While the social aspect of sustainable development comes into consideration, various methods can be used. The social aspect of sustainable development can be divided into two main groups: internal and community social aspects.

Raj et al. (2018) focused on the role of simultaneous green activities and corporate social responsibility (CSR) commitments of supply chain members for enhancing the level of sustainability in supply chains. Saberi et al. (2019) focused on the role of blockchains in finding and resolving the vast barriers that may potentially cause failures of a sustainable chain in the early steps of its life cycle. They listed four main barriers, including inter-organizational, intra-organizational, technical, and external barriers. Inter-organizational factors play a crucial role in increasing the performance of the supply chains.

Making appropriate teamwork in workstations is a critical internal social factor that can influence the performance of supply chains. Proper worker assignment can boost the profit of a supply chain (Norman et al. 2002) while optimizing the production rate. A data envelopment analysis has been done by Ertay and Ruan (2005) to make the correct decision of operator assignment and production rate. Lack of operators can be balanced effectively using correct work-sharing (Cesaní and Steudel 2005). Delgoshaei et al. (2016) showed how effective part scheduling could improve the system performance in the presence of system uncertainty.

Skill development is a crucial part of human resource development in sustainability studies (Suer and Cedeño 1996). There are significant research studies carried out to plan a manufacturing system considering the operator's skill. Considering different operator skill levels during worker assignment is an effective method for sustainable development (Aryanezhad et al. 2009). Developing appropriate team works can maximize labor flexibility and minimize training costs (Slomp and Suresh 2005). Manavalan and Jayakrishna (2019) argued that supply chains must increase their speed in developing sustainable systems. In their research, they emphasize the role Internet of Things (IoT) in achieving the goals of a supply chain.

Cross-training a philosophy of enhancing the operator's capability in a work environment (Van Oyen et al. 2001). Olorunniwo and Udo (2002) proved that cross-trained operators could raise the system performance noticeably. Kher (2000) discussed that re-learning programs must be considered in the development scheme of a manufacturing firm. For this purpose, they compared four policies for cross-training planning in manufacturing systems. Human resource scheduling is often complicated and; therefore, heuristic methods are applied for solving them. Askin and Huang (2001) used simulated annealing for simultaneous teamwork development and cross-training planning. Slomp and Molleman (2002) evaluated the impact of operator training on flow time, mean of tardiness, and square means of job tardiness. Sawyer (2011) focused on evaluating the role of leader behavior on creative performance. They showed that leaders have a significant role in problem defining and experts' self-confidence. Delgoshaei and Gomes (2016) proposed a new method using Artificial Neural Networks for scheduling manufacturing systems while uncertainty can affect operator's work assignments.

Using mathematical models is an effective way for human resource planning. Li et al. (2012) proposed a multi-objective model that can decide the correct number of cross-trained laborers and job assignments. Delgoshaei et al. (2017) focused on scheduling multi-skill operators and temporary workers in manufacturing systems. Koberg and Longoni (2019) reported that a better level of sustainability could be achieved by using the overall configuration of supply chains with a more confident connection between leading suppliers and multi-tier suppliers.

The creativity of human resources is an undeniable part of sustainable development in today's rivalry world. However, there are preconceptions that some factors can impose harmful effects on human resource creativeness. This study's main question is: what are the harmful factors in creativity? Furthermore, how deep can such factors affect creativity?

An in-depth survey in the history of HRM problems in SCM shows that staff creativeness in SCM studies is less developed (Barbosa-Póvoa et al. 2018). Moreover, no research was found that investigated the harms of negative factors in human resources in terms of inaction.

Therefore, in this paper, the focus is on individual and organizational factors that can affect the supply chain performance by reducing the personal innovation of human resources. An effort has been made to find out some of the most important individual and social factors, which can cause creative damage in designers and technical experts in supply chain companies. In this manner, a reliable statistical technique is used.

2 Research Methodology

In this section, we focus on the most critical researches about creativeness factors in manufacturing systems. In most scientific books, the word "creativeness" is defined as the power of presenting an idea. In today's rivalry world, successful companies focus on developing new ideas to improve their goods and services. Gardner discussed that user-driven innovation does not only have benefits for firms but has specific benefits for the involved users Gardner (2011). The prerequisite factor for this point of view is both creative personnel and innovative managers in the organization. However, in many companies, although managers are aware of the importance of creativeness, they cannot recognize creative human resources, develop their tendency to be creative, and use their new ideas in the workplace. Developing and using a creative mind requires multidisciplinary approaches. For instance, Blum-Kusterer and Hussain, in their study of 150 German and British pharmaceutical companies, found that regulation and technological progress can be considered as the two main drivers for sustainability innovations Blum-Kusterer and Hussain (2001).

It is clear that all human beings are interested in creativeness, but figuring out the intensity and value of creativeness depends on many individual and circumstance factors. In other words, the same person may show different degrees of creativeness throughout life depending on individual and social factors. It seems that the creativity may get harms which cause creativeness inaction period. Guilford designed a 3 part approach (operation, content, and product) and could create 120 mental factors by knowing subordinate categorizes and the relation between them Guilford (1982).

Traditionally, there were many theories beyond creativeness that mainly investigated psychological aspects of creativeness. Such ideas believed creativeness as a consequence of individual genius that is an inherent factor might be considered God's gift. However, other theories tried to realize cultural, social, historical, and individual facts that influence creativity. Shalley and Perry-Smith (2001) believed that there are attractive and persuasive myths with credence in ordinary minds. During the last four decades, it has been proven that creativity can be developed, directed, and improved despite incorporating social conditions that may damage it. Gardner (2011) argued that investigating people, products, and processes could be carried out in terms of 4 perspectives. They listed them as sub-individual, individual, social, and multi-individual approaches. Davis (2009) performed a field study on evaluating the positive mood enhances creativity factors. Their method is then applied to several actual case studies in the industry. Their outcomes indicated that using positive mood creativity level boosted.

Furthermore, the pattern of effect sizes supported a curvilinear relationship between affective intensity and creative performance. Choi and Thompson (2005) examined the impact of membership change on group creativity. Based on previous literature on the stimulus effects of group membership change, they hypothesized that membership change would increase group creativity. Membership change involved the random rotation of a subset of group members between groups during a series of creative tasks. In two experiments, they compared open group creativity (groups that experienced a change in their membership in performing tasks) with closed groups (groups whose membership was not fixed in performing tasks). In both experiments, it was found that open groups generate more ideas than closed groups and generate different types of ideas.

Shalley and Perry-Smith (2001) focused on psychological factors, named the controlling and informational aspects. In their method, individuals showed higher creativeness while anticipating an informational rather than a controlling evaluation. Besides, those individuals who received a creative example showed better creative performance than others did.

Considering the source of creativeness damage, individual creativeness damage emerging are divided into two main groups in this article: internal and external reasons. Internal reasons or "Individual factors" are based on an individual Oldham and Cummings (1996). Such reasons will cause a suitable or unsuitable reaction while confronting a situation. Hereditary and acquisitive psychological diseases, especially characteristic diseases, neurotic psychological diseases, social behaviors or habits (such as obedience mentality and not concentrating power), are reasons for internal-based creativeness inaction. In contrast, social factors or external creativeness harm factors, which are acquisitive issues from the environment, seem to have destructive effects on creativity power. Such harms have a vast range and may become different from organization to organization.

In this research, 24 individual and social factors are investigated using two separate questionnaires. We considered eight factors as individual factors in our survey, which are: Gender, age, Defeat fear, disappointment, inflexibility, work-accustom, shyness, obeying morals.

Besides, 16 factors are considered as social factors, which are: Boss Engagement, The Hierarchy Regulations in Organizations (written or unwritten), Organizational Position, Job Traditions, Job Structure, Traditions of Region (human beliefs), Organizational Customs, Loss of Persuasion and Supporting System, Improper Persuasion and Supporting System, Underground Networks and Mafia Gangs, External Standards and Regulations,

Existence of Other Creative Staff in Organization, Other's Scoff, Manager's Immediate Justification, Manager's Obstruction, Staff Grading.

In this research, we used a field study for data gathering. The questionnaires were distributed among 362 individuals (both males and females) in supply chain-based companies (mainly automotive and food manufacturing systems) located in IRAN. The number of samples is estimated using Cochran's sampling formula:

$$n = (Z^2.P.Q/d^2)/(1 + (1/N)(Z^2.P.Q/d^2 - 1)) = 361.1 \sim 362 \qquad (1)$$

p: q = 0.5 (whether an individual experienced creativity harms or not).
$\alpha = 0.05$.
Confidence level = 95%
N: approximately 6000 industrial firms which are located in 7 industrial zones.
$Z^2(\alpha/2) = 1.96$.
d = 0.05.

The primary presupposition in this research is that the sample members are chosen among the statistical population in various industries, primarily motor vehicle and food, follow the normal distribution. Such presupposition means that all members have some relative creativity tendencies.

2.1 Reliability Estimating

To evaluate the quality of the designed questionnaires before using them in actual cases, the Delphi technique was employed. All experts were selected from think tank rooms, R&D department staff of big companies, and senior lecturers of some Universities in Tehran. Internal consistency (alpha) coefficients were calculated using *SPSS*® 16.0 for both individual and social factors. Cronbach's Alphas showed 0.953 and o.797 for individual and social questionnaires, respectively, revealing that the designed questions are reliable enough to be used (Table 1 and Table 2).

Table 1. Reliability statistics for individual questionnaire

Number of items	Cronbach's alpha (Standardized)	Cronbach's alpha
8	.952	.953

Table 2. Reliability statistics for social questionnaire

Number of items	Cronbach's alpha (standardized)	Cronbach's alpha
16	.794	.797

The subjects are then asked to fill up the questionnaires where the intensity of each factor was asked as a question that follows the Likert Scale. The subjects are then asked to specify the period of inaction they experienced (or may experience) while encountering each factor. Afterward, the filled-up forms were gathered and analyzed using SPSS 16.0.

3 Results and Discussion

The results show that subjects can considerably understand differences of harm factors and their impact on creativity power. The results also demonstrated that subjects experienced all inaction periods during their work-life. Based on the data, the inaction period can fall into three-period sections: Short-term Period of Creativity Inertia, Long-term Creativity Inaction, and Organizational Creativity Death.

In the short-term period of individual creativeness inaction, which can start from one day to a few weeks, the employee may feel lost in creating new ideas or tend to create new ones due to some internal or external malformations. However, he/she may show some ideas in this period but not as frequently as before. 51% of subjects believed that continuing reasons of short time individual creativeness inaction could cause shifting to long-term inaction, preventing an employee from developing new ideas and can last from one month to one year.

Organizational Creativity Death of individual creativeness, which was experienced by 30% of subjects (both males and females), is caused by continuing such factors that make employee has no enthusiasm about the workplace. Consequently, he/she will avoid finding new ways of solving problems. This period will continue until the organization ends organization and will be removed by transmitting to another organization. The following two tables show how individual and social factors influence people's creative power and how the population evaluated and ranked these factors.

Table 3. Results gained for causes of individual creativity harm and their impact on creativity inertia (significant level: 5%)

Abbr	Factor	$\overline{X}$	Std.Div.	Confidence interval		Inertia (%)		
				L-bound	U-bound	STP*	LTP**	OCD***
01	Gender	2.446	1.677	2.318	2.574	68.99	29.11	1.90
02	Age	3.367	1.561	3.248	3.486	44.30	52.53	3.16
03	Defeat fear	4.383	1.547	4.265	4.501	32.91	40.51	26.58
04	Inflexibility	5.584	1.528	5.468	5.701	10.13	37.34	52.53
05	Disappointment	5.961	1.643	5.836	6.087	10.76	25.32	63.92
06	Work-accustom	8.006	1.256	7.910	8.102	1.90	0.00	98.10
07	Shyness	5.761	1.702	5.631	5.891	15.82	24.05	60.13
08	Obeying moral	6.637	1.194	6.546	6.728	0.63	17.72	81.65

Table 3 shows the intensity of individual harm factors and their results to the inertia of creativity. It is noticeable that subjects chose Work Accustom as the top-ranked individual factor, which causes creativity inertia, but at the same time, they did not believe that Gender and Age can play a significant role in destroying the creative mind. Moreover, most of them believed that individual factors usually set them to short-term creativity in action, and 98.1% of studied cases; they believed that Work Accustom

Table 4. Results gained for causes of social creativity harm and their impact on creativity inertia (significant level: 5%)

Abbr	factor	$\overline{X}$	Std. Div.	Confidence interval		Inertia (%)		
				L-bound	U-bound	STP*	LTP**	OCD***
09	Boss engagement	5.873	1.583	5.752	5.994	1.27	67.09	31.65
10	Hierarchy regulations	5.095	1.409	4.987	5.203	15.19	68.35	16.46
11	Organizational position	5.076	1.285	4.978	5.174	8.86	77.85	13.29
12	Job traditions	3.525	1.133	3.439	3.612	39.24	60.13	0.63
13	Job structure	4.443	1.062	4.362	4.524	14.56	82.28	3.16
14	Traditions of the region	3.367	1.191	3.276	3.458	36.71	62.03	1.27
15	Other's scoff	4.614	1.538	4.496	4.731	27.22	57.59	15.19
16	Organizational customs	4.525	0.935	4.454	4.597	12.03	85.44	2.53
17	Loss of Persuasion and supporting system	4.804	0.947	4.731	4.876	7.59	87.34	5.06
18	Improper persuasion and supporting system	6.614	1.188	6.523	6.705	0.00	43.67	56.33
19	Underground networks and mafia gangs	5.975	1.450	5.864	6.085	0.63	76.58	22.78
20	External standards and regulations	5.462	1.439	5.352	5.572	8.23	70.89	20.89
21	Other creative staff	3.854	1.475	3.742	3.967	50.00	46.20	3.80
22	Immediate justification	8.057	1.283	7.959	8.155	0.00	9.49	90.51
23	Manager's obstruction	4.633	1.030	4.554	4.712	13.29	84.81	1.90
24	Staff grading	7.133	1.089	7.050	7.216	0.00	25.95	74.05

*STP: Short term of inertia creativeness
**LTP: Long term period of individual inertia creativeness.
***ODC: Organizational creativity death.

could shift them directly to the Organizational Creativity Death. Table 3 also suggests a two-tail confidence interval of each factor's assessment while the confidence coefficient is supposed 95%. ($\alpha = 0.05$).

Table 4, in contrast, reveals the results of social factors that can influence individual creativity. It is observed that most of the subjects believed that Immediate Justification is the most significant reason that affects their sense of creativity through their work-life. The responders also believed that their managers do not show enough gratitude for their new ideas but also reject them immediately without letting the owner of the idea develop it. Six other factors were considered important factors: Staff Grading, Improper Persuasion and Supporting System, Underground Networks and Mafia Gangs, Boss Engagement, External Standards and Regulations, and Hierarchy Regulations, respectively.

Despite individual factors, the statistical population believed that almost all the social harm factors, except Immediate Justification and Staff Grading, will cause long-term creativity in action.

As another question, we asked both groups of males and female volunteers to answer this question: "when did you present your last new idea to your company?" which reveals that in what category of creativeness they are located. Figure 1 compares results of creativity in action in both males and females:

Table 6 illustrates that the mean of males and females groups are significantly different in some factors such as organizational position, job structure, other's scoff, and external standards and regulations. Such difference means that males and females have different experiences (or feelings) in some factors. In contrast, other factors show no significant mean differences among male and female groups. In addition, those factors that have a significant value of homogeneity of variances between males and females will allow us to have close ideas regarding these factors, which means population member's experiences are similar.

Table 5. ANOVA test of creativeness inaction between males and females

Row	Factor	F	Sig.	Result	Homogeneity of variances		Variance equality
					Levene statistics	Sig	
1	Gender	.883	.502	N*	2.242	.083	Rejected
2	Age	.882	.538	N	1.357	.267	Rejected
3	Defeat fear	1.104	.375	N	.318	.864	Equal Var
4	Inflexibility	.807	.529	N	.864	.505	Rejected
5	Disappointment	1.263	.300	N	1.073	.397	Rejected
6	Work-accustom	3.667	.094	N	2.232	.075	Rejected
7	Shyness	1.205	.327	N	2.195	.077	Rejected
8	Obeying moral	1.098	.362	N	1.278	.296	Rejected
9	Boss engagement	.828	.556	N	1.053	.402	Rejected
10	Hierarchy regulations	.724	.610	N	1.248	.307	Rejected
11	Organizational position	.386	.858	SD**	1.383	0.253	Rejected
12	Job traditions	1.423	.294	N	.874	.488	Rejected
13	Job structure	.266	.949	SD	1.669	.167	Rejected
14	Traditions of the region	1.699	0.170	N	.122	.946	Equal Var
15	Other's scoff	.428	.855	SD	.783	.543	Rejected
16	Organizational customs	1.499	.230	N	1.571	.212	Rejected
17	Loss of Persuasion/supporting system	2.081	.102	N	2.648	.063	Rejected
18	Improper Persuasion/supporting system	.696	.599	N	2.061	.105	Rejected
19	Underground networks and mafia gangs	2.686	.029	N	4.058	.050	Rejected
20	External standards and regulations	.606	.695	SD	2.241	.071	Rejected
21	Other creative staff	1.681	.146	N	.462	.763	Equal Var
22	Immediate justification	.770	.587	N	.695	.600	Rejected
23	Manager's obstruction	.794	.561	N	1.038	.387	Rejected
24	Staff grading	1.610	.182	N	2.190	.106	Rejected

* SD: significantly different ** N: No significance difference

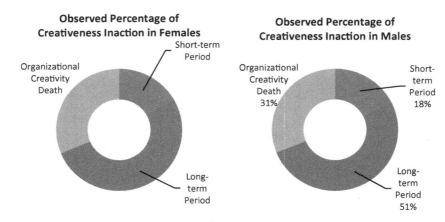

Fig. 1. Results of comparing male and female's creativity inertia

Since most of the mentioned challenges are influenced by other factors, it is impossible to ignore the interaction between factors. For this purpose, a two-tailed Pearson correlation of factors is calculated (Table 7 and Table 8). Results showed that some of the factors, which are effective on individual creativeness inaction, are relative to each other. Increasing the intensity of these factors may increase the intensity of the other and cause strengthen in violence and a period of individual creativeness inaction. Table 6 and Table 7 show correlations and reciprocal effects of individual and social factors ($p \leq 0.01$; $p \leq 0.05$).

The findings reveal that all of the individual factors have interrelationships at the significant level of 0.01. A similar trend is observed for social harm factors, but not in all cases making this paradigm strengthen that simultaneous happening of these factors causes more significant trouble for creative minds (like Organizational Position and Other's Scoff). Only a few factors are found independent (Job Structure and Existence of Underground Mafia Networks).

Table 6. Interaction among individual factors

Factor		01	02	03	04	05	06	07	08
01	Pearson Corrl	1	.914**	.847**	.720**	.865**	.773**	.926**	.647**
	Sig. 2-tailed		.000	.000	.000	.000	.000	.000	.000
	N	360	360	357	357	358	358	358	156
02	Pearson Corrl	.914**	1	.805**	.689**	.793**	.703**	.836**	.630**
	Sig. 2-tailed	.000		.000	.000	.000	.000	.000	.000
	N	360	362	358	358	359	359	359	360

(continued)

Table 6. (*continued*)

Factor		01	02	03	04	05	06	07	08
01	Pearson Corrl	1	.914**	.847**	.720**	.865**	.773**	.926**	.647**
03	Pearson Corrl	.847**	.805**	1	.633**	.771**	.676**	.806**	.544**
	Sig. 2-tailed	.000	.000		.000	.000	.000	.000	.000
	N	357	358	358	355	356	356	356	357
04	Pearson Corrl	.720**	.689**	.633**	1	.673**	.609**	.712**	.472**
	Sig. 2-tailed	.000	.000	.000		.000	.000	.000	.000
	N	357	358	355	358	356	356	356	357
05	Pearson Corrl	.865**	.793**	.771**	.673**	1	.674**	.799**	.611**
	Sig. 2-tailed	.000	.000	.000	.000		.000	.000	.000
	N	358	359	356	356	359	357	357	358
06	Pearson Corrl	.773**	.703**	.676**	.609**	.674**	1	.757**	.537**
	Sig. 2-tailed	.000	.000	.000	.000	.000		.000	.000
	N	358	359	356	356	357	359	357	358
07	Pearson Corrl	.926**	.836**	.806**	.712**	.799**	.757**	1	.611**
	Sig. 2-tailed	.000	.000	.000	.000	.000	.000		.000
	N	358	359	356	356	357	357	359	358
08	Pearson Corrl	.647**	.630**	.544**	.472**	.611**	.537**	.611**	1
	Sig. 2-tailed	.000	.000	.000	.000	.000	.000	.000	
	N	362	360	357	357	358	358	358	360

[*] Correlation is significant at the 0.05 level (2-tailed) ** Correlation is significant at the 0.01 level (2-tailed)

Table 7. Interaction among social factors

Factor		09	10	11	12	13	14	15	16	17	18	19	20	21	22	23	24
09	Pearson Crrl	1	.702**	.622**	.087	.375**	.032	.618**	.299**	.514**	.177*	.054	.856**	-.032	.465**	.174*	.320**
	Sig. 2-tailed		.000	.000	.277	.000	.694	.000	.000	.000	.026	.500	.000	.685	.000	.028	.000
	N	362	362	362	362	362	362	362	362	362	362	362	362	362	362	362	362
10	Pearson Crrl	.702**	1	.309**	.016	.321**	-.013	.408**	.160*	.382**	.209**	.051	.625**	-.082	.307**	.068	.241**
	Sig. 2-tailed	.000		.000	.837	.000	.868	.000	.045	.000	.009	.524	.000	.305	.000	.396	.002
	N	362	362	362	362	362	362	362	362	362	362	362	362	362	362	362	362
11	Pearson Crrl	.622**	.309**	1	.038	.083	-.010	.527**	.141	.347**	.136	.138	.570**	-.004	.357**	.223**	.321**
	Sig. 2-tailed	.000	.000		.635	.302	.901	.000	.076	.000	.088	.084	.000	.958	.000	.005	.000
	N	362	362	362	362	362	362	362	362	362	362	362	362	362	362	362	362
12	Pearson Crrl	.087	.016	.038	1	.218**	.791**	-.029	.050	.031	-.085	.024	.049	.603**	-.051	.041	-.016
	Sig. 2-tailed	.277	.837	.635		.006	.000	.717	.529	.696	.288	.768	.537	.000	.521	.611	.845
	N	362	362	362	362	362	362	362	362	362	362	362	362	362	362	362	362
13	Pearson Crrl	.375**	.321**	.083	.218**	1	.173*	.051	.412**	.296**	.142	-.026	.286**	.143	.117	.027	.031
	Sig. 2-tailed	.000	.000	.302	.006		.030	.526	.000	.000	.076	.748	.000	.073	.143	.733	.696
	N	362	362	362	362	362	362	362	362	362	362	362	362	362	362	362	362
14	Pearson Crrl	.032	-.013	-.010	.791**	.173*	1	-.058	.020	.064	-.021	.087	.027	.571**	-.047	.017	-.018
	Sig. 2–tailed	.694	.868	.901	.000	.030		.471	.801	.423	.796	.279	.739	.000	.557	.831	.820
	N	362	362	362	362	362	362	362	362	362	362	362	362	362	362	362	362

(continued)

Table 7. (*continued*)

Factor		09	10	11	12	13	14	15	16	17	18	19	20	21	22	23	24
15	Pearson Crrl	.618**	.408**	.527**	−.029	.051	−.058	1	.257**	.498**	.092	.138	.539**	−.126	.327**	.119	.282**
	Sig. 2−tailed	.000	.000	.000	.717	.526	.471		.001	.000	.249	.083	.000	.115	.000	.136	.000
	N	362	362	362	362	362	362	362	362	362	362	362	362	362	362	362	362
16	Pearson Crrl	.299**	.160*	.141	.050	.412**	.020	.257**	1	.563**	.103	−.056	.230**	−.004	.102	−.017	.050
	Sig. 2−tailed	.000	.045	.076	.529	.000	.801	.001		.000	.196	.486	.004	.958	.201	.835	.534
	N	362	362	362	362	362	362	362	362	362	362	362	362	362	362	362	362
17	Pearson Crrl	.514**	.382**	.347**	.031	.296**	.064	.498**	.563**	1	.085	.043	.473**	−.039	.287**	.050	.211**
	Sig. 2−tailed	.000	.000	.000	.696	.000	.423	.000	.000		.288	.594	.000	.628	.000	.535	.008
	N	362	362	362	362	362	362	362	362	362	362	362	362	362	362	362	362
18	Pearson Crrl	.177*	.209**	.136	−.085	.142	−.021	.092	.103	.085	1	.024	.176*	−.021	.353**	.175*	.188*
	Sig. 2−tailed	.026	.009	.088	.288	.076	.796	.249	.196	.288		.766	.027	.790	.000	.028	.018
	N	362	362	362	362	362	362	362	362	362	362	362	362	362	362	362	362
19	Pearson Crrl	.054	.051	.138	.024	−.026	.087	.138	−.056	.043	.024	1	.058	.046	.299**	.113	.228**
	Sig. 2−tailed	.500	.524	.084	.768	.748	.279	.083	.486	.594	.766		.473	.567	.000	.157	.004
	N	362	362	362	362	362	362	362	362	362	362	362	362	362	362	362	362
20	Pearson Crrl	.856**	.625**	.570**	.049	.286**	.027	.539**	.230**	.473**	.176*	.058	1	.005	.427**	.201*	.278**
	Sig. 2−tailed	.000	.000	.000	.537	.000	.739	.000	.004	.000	.027	.473		.951	.000	.011	.000
	N	362	362	362	362	362	362	362	362	362	362	362	362	362	362	362	362

(*continued*)

Table 7. (*continued*)

Factor		09	10	11	12	13	14	15	16	17	18	19	20	21	22	23	24
21	Pearson Crrl	-.032	-.082	-.004	.603**	.143	.571**	-.126	-.004	-.039	-.021	.046	.005	1	-.006	.065	-.035
	Sig. 2–tailed	.685	.305	.958	.000	.073	.000	.115	.958	.628	.790	.567	.951		.943	.416	.658
	N	362	362	362	362	362	362	362	362	362	362	362	362	362	362	362	362
22	Pearson Crrl	.465**	.307**	.357**	-.051	.117	-.047	.327**	.102	.287**	.353**	.299**	.427**	-.006	1	.536**	.592**
	Sig. 2–tailed	.000	.000	.000	.521	.143	.557	.000	.201	.000	.000	.000	.000	.943		.000	.000
	N	362	362	362	362	362	362	362	362	362	362	362	362	362	362	362	362
23	Pearson Crrl	.174*	.068	.223**	.041	.027	.017	.119	-.017	.050	.175*	.113	.201*	.065	.536**	1	.362**
	Sig. 2–tailed	.028	.396	.005	.611	.733	.831	.136	.835	.535	.028	.360	.011	.416	.000		.000
	N	362	362	362	362	362	362	362	362	362	362	362	362	362	362	362	362
24	Pearson Crrl	.320**	.241**	.321**	-.016	.031	-.018	.282**	.050	.211**	.188*	.228**	.278**	-.035	.592**	.362**	1
	Sig. 2–tailed	.000	.002	.000	.845	.696	.820	.000	.534	.008	.018	.004	.000	.658	.000	.000	
	N	362	362	362	362	362	362	362	362	362	362	362	362	362	362	362	362

4 Conclusions

In this article, the impacts of individual and social factors on creativity inaction in supply chain systems are investigated. The findings show that the intensity of the factors significantly impacts the period of individual creativeness inaction in employees. It is also observed that focusing on factors that can cause individual creativeness inaction; creativity inertia can be categorized into three time series: short-term individual creativeness inertia, long-term individual creativeness inaction, and organizational creativity death. Most employees believed that accustoming to work circumstances and managers' negative points of view negatively impacted creativity inertia. It is observed that in such circumstances, in studied subjects, most of the creative staff believed that they would quickly pass from short-term to a long-term level of creativity inaction or even to organizational creativity death levels just before they stay enough in this level to find a shelter.

In some cases, effective factors of individual creativeness harm have strong interactions. Such interactions caused more influences to create harm and shifted victims from lower inaction to higher ones. Hence, finding ways to persuade creative minds to prevent suffering from such harmful factors and find shelters to relieve them from creativity inactions. Further expansion of this research in the mentioned areas (both theoretical and empirical) is suggested for future studies.

Acknowledgments. The authors would like to sincerely appreciate the editor and anonymous reviewers for their constructive comments during publication.

References

Aryanezhad, M., Deljoo, V., Mirzapour Al-e-hashem, S.: Dynamic cell formation and the worker assignment problem: a new model. Int. J. Adv. Manuf. Technol. **41**(3–4), 329–342 (2009)

Askin, R., Huang, Y.: Forming effective worker teams for cellular manufacturing. Int. J. Prod. Res. **39**(11), 2431–2451 (2001)

Barbosa-Póvoa, A.P., da Silva, C., Carvalho, A.: Opportunities and challenges in sustainable supply chain: an operations research perspective. Eur. J. Oper. Res. **268**(2), 399–431 (2018)

Blum-Kusterer, M., Hussain, S.S.: Innovation and corporate sustainability: an investigation into the process of change in the pharmaceuticals industry. Bus. Strateg. Environ. **10**(5), 300–316 (2001)

Cesaní, V.I., Steudel, H.J.: A study of labor assignment flexibility in cellular manufacturing systems. Comput. Ind. Eng. **48**(3), 571–591 (2005)

Choi, H.-S., Thompson, L.: Old wine in a new bottle: impact of membership change on group creativity. Organ. Behav. Hum. Decis. Process. **98**(2), 121–132 (2005)

Davis, M.A.: Understanding the relationship between mood and creativity: a meta-analysis. Organ. Behav. Hum. Decis. Process. **108**(1), 25–38 (2009)

Delgoshaei, A., Ali, A., Ariffin, M.K.A., Gomes, C.: A multi-period scheduling of dynamic cellular manufacturing systems in the presence of cost uncertainty. Comput. Ind. Eng. **100**, 110–132 (2016)

Delgoshaei, A., Ariffin, M.K.A., Ali, A.: A multi-period scheduling method for trading-off between skilled-workers allocation and outsource service usage in dynamic CMS. Int. J. Prod. Res. **55**(4), 997–1039 (2017)

Delgoshaei, A., Gomes, C.: A multi-layer perceptron for scheduling cellular manufacturing systems in the presence of unreliable machines and uncertain cost. Appl. Soft Comput. **49**, 27–55 (2016)

Ertay, T., Ruan, D.: Data envelopment analysis based decision model for optimal operator allocation in CMS. Eur. J. Oper. Res. **164**(3), 800–810 (2005)

Gardner, H.: Creating Minds: An Anatomy of Creativity Seen Through the Lives of Freud, Einstein, Picasso, Stravinsky Graham, and Gandhi. Basic Books, Eliot (2011)

Guilford, J.: Cognitive psychology's ambiguities: some suggested remedies. Psychol. Rev. **89**(1), 48 (1982)

Kher, H.V.: Examination of flexibility acquisition policies in dual resource constrained job shops with simultaneous worker learning and forgetting effects. J. Operat. Res. Soc. **51**, 592–601 (2000)

Koberg, E., Longoni, A.: A systematic review of sustainable supply chain management in global supply chains. J. Clean. Prod. **207**, 1084–1098 (2019)

Li, Q., Gong, J., Fung, R.Y., Tang, J.: Multi-objective optimal cross-training configuration models for an assembly cell using non-dominated sorting genetic algorithm-II. Int. J. Comput. Integr. Manuf. **25**(11), 981–995 (2012)

Manavalan, E., Jayakrishna, K.: A review of Internet of Things (IoT) embedded sustainable supply chain for industry 4.0 requirements. Comput. Ind. Eng. **127**, 925–953 (2019)

Norman, B.A., Tharmmaphornphilas, W., Needy, K.L., Bidanda, B., Warner, R.C.: Worker assignment in cellular manufacturing considering technical and human skills. Int. J. Prod. Res. **40**(6), 1479–1492 (2002)

Oldham, G.R., Cummings, A.: Employee creativity: personal and contextual factors at work. Acad. Manag. J. **39**(3), 607–634 (1996)

Olorunniwo, F., Udo, G.: The impact of management and employees on cellular manufacturing implementation. Int. J. Prod. Econ. **76**(1), 27–38 (2002)

Raj, A., Biswas, I., Srivastava, S.K.: Designing supply contracts for the sustainable supply chain using game theory. J. Clean. Prod. **185**, 275–284 (2018)

Saberi, S., Kouhizadeh, M., Sarkis, J., Shen, L.: Blockchain technology and its relationships to sustainable supply chain management. Int. J. Prod. Res. **57**(7), 2117–2135 (2019)

Sawyer, R.K.: Explaining Creativity: The Science of Human Innovation. Oxford University Press, Oxford (2011)

Shalley, C.E., Perry-Smith, J.E.: Effects of social-psychological factors on creative performance: the role of informational and controlling expected evaluation and modeling experience. Organ. Behav. Hum. Decis. Process. **84**(1), 1–22 (2001)

Slomp, J., Molleman, E.: Cross-training policies and team performance. Int. J. Prod. Res. **40**(5), 1193–1219 (2002)

Slomp, J., Suresh, N.C.: The shift team formation problem in multi-shift manufacturing operations. Eur. J. Oper. Res. **165**(3), 708–728 (2005)

Suer, G.A., Cedeño, A.A.: A configuration-based clustering algorithm for family formation. Comput. Ind. Eng. **31**(1), 147–150 (1996)

Tompkins, J., White, J., Bozer, Y., Tanchoco, J.: Facilities Planning. Wiley, New York (2003)

Van Oyen, M.P., Gel, E.G., Hopp, W.J.: Performance opportunity for workforce agility in collaborative and noncollaborative work systems. IIE Trans. **33**(9), 761–777 (2001)

Mapping the Supply Chain Resilience Enablers (SCRE) Model with a Hybrid (FDelphi-ISM-DEMATEL) Approach

Ammar Feyzi[✉]

Industrial Management, Islamic Azad University, Saveh, Iran

Abstract. Resilience refers to the ability to deal with unexpected disturbances or the ability of the system to return to its original state or to a more favorable state than before, after experiencing a disturbance and avoiding the occurrence of failure states. This study aimed to improve the level of resilience of Iran Khodro Company (IKCO) by identifying Resilience Supply Chain Enablers (RSCE) with Fuzzy Delphi technique (FD), developing a RSC model and classifying identified factors using Interpretive Structural Modeling (ISM) and investigating the intensity of their relations with Decision Making Trial And Evaluation Laboratory (DEMATEL) technique. The innovation of this study is to identify RSCE using FD technique, leveling and model design using ISM model and investigating the intensity of communication between these components using DEMATEL technique. Following the literature review, RSCE were identified and screened using FD technique and expert opinions. ISM and influence-dependence matrix were then used. The variables were classified in six different levels and the ISM graph was depicted based on the available relations. The DEMATEL technique was then used to investigate the causal relations between all RSCE. The results help automotive industry managers analyze resilience and select effective strategies to reduce supply chain risks. The results can also facilitate decision making and decision taking processes.

Keywords: Enablers · Supply Chain Resilience (SCR) · Fuzzy Delphi (FD) · Interpretative Structural Modeling (ISM) · Influence-dependency power matrix · Decision Making Trial and Evaluation Laboratory (DEMATEL)

1 Introduction

Today, competition between individual companies is replaced by competition between supply chains; in other words, a network of companies interact to convert raw materials into final products and to deliver them to the customer. This network of entities responsible for various processes, from supplying, production, storage and distribution, is known as supply chain [1]. In a supply chain, any activity is associated with an inherent risk that may cause some disruptions. There are different kinds of risks that are mainly caused either by external factors (e.g. natural disasters) or by internal factors (e.g. equipment failures). Events such as loss of vital suppliers, fire in a manufacturing plant or terrorist attacks can significantly affect costs and income levels [2]. In order to cope with various

© Springer Nature Switzerland AG 2021
Z. Molamohamadi et al. (Eds.): LSCM 2020, CCIS 1458, pp. 319–338, 2021.
https://doi.org/10.1007/978-3-030-89743-7_18

risks, a supply chain must be designed in a way to be prepared to face such disruptions and to respond them efficiently and effectively. In addition, it should be capable of recovering from disruption and moving toward a normal or even better state. This feature represents the importance of resilience in supply chains [3]. Automobile manufacturing, as the world's largest manufacturing activity, has been subject to profound changes in recent years. These changes began in Japan (as a leading country in the automotive industry) by providing superior techniques in the areas of quality, repair, maintenance and inventory control. These changes forced all automobile manufacturers around the world to rebuild themselves in order to survive [4]. Due to its deep connections with other key economic sectors, the automotive industry has a special place in the national economy of Iran.

This industry has the potential to become a driving force for economic growth. The gradual opening of the Iranian economy to foreign competitors in recent years has forced Iran's domestic automobile manufacturers to compete with international manufacturers [5]. In this research, IKCO was the case study. This company is the largest automaker in the Middle East and the livelihood of more than one million people depends directly or indirectly on this company. Despite its long (half a century) history, this company has often teetered on the brink of bankruptcy. Frequent stops have imposed huge losses or lost profits on shareholders in recent years. The main issues addressed by the present research include poor knowledge of industrial and manufacturing managers of IKCO about supply chain resilience enablers as well as about the intensity of their relations. The main research question is that what resilience enablers should be taken into consideration in the supply chain of IKCO. This study aimed to improve the level of resilience of Iran Khodro Company (IKCO) by identifying supply chain resilience enablers with Fuzzy Delphi technique, developing a resilient supply chain model and classifying identified factors using Interpretive Structural Modeling (ISM) and investigating the intensity of their relations with Decision Making Trial And Evaluation Laboratory (DEMATEL) technique.

The innovative aspect of the present study is utilizing a hybrid qualitative and quantitative approach, Fuzzy Delphi technique, ISM and DEMATEL technique. So far, many researchers have investigated resilience of various supply chains [6–18]; however, no study has been conducted to investigate the supply chain resilience enablers in IKCO and this indicates the novelty of the present research.

2   Research Theoretical Fundamentals

Resilience studies are rooted in the social psychology theory, which is an emerging theory. The concept of resilience is directly associated with important issues such as social and environmental vulnerability, disaster recovery policies and psychology and risk management in critical situations. Numerous definitions of resilience have been presented in different disciplines; however, no explicit definition has been presented so far [8].

2.1   Concept and Definitions of Supply Chain Resilience

Many organizations have designed their supply chain with the aim of cost minimization or service optimization. Today, markets are continuously influenced by environmental and external activities and the concept of resilience has been introduced as an approach for coping with high levels of instability and turbulence [9]. The literal definition of resilience is as follows: "the ability of a substance to spring back into shape after deformation (bending, pulling and compressing)". In another definition, resilience is defined as: "the ability of a system to return to its original or even better state after being disturbed or its ability to recover from large-scale disruptions". Various definitions have been presented for the term "resilience" in different disciplines [10]. Table 1 presents these definitions.

Table 1. Resilience definitions

Rsearcher	Field	Definition of resilience
[10]	Environmental systems	How and speed of initial structure recovery
[11]		The system's sustainability criteria, and the ability to absorb changes and disruptions as well as maintain relationships between variables
[13]	Social systems - environmental	The degree of tolerance is the severity of disorder, before transformation transforms the system into a different state
[14]	Engineering	The body's ability to recover after any form of change
[7]	Psychology	The ability to return from the troubles
[15]		Expandable capacity to return from a difficult situation or disadvantage
[11]	Supply chain	Inhibition and recovery of disturbance
[12]		The ability of the system to return to the initial state or a more favorable and newer state of the past after the disruption
[3]		The capacity of complex industrial systems to survive, adapt and grow in turbulent and changing conditions
[16]		The ability to maintain and restore operations after a disturbance
[13]		The supply chain ability to reduce the likelihood of disruption and its consequences in the event of disruption and reduce the recovery time of the normal supply chain performance
[9]		Compatibility of the supply chain to prepare for unwanted events, respond to disruptions and recover them by maintaining the continuity of operations at the optimal level and controlling structure and operation
[18]		Ability to respond to negative impacts of disturbance to maintain supply chain goals

2.2 Principles of Resilient Supply Chain Design

One of the principles that need to be taken into account during the design and reengineering of a supply chain is adopting strategies to keep different options open in order to create and improve resilience. These strategies impose costs on systems; however, they can reduce the consequences of disruptions as well. Obviously, despite reducing costs, centralized strategies increase vulnerability. Excess inventory and capacities are often regarded as waste and undesirable; however, strategic excess inventory and capacities are very important in creating supply chain resilience at bottlenecks [14]. Therefore, it is vital to analyze the benefits and costs of both inventory shortages and excesses. It is not recommended to create buffer inventory at all supply chain stages through excess capacity and precautionary inventory; however, it is essential to use them strategically and selectively to create supply chain resilience [15]. Figure 1 illustrates the principles of designing a resilient supply chain.

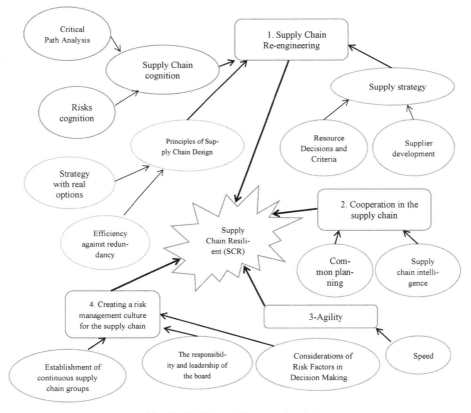

Fig. 1. Build a resilient supply chain

2.3   Research Background

Despite the increasing number of studies in the area of supply chain resilience, a review on the research literature shows that unfortunately the issue of resilience enablers has not been adequately addressed. The major studies conducted in this area are presented in Table 2.

Table 2. Supply chain resilience research

Authors/Year	Title of research	Variables/Results and findings
[34]	Empirical investigation of data analytics capability and organizational flexibility as complements to supply chain resilience	The supply chain resilience and data analytics capability has generated increased interest in academia and among practitioners. The findings of themr study offer a uniquecontribution to information systems (IS) and operations management (OM) literature
[36]	Integrating sustainability and resilience in the supply chain: A systematic literature review and a research agenda	This study completes a systematic literature review that critically examines several major observations and directions. They find the concept of sustainable supply chains is more established, and general agreement on its theoretical foundations exists. Supply chain resilience is relatively less mature
[18]	Design framework for performance improvement resilience supply chain	This study discussed the concept of supply chain resilience and investigated supply chain risks, disruptions, measures, enablers and characteristics. The results showed that a profound understanding of resilience can help managers maintain their competitiveness and improve the performance of their supply chain
[5]	Designing a resonance model in Iran Khodro supply chain with the approach of structural equation modeling and Delphi techniques	This study used the FMEA technique to determine the supply chain resilience strategies before failure. It also used a new technique, called FAAO, to determine the supply chain resilience strategies after failure
[17]	Resiliency measurement of supply chain with systems approach compliant complaint; case study: Iran's pharmaceutical industry	This research aimed to measure the Supply Chain Resilience of pharmaceutical industries (Iran, Daru and Ghazi pharmaceutical companies) using a Complex Adaptive Systems approach as the main research theory. In addition, the levels of resiliency of the two supply chains were assessed using interpretive structural modeling (ISM), DEMATEL, graph theory and matrix approach (GTMA) and importance-performance analysis (IPA)

(continued)

Table 2. (*continued*)

Authors/Year	Title of research	Variables/Results and findings
[24]	Mapping sustainable production model with interpretive structural modelingapproach Fuzzy DEMATEL	In this study, Azar et al. (2017), modeling sustainable production using an interpretive structural approach and studying the intensity of the relationship between variables with fuzzy DMF. The results of his research showed that monitoring and controlling, using high-efficiency resources, using high-tech technology, optimizing the production plan to improve the efficiency of the basic indicators for achieving sustainable production in the refinery improves
[26]	Production model with interpretive structural modeling important factors in development green products	The main purpose of the present research is to provide a method for assessing the environmental impacts of health products and also to identify and develop green products. In this study, interpretative structural modeling, fuzzy multidimensional decision-making, Chang's fuzzy development analysis, and fuzzy hierarchical TOPSIS were used. The results showed that the selection of raw materials for the production of carbide and the end of life were ranked first to fifth respectively
[27]	Identify and ranking criteria for select supply chain in LARG supply chain (case study: kaleh food industrials)	In the present study, the indicators affecting the selection of suppliers and the degree of their importance in Kalleh Dairy Company were evaluated using the lean, agile, resilient, and green approaches
[28]	Sustainable procurement and logistics for disaster resilient supply chain	The proposed model in the paper is referred as SPL_DRSCM (sustainable procurement and logistics for disaster resilient supply chain management) and is validated through illustrations. Important useful managerial and practical insights are obtained from set of five different deterministic data sets. Proposed model for SPL_DRSCM MINLP shows significant cost saving while optimizing procurement and its logistics under carbon emissionconstraint. Comparative analysis is conducted and detailed numerical results are presented

(*continued*)

Table 2. (*continued*)

Authors/Year	Title of research	Variables/Results and findings
[31]	Developing a resilient supply chain through supplier flexibility and reliability assessment	Findings generally show that highly flexible suppliers receive less allocation, and their flexible capacity is reserved for disruptions. This study highlights the supply chain risk management strategy of regionalising as ameans for minimising the impact of environmental disruptions
[32]	Firm's resilience to supply chain disruptions: scale development and empirical examination	Supply chain disruption oriented firms require the ability to reconfigure resources or have a risk management resource infrastructure to develop resilience. The way in which supply chain disruption oriented firms develop resilience through resource reconfiguration or risk management infrastructure depends on the context of the disruption as high impact or low impact. In a high impact disruption context, resource reconfiguration fully mediates the relationship between supply chain disruption orientation and firm resilience. In a low impact disruption context, supply chain disruption orientation and risk management infrastructure have a synergistic effect on developing firm resilience

2.4 Supply Chain Resilience Enablers

Considering the increased risk in the business environment, the study of resilient supply chain is crucially important. Moreover, the resilience enablers of this area have not been yet fully identified [12, 18]. Table 3 presents the dimensions of resilience enablers in the supply chain.

2.5 Fuzzy DELPHI

The Delphi technique is performed on the basis of respondent opinions. In this technique, verbal expressions are used to measure opinions and viewpoints. These expressions cannot fully reflect latent mentalities of respondents. For example, the expression "high" for individual A, who is a strict person, is different from "high" expressed by individual B. Using absolute values for quantification of the attitudes of these two individuals will lead to skewed results. Therefore, developing a suitable fuzzy spectrum will help a researcher overcome the problem. Ishikawa et al. (2012) introduced and further developed the application of fuzzy theory in the Delphi method and developed the fuzzy integration algorithm to predict the future penetration of computers in organizations. In this study, the Fuzzy Delphi technique algorithm was used to screen the enablers of resilient supply

Table 3. The SCRE and the supporting literature references

Row	Enablers	Supporting literature references
1	Sustainability (Sus)	[27–29]
2	Flexibility (F)	[15, 20, 21]
3	Information Sharing (IS)	[14, 31, 32]
4	Supply Chain Recovery Speed (SCRS)	[27–29]
5	Structure (S)	[28, 29, 32]
6	Adaptive Capability (AC)	[27–29]
7	Agility (A)	[31–33]
8	Risk and Revenue Sharing (RRS)	[27–29]
9	Collaboration (C)	[26, 30, 34]
10	Visibility (V)	[27–29]
11	Trust (T)	[31, 32, 34]
12	Risk Management Culture (RMC)	[13, 31, 33]
13	Responsiveness (R)	[30, 33, 34]

chains. In summary, various stages of the fuzzy Delphi method include: identification of proper spectrum for fuzzification of verbal expressions; fuzzy aggregation of fuzzified values, defuzzification of values, selecting the threshold intensity and screening the criteria. Table 3 presents the triangular fuzzy numbers with a five-point Likert scale.

The defuzzy numbers in Table 4 are calculated using Eq. (1). If $\tilde{N} = (1, m, u)$ it's. (A fuzzy number).

$$\text{Crisp}(\tilde{N}) = \frac{2m + 1 + u}{4} \qquad \textbf{Relationship} \qquad (1)$$

Table 4. Triangular fuzzy numbers [23]

Verbal variables	Very low	Low	Average	Much	Very much
Triangular fuzzy number	(0, 0, 0.25)	(0, 0.25, 0.5)	(0.25, 0.5, 0.75)	(0.5, 0.75, 1)	(0.75, 1, 1)
Defuzzy number	0.063	0.25	0.5	0.75	0.94

Step1: Select SCRE with Fuzzy DELPHI

In this section, according to the experts' opinion and the questionnaire with the spectra of Table 4, Them view on the dimensions of SCRE was acquired and analyzed in two phases with a fuzzy Delphi technique. Due to the limited number of pages in the article, mention are not made of the details. Table 5 shows the de-intermediate difference between the first and second stages of the survey.

Table 5. Departed average deviation of first and second round of polling

Variables	Defuzzy average of the first stage	Defuzzy average of the second stage	Average difference Defuzzy first & second stage	Variables	Defuzzy average of the first stage	Defuzzy average of the second stage	Average difference Defuzzy first & second stage
Sustainability	0.547	0.573	0.026	Risk and revenue sharing	0.526	0.594	0.068
Flexibility	0.663	0.671	0.008	Collaboration	0.632	0.707	0.075
Information sharing	0.529	0.568	0.039	Visibility	0.650	0.650	0
Supply chain recovery speed	0.481	0.486	0.005	Trust	0.539	0.539	0
Structure	0.647	0.647	0	Risk management culture	0.701	0.719	0.018
Adaptive capability	0.713	0.747	0.034	Responsiveness	0.608	0.637	0.029
Agility	0.718	0.739	0.021				

Due to the fact that divergent divergence of the defuzzy are, the opinion of experts is less than (0.1) in two steps, The experts reached a consensus, So the poll stops at this stage and the identified dimensions of the supply chain resilience enablers are confirmed.

3   Research Methodology

The present study was conducted in three stages. In the first stage, the supply chain resilience enablers were identified with regard to the research literature and by using Fuzzy Delphi technique. In the second stage, ISM technique was used to determine the relationships between variables and the types of variables and MICMAC analysis was used to determine the type of criteria. Finally, in the third stage, the intensity of the relationships between the resilience enablers was examined using DEMATEL technique. The study population included 20 senior managers of IKCO, with at least 10 years of experience in the production and supply chain management sectors. Due to the small size of the study population, complete enumeration sampling method was used and all the managers were interviewed to screen resilience enablers and were asked to complete the questionnaires. The validity of the questionnaires was confirmed by the experts and Gogus and Boucher's consistency test was used to test its reliability. The inconsistency rate of the DEMATEL questionnaire was (0.03) that was less than (0.1); thus, the questionnaire has a desirable reliability.

3.1   Interpretive Structural Modeling (ISM)

The interpretive structural modeling was presented by Andrew Sage in 1977. ISM is an interpretive structural method that was introduced by Agarwal in 2006 and used in an article in 2007 by Kannan. This method first identifies effective and fundamental factors and then investigates the relationships between these factors and finds strategies to achieve progress via these factors. This method breaks down criteria to several different levels in order to analyze their relations. ISM can determine the relationships between some indices that are dependent on each other individually or aggregately. This method can be used to analyze the relationships between the properties of some variables. In summary, various steps involved in ISM modeling are as follows: 1-identifying the variables associated with the problem; 2-developing a structural self-interaction matrix (SSIM) of variables; 3-establishing an initial reachability matrix; 4-establishing the final reachability matrix; 5-classifying final reachability matrix into different levels; 6-drawing the ISM graph and 7-clustering the indices using MICMAC analysis (influence-dependence matrix) [24].

3.2   Decision Making Trial and Evaluation Laboratory (DEMATEL)

DEMATEL technique is a scientific and useful tool, especially for illustrating the complex structure of causal relationships by graph or matrix. Matrices or graphs show the relationships established based on the system elements and the numbers displayed on graphs represent the intensity of the effect of each element. In 2013, Wu et al. presented the following four steps for the implementation of DEMATEL technique based on the Fontela and Gabus' method: step 1: finding the average matrix; step 2: normalizing the direct-relation matrix; step 3: calculating the total-relation matrix and step four: calculating the threshold of relations and depicting the Cartesian graphs of causal relationships [25].

3.3   Execution Model

Chart (1) shows the executive model of the present study.

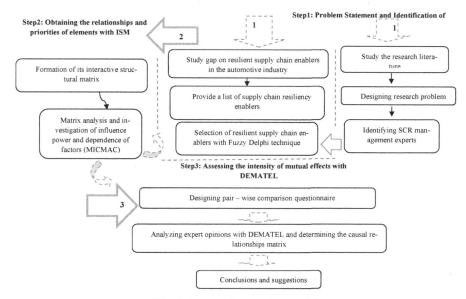

Fig. 2. Research execution model

4 Findings

4.1 Step2: Designing Interpretative Structural Model (ISM) in Iran Khodro Resilient Supply Chain

Stage 1 and 2: Identify Related Indexes and Formulate Interactive Structural Matrices

After identifying the dimensions of supply chain resilience with Fuzzy Delphi technique in the first Step, ISM was used to illustrate the structural interpretation modeling. The research dimensions were identified and in the next step, a matrix was used to investigate the dimensions of the research variables in the first row and column, respectively. Then the pairwise relations of the variables were determined by symbols. The self-interaction matrix was formed based on the expert opinions. Expert opinions and various managerial techniques, including brainstorming and nominal group technique are recommended to be used to determine the type of relations [24]. Table 6 shows the symbols used to determine the relations.

In the first step, a questionnaire was designed to form the self-interaction matrix. The questionnaire is briefly presented in Table 7. In this regard, 13 supply chain resilience enablers selected in the table rows and columns were displayed and the experts were asked to use the symbols of Table 6 to determine the types of their pairwise relations. Then, the experts were interviewed to eliminate disagreements and finally, the relations presented in Table 7 were obtained.

Table 6. Conceptual relationships in the formation of structural self-correlation matrix [7]

Symbol	Symbolic concept
V	i leads to j (line leading to the column)
A	j leads to i (column leading to row)
X	There is a two-way relationship between i and j
O	There is no valid relationship

Table 7. Self-interactive matrix

Factors	13	12	11	10	9	8	7	6	5	4	3	2
1	O	O	O	A	A	A	O	O	A	A	A	A
2	X	A	V	A	V	V	O	A	A	O	X	-
3	O	O	V	O	O	V	A	V	V	V	-	
4	X	X	V	O	A	A	V	O	A	-		
5	V	V	O	O	O	A	A	X	-			
6	X	X	A	A	O	V	V	-				
7	O	O	O	A	V	X	-					
8	A	O	A	V	V	-						
9	V	O	X	A	-							
10	A	O	O	-								
11	X	V	-									
12	O	-										

Stage Three: Create an Access Primary Matrix

The rules used for converting these symbols to numbers zero and one are shown in Table 8. Table 9 shows the initial reachability matrix of the present study.

Table 8. Way to convert conceptual relationships to numbers [7]

Symbol	i to j	j to i
V	1	0
A	0	1
X	1	1
O	0	0

Stage Four: Create the Final Access Matrix

The results are presented in Table 10. In this table, numbers marked with asterisk symbol (*) have been zero in the initial reachability matrix and they have been converted to one, following the compatibility process.

Table 9. Primary access matrix

Factors	1	2	3	4	5	6	7	8	9	10	11	12	13
Sustainability	1	0	0	0	0	0	0	0	0	0	0	0	0
Flexibility	1	1	1	0	0	0	0	1	1	0	1	0	1
Information Sharing	1	1	1	1	1	1	0	1	0	0	1	0	0
Supply Chain Recovery Speed	1	0	0	1	0	0	1	0	0	0	1	1	1
Structure	1	1	0	1	1	1	0	0	0	0	0	1	1
AdaptiveCapability	0	1	0	0	1	1	1	1	0	0	0	1	1
Agility	0	0	1	0	1	0	1	1	1	0	0	0	0
Risk and RevenueSharing	1	0	0	1	1	0	1	1	1	1	0	0	0
Collaboration	1	0	0	1	0	0	0	0	1	0	1	0	1
Visibility	1	1	0	0	0	1	0	0	1	1	0	0	0
Trust	0	0	0	0	0	1	0	1	1	0	1	1	1
Risk Management Culture	0	1	0	1	0	1	0	0	0	0	0	1	0
Responsiveness	0	1	0	1	0	1	0	1	0	1	1	0	1

Table 10. Matrix ultimate access

Factors	1	2	3	4	5	6	7	8	9	10	11	12	13
Sustainability	1	0	0	0	0	0	0	0	0	0	0	0	0
Flexibility	1	1	1	0	1*	0	1*	1	1	1*	0	1	0
Information Sharing	1	1	1	1	1	1	1*	1	1*	0	1	0	0
Supply Chain Recovery Speed	1	0	1*	1	0	1*	1	1*	1*	1*	1	1	1
Structure	1	1	1*	0	1	1	1*	1*	1*	1*	0	1	1
AdaptiveCapability	1*	1	1*	1*	1	1	1	1	0	0	0	1	1
Agility	1*	1*	1	1*	1	1*	1	1	1	0	0	0	0
Risk and RevenueSharing	1	1*	1*	1	1	0	1	1	1	1	0	0	0
Collaboration	1	0	1*	1	1*	0	1*	0	1	0	1	0	1
Visibility	1	1	1*	1*	1*	1	0	0	1	1	0	0	0
Trust	1*	0	1*	0	0	1	0	1	1	0	1	1	1
Risk Management Culture	1*	1	0	1	1*	1	0	0	0	0	0	1	0
Responsiveness	1*	1	0	1	0	1	0	1	0	1	1	0	1

Stage Five: Level Partitioning
Table 11 shows the levels of identified factors.

Stage Six: Draw a Model
The present research, the variables were classified into 6 levels. Figure 2 shows that the ISM model of resilience for Iran Khodro supply chain. Risk management culture, flexibility, accountability, and trust are at the lowest level of this model. These factors act as the cornerstone of this model and are necessary for the promotion of resilience in Iran Khodro supply chain. The fifth level includes the transparency and speed of supply chain recovery which is in interaction with the four factors of the sixth level. Interconnectedly, collaboration and sharing of risk and income are in the fourth place. The third level consists of agility and compatibility which affect the structure and sharing

Table 11. Determining the level of variables

Variables	Entrance set	Output set	Common set	Level
Sustainability	13	13	2,3,4,5,6,7,8,9,10,11,12,13	1
Information sharing	2,3,4,5,6,7,8,9,10,11	1,2,3,4,5,6,7,8,9,11	2,3,4,5,6,7,8,9,11	2
Supply chain recovery speed	3,4,7,8,9,10,12,13	4,7,11,12,13	4,7,12,13	5
Structure	2,3,5,6,7,8,9,10,12	1,2,3,5,6,7,8,9,10,12,13	2,3,5,6,7,8,9,10,12	2
Responsiveness	4,5,6,9,11,13	1,2,10,11,13	10,11,13	6
Adaptive capability	1,2,3,4,5,6,7,10,11,12,13	3,4,5,6,7,12,13	3,4,5,6,7,12,13	3
Agility	2,3,4,5,6,7,8,9	1,2,3,4,5,7,8,9	2,3,4,5,7,8,9	3
Flexibility	2,3,5,6,7,8,10,12,13	1,2,9,12,13	2,12,13	6
Risk and revenue sharing	2,3,4,5,6,7,8,11,13	1,2,3,4,5,7,8,9,10	2,3,4,5,7,8	4
Collaboration	2,3,4,5,7,8,9,10,11	1,3,4,5,7,9,11,13	3,4,5,7,9,11	4
Risk management culture	2,4,5,6,11,12	1,2,4,12	2,4,12	6
Visibility	2,4,5,8,10,13	1,2,3,4,5,6,9,10	2,4,5,10	5
Trust	3,4,11,13	1,3,6,8,9,11,12,13	3,11,13	6

of information in the second level. Finally, the first level includes sustainability which is influenced by factors of the second level (Fig. 3).

Stage Seven: Analysis of MICMAC

In the analysis of MICMAC, factors (enablers) are classified into 4 groups of linkage, independent, dependent and autonomous factors based on driving and dependence power [24]. Table 12 shows the degree of drive power and dependence power of the research model factors, and Fig. 2 displays factors clustering using MICMAC analysis method.

4.2   Step3: Investigating the Causal-Causal Relationships of Resiliency Enablers of Supply Chain with DEMATEL

The results are according to Table 13.

Also, We have Fig. 4 shown result of DEMATEL technique.

According to Fig. 4, the net effect intensity of enablers which are placed above the horizontal line is positive, and they are classified as causal, stimulating, or influential enablers (trust (T), responsiveness (R), flexibility (F), risk management culture (RMC), collaboration (C), supply chain recovery speed (SCRS), and visibility (V)). In addition, the net effect intensity of enablers which are placed below the horizontal line is negative; these enablers are classified as dependent enablers (sustainability (SUS), information sharing (IS), structure (S), adaptability (A), risk and revenue sharing (RRS)). The higher

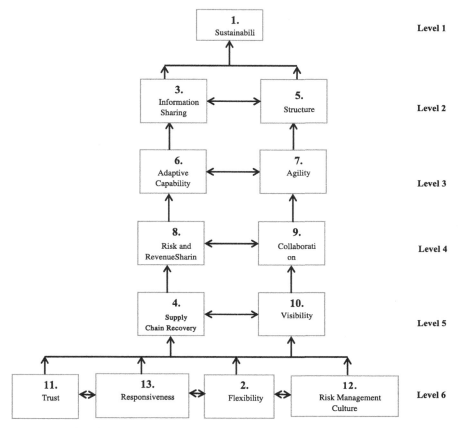

Fig. 3. Model ISM enablers Iran Khodro supply chain resilience

Table 12. Degree of influence and affiliation factors

Variable	1	2	3	4	5	6	7	8	9	10	11	12	13
Power of influence	1	10	4	9	4	6	6	8	8	9	10	10	10
Dependency power	10	1	9	4	9	7	7	6	6	4	1	1	1

the enablers are placed, the greater degree of effectiveness they have, and the lower the enablers are placed, the greater degree of affectivity they show. Additionally, as enablers move toward the right side of the diagram, their importance increases, because the sum of their effectiveness and affectivity is greater. In the other words, the enabler with greater interaction with other enablers is more important (Fig. 5).

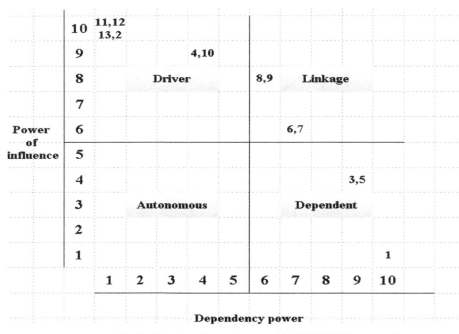

Fig. 4. Intrusion-dependency matrix (MICMAC)

Table 13. The influence order of enablers affecting the SCR with DEMATEL

The factors	Based on the maximum total row (R)	The factors	Based on the maximum total column (D)	The factors	Based on R+D	The factors	Based on R-D
T	8.964	Sus	7.514	T	9.778	T	8.15
R	7.583	S	6.148	R	8.541	R	6.625
F	6.541	A	5.217	RMC	8.532	F	5.294
RMC	6.214	RRS	4.283	RRS	8.462	RMC	3.896
SCRS	5.598	AC	4.173	A	8.227	SCRS	3.057
V	5.139	C	3.015	SCRS	8.139	V	2.492
RRS	4.179	IS	2.984	AC	8.063	RRS	−0.104
C	4.028	V	2.647	V	7.786	C	1.013
AC	3.89	SCRS	2.541	C	7.043	AC	−0.283
A	3.01	RMC	2.318	F	7.788	A	−2.207
IS	2.47	F	1.247	IS	8.16	IS	−0.514
S	1.348	R	0.958	S	7.496	S	−4.8
Sus	0.318	T	0.814	Sus	7.832	Sus	−7.196

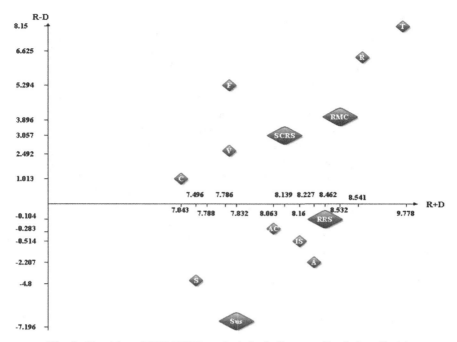

Fig. 5. Resulting of DEMATEL analysis for indicators of banks' credit risk

5   Conclusion

Moving toward resilient supply chain in automotive industry requires identifying main and effective enablers of the resilient chain as well as determining the relationship between them and the intensity of the effect of them on each other. In the current research, after reviewing the theoretical principles, and using fuzzy Delphi method, finally, 13 enabler factors of the resilient supply chain in the automotive industry were identified through experts' opinions. Results obtained from interpretive structural modeling (ISM) showed that enablers of risk management culture, flexibility, responsiveness, and trust are placed in the lowest hierarchical levels of ISM. This means that these factors have affected other factors, but are not affected by them.

According to the results of the MICMAC analysis, no indicators are classified in the Autonomous region, and the factors mentioned are those factors that have a small effect on other factors and are little affected by them. The lack of these types of factors can be interpreted so that all the factors used in this study play a significant role in moving towards resilient production. Risk and Revenue Sharing (RRS), Collaboration (C), Agility (A), Adaptation Capability (AC), are classified in the third section of the chart, linkage agents. These factors are among the most important factors because they have a significant impact, and they are affected significantly too. These factors are unstable because every small event that happens to them can affect the performance of other factors or even themselves.

Risk Management Culture (RMC), Flexibility (F), Responsiveness (R), Trust (T), Transparency (T), and Supply Chain Recovery Speed (SCRS) are categorized in the fourth section of the chart, independent criteria. All of the factors that are in the empowerment group of the resilient supply chain of Iran Khodro Company- are among the "most important" independent factors. As a result, organization management should take the basic steps in order to strengthen these empowerers, as these empowerers have a significant power in influencing other factors.

In the second part of this study, the DEMATEL method was used to obtain a deeper insight into the resilient supply chain enablers. According to DEMATEL cause-effect relationships, the empowerers of Trust (T), Responsiveness (R), Flexibility (F), risk management culture (RMC), are ranked first in terms of severity of net effect, it means that they have the highest impact on other influencers and are little affected by them.

According to the results of this study, four mentioned factors have been identified as strong effective factors in this method. Sustainability (SUS), structure (S), and agility (A) are also identified as the most powerful empowerers. Furthermore, the empoweres of flexibility (F), responsiveness (R), and trust (T) are identified as the most important empowerers.

References

1. Jafarnejad, A., Hashemi, H., Talaei, H.R.: New Approaches to Supply Chain Management, 1st edn. Negahedanesh Publishers (2014). (in Persian)
2. Rahimian, M.M., Rajabzadeh, A.: Resiliency measurement of supply chain with systems approach compliant complaint; case study: Iran's pharmaceutical industry. J. Modern Res. Decis. Making (MRDM) **2**(2), 155–195 (2017). (in Persian)
3. Ponomarov, S.Y., Holcomb, M.C.: Understanding the concept of supply chain resilience. Int. J. Log. Manage. **20**(1), 124–143 (2009)
4. Hoseyni, M., Marei, S.P.: Provide a conceptual framework for selecting new and innovative product ideas (Case Study: Iran Khodro Co.). New Res. Manage. Account. **8**(3), 67–88 (2016). (in Persian)
5. Ravanestan, K., Hassanali, A., Safaee Ghadikelae, A.H., Yahyazadehfar, M.: Designing a resonance model in iran khodro supply chain with the approach of structural equation modeling and Delphi techniques. J. Ind. Manage. Sanandaj **12**(40), 49–62 (2017). (in Persian)
6. Thun, J.H., Hoenig, D.: An empirical analysis of supply chain risk management in the German automotive industry. Int. J. Prod. Econ. **131**(1), 242–249 (2011)
7. Soni, U., Jain, V., Kumar, S.: Measuring supply chain resilience using a deterministic modeling approach. Comput. Ind. Eng. **74**, 11–25 (2014)
8. Rezapour, S., Farahani, R.Z., Pourakbar, M.: Resilient supply chain network design under competition: a case study Eur. J. Oper. Res. **259**(3), 1017–1035 (2017)
9. Levalle, R., Nof, S.Y.: Resilience in supply networks: definition, dimensions, and levels. Ann. Rev. Contr. **43** 224–236 (2017)
10. Rajesh, R.: Technological capabilities and supply chain resilience of firms: a relational analysis using Total Interpretive Structural Modeling (TISM). Technol. Forecast. Soc. Change **118**, 161–169 (2017)
11. Rajesh, R.: Forecasting supply chain resilience performance using grey prediction. Electron. Commerce Res. Appl. **20**, 42–58 (2016)

12. Liu, C.L., Shang, K.C., Lirn, T.C., Lai, K.H., Lun, Y.V.: Supply chain resilience, firm performance, and management policies in the liner shipping industry. Transp. Res. Part A Policy Pract.**110**, 202–219 (2017)

13. Kamalahmadi, M., Parast, M.: A review of the literature on the principles of enterprise and supply chain resilience: major findings and directions for future research. Int. J. Prod. Econ. **171**, 116–133 (2016)

14. Chowdhury, M.M.H., Quaddus, M.: Supply chain resilience: conceptualization and scale development using dynamic capability theory. Int. J. Prod. Econ. **188**, 185–204 (2017)

15. Azevedo, S.G., Govindan, K., Carvalho, H., Cruz-Machado, V.: Ecosilient Index to assess the greenness and resilience of the upstream automotive supply chain. J. Clean. Prod. **56**, 131–146 (2013)

16. Ambulkar, S., Blackhurst, J., Grawe, S.: Firm's resilience to supply chain disruptions: scale development and empirical examination. J. Oper. Manage. **33**(34), 111–122 (2015)

17. Rahimian, M.M., Rajabzadeh, A.: Resiliency measurement of supply chain with systems approach compliant complaint; case study: Iran's pharmaceutical industry. J. Modern Res. Decis. Mak. (MRDM) **2**(2), 155–195 (2017). (in Persian)

18. Jafarnejad, A., Mohseni, M.: Design framework for performance improvement resilience supply chain. Supply Chain Manage. **48**(17), 38–51 (2015). (in Persian)

19. Rice, J.B., Caniato, F.: Building a secure and resilient supply network loop. Disaster Prev. Manage. **3**(3), 59–80 (2003)

20. Christopher, M., Peck, H.: Building the resilient supply chain. Int. J. Log. Manage. **15**(2), 1–13 (2004)

21. Tang, C.: Robust strategies for mitigating supply chain disruptions. Int. J. Log. Manag. **9**(1), 33 (2006)

22. Iakovou, E., Vlachos, D., Xanthopoulos, A.: An analytical methodological framework for the optimal design of resilient supply chains. Int. J. Logist. Econ. Global. **1**(1), 1–20 (2007)

23. Sepahvand, R., Arefnejad, M., Shareatnejad, A.: Identify and prioritize inertia factors Organization Using Fuzzy Delphi Method. J. Modern Res. Decis. Making (MRDM) **2**(1), 95–117 (2017). (in Persian)

24. Azar, A., Rajabzadeghatari, A., Akhavan, A.: Mapping sustainable production model with interpretive structural modeling approachFuzzy DEMATEL. J. Ind. Manage. Stud. **15**(46), 1–26 (2017). (in Persian)

25. Jokar, A., Maleki, M.H.: Providing a framework for assessing the readiness of companies from the standpoint of successor with the combined approach of gray and DEMATEL network analysis process. J. Modern Res. Decis. Making (MRDM) **2**(1), 73–94 (2017). (in Persian)

26. Parhizgari, R., Fazli, S.: Production model with interpretive structural modeling important factors in development green products. J. Modern Res. Decis. Making (MRDM) **1**(4), 25–44 (2016). (in Persian)

27. Mohammadnejad, F., Safaee, A.H.: Identify and ranking criteria for select supply chain in LARG supply chain (case study: kaleh food industrials). J. Oper. Res. **51**, 103–120 (2016). (in Persian)

28. Kaur, H., Singh, S.P.: Sustainable procurement and logistics for disaster resilient supply chain. Int. Appl. OR Disas. Relief Oper. **283**, 1–46 (2016)

29. Kamalahmadi, M., Mellatparast, M.: A review of the literature on the principles of enterprise and supply chain resilience: major findings and directions for future research. Int. J. Prod. Econ. **171**, 1–59 (2015)

30. Rezapour, S., Zanjiran Farahani, R., Pourakbar, M.: Resilient supply chain network design under competition: a case study. Eur. J. Oper. Res. **259**, 1017–1035 (2016)

31. Kamalahmadi, M., Mellat- Parast, M.: Developing a resilient supply chain through supplier flexibility and reliability assessment. Int. J. Prod. Res. **54**, 1–20 (2015)

32. Ambulkar, S., Blackhurtst, J., Grawe, S: Firm's resilience to supply chain disruptions: scale development and empirical examination. J. Oper. Manage. **33**, 111–122 (2015)
33. Reyes Levalle, R., Nof, S.: Resilience in supply networks: definition, dimensions, and levels. Ann. Rev. Control **43**, 224–236 (2017)
34. Dubey, R., Gunasekaran, A., Childe, S., Fosso Wamba, S., Roubaud, D., Foropon, C.: Empirical investigation of data analytics capability and organizational flexibility as complements to supply chain resilience. Int. J. Prod. Res. **59**(1), 110–128 (2021)
35. El Baz, J., Ruel, S.:. Can supply chain risk management practices mitigate the disruption impacts on supply chains' resilience and robustness? Evidence from an empirical survey in a COVID-19 outbreak era. Int. J. Prod. Econ. **233**, 107972 (2021)
36. Negri, M., Cagno, E., Colicchia, C., Sarkis, J.: Integrating sustainability and resilience in the supply chain: a systematic literature review and a research agenda. Bus. Strategy Env. (2021)

Humanitarian Supply Chain Management

Multi-objective Model to Distribute Relief Items After the Disaster by Considering Location Priority, Airborne Vehicles, Ground Vehicles, and Emergency Roadway Repair

Behnam Momeni[1]([✉]) [iD], Samira Al-sadat Salari[1], Amir Aghsami[2], and Fariborz Jolai[1]

[1] School of Industrial Engineering, College of Engineering, University of Tehran, Tehran, Iran
{behnammomeni,samira.salari.sa,fjolai}@ut.ac.ir
[2] School of Industrial Engineering, K. N. Toosi University of Technology, Tehran, Iran
a.aghsami@ut.ac.ir

Abstract. The most important aim in humanity's relief problems is the minimization of response time and reducing the rate of casualty in disasters. If paths are inappropriate for transferring relief items to the disaster areas, response time and the number of casualties will be increased. Therefore, airborne vehicles like helicopters are used in this study to transfer equipment to the disaster areas to build a "Field hospital." Each path is assigned a weight for illustrating priority and humanitarian operation starts its operation from the highest priority. What the instances' evaluations conclude is that most of the injured individuals are rescued by this strategy because the humanitarian operation is independent of paths' situation, and humanitarian operation time will be reduced. We wish to present the innovation of airborne and ground vehicles by reviewing demand priority in the disaster literature to show its efficiency in humanitarian operations. This study shows that the airborne vehicle strategy is cost-efficient and reduces the response time to extend possible; thus, the casualty rate could be decreased.

Keywords: Humanity relief · Airborne vehicles · Ground vehicles · Routing problem · Priority relief

1 Introduction

According to the occurrence of natural disasters, such as floods, earthquakes, and hurricanes worldwide, financial and casualty damages hurt human beings' lives. Referring to some criteria or factors, like population growth and climate changes, we can see an increase in natural disaster happenings. It is predicting that the present help is insufficient to respond to all the demands. The accidental and unpredictable nature of natural crises necessitates comprehensive plans to decrease the danger. In this regard, one of the most critical decisions is to have essential facilities distribution plans: relief team, tent, drug, ambulance.

© Springer Nature Switzerland AG 2021
Z. Molamohamadi et al. (Eds.): LSCM 2020, CCIS 1458, pp. 341–361, 2021.
https://doi.org/10.1007/978-3-030-89743-7_19

The destruction of roads in a natural disaster occurrence entails traffic and disorder in transportation systems, also cease the relief operations, rescue teams, and the ambulance movements. So, road reliability is defined according to road damage percentage. Ability and high capability in distribution result in a quicker launching of roads for land vehicles to deliver required resources at the desired location on time. Hence, how to repair damaged roads in the least possible time is an important issue after a catastrophe.

On the one hand, the main concern of solving these issues is saving the lives of injured people and preventing more casualties, and on the other hand, much time is needed to reopen blocked land routes to cities and villages, so lots of injured people will lose their lives. Consequently, the best solution is to provide airborne assistance to different areas at a time when the landing path is reopening. Drugs, medical teams, and first aid depots can be delivered by helicopters, and after the reopening of the land route, food, and other necessities -like tents- will also be sent.

Based on the high limitations in relief supplies and equipment in the widespread catastrophic occurrence, we cannot respond to all affected cities and areas at the same time. So, authorities prioritize them based on criteria in the way to rescue the maximum population, who are likely to be injured and killed, in every relief operation. Accordingly, we have also taken our interest in maximizing the satisfaction of the community.

In this research, we attempt to choose the most effective strategy to minimize the number of people killed during the disaster, according to available facilities: air and ground vehicles, medical supplies, hygienic accessories, and heating for the disaster-stricken people. This paper reviews both airborne and ground vehicles in disaster literature and considers location's priority and repair groups. Due to the sensitivity of time, we suggest the strategy of delivery in the fastest possible way. For this reason, medical assistance is sent using helicopters to save the injured people, and then health accessories and food will be sent through ground vehicles. The importance of all paths is various, and in the following chapters, we will discuss the method for determining this factor.

2 Literature Review

In recent years, ambulance routing issues have tried to make the model work. The goals of research in this area are high convergence and coherence; for example, Chang et al. [1] examined the logistics' timing in crises and assumed resource diversion (nurses, transportation car for food, ambulances) for limited resource conditions. The goals achieved in this research are: 1) minimizing unsatisfied applications 2) minimizing the time to reach the region where the disaster has occurred 3) Minimizing the cost of logistics in the transportation sector.

From other articles in the field of transportation networks, research of [2] describes the effect of stored equipment, which will ultimately lead to a decrease in the number of deaths. This paper includes restrictions on the way, time, the number of expedition cars, the division of each area into sub-section, and it considers a rescue bus-including a doctor, five nurses, three dogs, and some food-that can save 15 people.

In the field of ambulance allocation for quick services, we can mention [3] that increased the performance of the model to a desirable level for various areas (urban and rural areas) by several ambulances. These assumptions helped to cover many natural disasters that affect different areas and provide proper relief. The article [4] presents the model by considering the blocking path and then opening the path by specific means. The research has a multi-objective problem, and its purposes are: 1) minimizing the time to arrive at the disaster area, 2) minimizing the total costs, and 3) maximizing the reliability level in serving the victims of the incident. Also, it solved the model using two meta-heuristic algorithms and provided acceptable numerical answers. According to reducing the number of deaths in disaster areas, we can use various technologies, including different relief vehicles. Helicopters are one of the most used vehicles in post-disaster situations. Due to the difficulties of ground transportation in case of a catastrophic event, the article [5] considered a medium-scale UAV helicopter for commodity transportation in the first few hours after the earthquake.

The prioritizing process for multiple patients when resources are insufficient for the rapid treatment of all patients is the topic of the study [6]. It develops an integrated criterion for mass casualty categorization to ensure interoperability and standardization to respond to an incident that has four categories: general considerations, global sorting, interventions life-saving, and individual assessment of patient classification. Further studies can be referred to the article [7] that prioritizes patients to get services. It divides the ambulance cost into two fixed and variable costs. The variable costs are proportional to the distance traveled by ambulance. In the research, two objective functions are considered, the first objective function is to minimize the response time, and the second is to minimize the total cost.

Another study was [8], in which a cost for an ambulance to move is considered, and it provides a model for balancing cost and responsibility. As expected, this study considers several objectives and seeks an ideal way to satisfy constraints and optimize answers. To this end, minimizing the total transportation costs and the sum of the earliness or tardiness of the transferring times is considered in the paper [9], and a multi-objective mathematical model and fuzzy parameters have been considered to make the modeling results more realistic.

Since the adequate response to public demands, through the first aid providing and transporting injured to the hospital's emergency department is a vital process, researchers are interested in modeling with the constraints of reality as closely as possible and using different methods to solve models. Therefore, [10] used existing models in their research and presented new results using uncertainty-based simulation. Whereas the article [11] modeled the problems related to cost, dynamic demand, and real-time service, using OR/MS models, and in this regard, uncertain data, the dynamic systems, and soft OR are used.

Patients' triage after entering the hospital ward is one of the issues in the field of health. This can be done in many different categories, including the severity of the illness or injury, prioritizing patients for treatment, and making the most of emergency department operations. In the article [12] the reliability and validity of the triage of patients in 3 and 5 categories are evaluated by nurses' experience in patient classification and evaluated in terms of managerial values. This study shows that the five triages category is safer and more highly reliable than the 3-category system. In the paper [13], the objective is to determine the efficiency of the simple classification and rapid treatment model, which was evaluated in a 2003 train crash disaster. It obtained the arrival time of the wounded people to the hospital and determined the correct patient triage values using a combination of modified Baxt criteria and hospital admission.

Systematic perspectives on technological disasters and incidents, especially human disasters [14] have been addressed. This study discusses the impact of two common barriers to learning from natural disasters: (1) information disruption; and (2) organizational policies, as well as ways to address these barriers.

One of the challenges faced at the regional and country-level is identified standards and criteria for measuring flexibility in events. The paper [15] provided a new framework, the disaster flexibility model, to improve the relative assessment of disaster resilience locally or socially.

Mathematical models, along with other optimization methods, have been used in many studies, including the article [16]. This paper presented two mathematical models and simulation methods for locating and enhancing the efficiency of ambulance services with the aims of 1) minimizing the maximum response time in emergency medical care, 2) minimizing the number of ambulances, and 3) optimizing the minimum costs involved. Initially, the problem constraints are defined by solving the problem of location regardless of the number of ambulances in each center and then allocating ambulances. In another paper with the linear mathematical model and simulation method presented by [17], the authors considered a complex nonlinear problem involving decision making with uncertain parameters and an iterative optimization algorithm. In this study [18] a multi-objective MINLP mathematical model with several uncertain parameters is developed. The authors in this article [19] focused on assessing the conditions and requirements after a disaster. A humanitarian relief supply chain is recently developed by [20] and they analyzed repair groups, reliability of routes, and monitoring operation before distributing relief items in various stages (Table 1).

Table 1. Related recent papers in humanitarian relief supply chain

reference	Features					objective			
	Relief items	Quick response	Reliability of route	Repair groups	Air-borne system	Reliability	Response time	Total cost	Air-borne completion time
This study	√	√	√	√	√	√	√	√	√
[1]	√	√	-----	-----	-----	-----	√	-----	-----
[2]	√	√	√	-----	-----	√	-----	-----	-----
[3]	-----	√	√	√	-----	-----	√	-----	-----
[4]	√	√	√	√	-----	√	√	√	-----
[5]	-----	√	-----	-----	√	-----	√	-----	√
[6]	-----	√	-----	-----	-----	-----	√	-----	-----
[7]	√	√	-----	-----	-----	-----	√	√	√
[8]	√	√	-----	-----	-----	-----	√	√	-----
[9]	-----	-----	-----	-----	-----	-----	√	√	-----
[10]	-----	√	-----	-----	-----	-----	√	√	-----
[11]	-----	√	-----	-----	-----	-----	√	-----	-----
[12]	-----	√	-----	-----	-----	√	√	-----	-----
[13]	-----	√	-----	-----	-----	-----	√	-----	-----
[14]	-----	-----	-----	-----	-----	-----	-----	√	-----
[15]	-----	√	-----	-----	-----	-----	√	-----	-----
[16]	-----	√	-----	-----	-----	-----	√	√	-----
[17]	-----	√	√	-----	-----	√	√	-----	-----

3 Problem Definition

In this research, we consider a relief network distribution like [4], comes with two types of relief operation: airborne and ground relief, specific plan and routing for increasing efficiency. Since not all disaster areas have the same degree of priority, we adopt a formula to determine the priority of the zones, inspired by the modeling of the problem in the article [2]. For this purpose, Formula I indicate the degree of priority:

$$l_{ij} = \frac{population\ of\ a\ disaster\ district}{total\ population\ of\ disaster\ zone}$$

In the first step, the information of disaster areas, which considers the number of injured people and their locations, is sent to relief centers. Relief centers allocate a priority number to each disaster area, and relief operations start with the biggest priority number and continue until the last disaster area. In this method, the disaster area with more injured people or has the worst situation is served sooner than disaster areas with more appropriate situations. Because the roads are in an inappropriate situation for transferring vital items to disaster areas and time is so important in this case, helicopters are used to transfer vital items to disaster areas in the shortest time for "Field hospital". The helicopters transport initial equipment medicals and medical teams to disaster areas; then medical teams start building field hospitals and treating injured people to save their lives. With this method, response time is decreased to the extent. Helicopters come back to distributions, pick up other medical team items, and transfer them to another disaster area. This process continues until all the disaster areas have field hospitals. So, we utilize helicopters to satisfy humanity's main objective and decrease the dead people rate.

While helicopters transfer medical equipment and medical teams, repair groups start their duty to increase the reliability of each road and make a good situation for sharing relief items (for example, food, blanket, and tents) to disaster areas. Similar to the previous step, all relief operations start from the biggest priority number. This step boosts the speed of response time for relief items and helps people in disaster areas. Ground vehicles are sent to disaster areas in the last step, and they deliver relief items. The ground vehicles do not come back to distributions and stay at disaster areas. Each disaster area can receive relief items from different ground vehicles.

3.1 Assumptions

1. Information is collected completely and sent to distributions.
2. Each disaster area could be visited several times.
3. Ground vehicles do not come back to distributions.
4. Field hospitals locate next to disaster areas.
5. Both vehicles are capacitated.
6. Airborne vehicles can transfer appropriate equipment, nurses, and doctors to establish field hospitals.
7. The shortage is not considered.
8. Distributions are not allowed to send vehicles to each other.
9. An appropriate situation is considered, so sufficient vehicles, equipment, nurses, and doctors are available.
10. Loading and discharging time of ground vehicles are not considered since repairing roads needs time and at this time the ground vehicles prepare themselves for fast operation.

Now notations are introduced:

Sets

N Set of disaster areas $\{1;...; n\}$
M Set of candidate distribution nodes or centres $\{n + 1,..., n + m\}$
V Set of nodes $\{1;...; n + m\}$
K Set of vehicles $\{1;...; k\}$
T Set of time periods $\{1;...; t\}$
L Set of relief $\{1;...; l\}$

E Set of available traffic links $\{(i,j), i.j \epsilon V . i \neq j$

i,j Indices to nodes $i.j\epsilon V$

l Indices to relief

k Indices to aerial vehicles

t Indices to time periods

G Set of repair groups

H Set of helicopters

O Set of candidate hospitals

Parameters

f_j Fixed cost of constructing the distribution nodes j,$\forall j\epsilon M$

W Cost of each repairing work teams in each period

d_{ij} Distance of link (i,j),$\forall (i.j)\epsilon E$

h_{ij} Road damage percentage between nodes i and j,$i.j\epsilon V$

e_{ij} Number of time periods needed for repairing road between nodes i and j, $(i.j)\epsilon V$

D_{ilj} Quantity of relief l demanded by disaster area i, at any period

uv_l Unit volume of relief l,$\forall l\epsilon L$

c_k Transportation cost per kilometer of vehicle k,$\forall k\epsilon K$

Q_l Amount of relief l available in a traffic network,$\forall l\epsilon L$

v_k The normal speed of vehicle k,$\forall k\epsilon K$

l_k Loading capacity of aerial vehicle k,$\forall k\epsilon K$

G Number of available work teams

I_j Weight of node

B_j Duration of discharge from a helicopter

B_i Duration of loading

N_h Number of helicopters

Decision variables

x_{jt} Equal = 1 if candidate DC j is opened at period t, 0, else,$\forall i\epsilon j.t\epsilon T$

y_{ijkt} Equal = 1 if i precedes j in route of aerial vehicle k at period t, 0, else

y'_{ijht} Equal = 1 If helicopter h does its mission at ij in period t, 0, else

z_{ijkt} Equal = 1 if i is on the route of aerial vehicle k in period t, 0, else,$\forall k\epsilon K .t\epsilon T .(i.j)\epsilon E$

vf_{ikt} Equal = 1 if the last demand point serviced by aerial vehicle k is a node ϵN; 0, else

p_{ijt} Equal = 1 if a work team starts repairing road between nodes i and j at period t

r_{ijt} Reliability of road between nodes i and j in period t

dev_{ilt} Amount of unsatisfied demand relief type l at node

I At the end of the operation at period t

q_{ilkt} Quantity of relief l distributed by K to demand point i at period t,$\forall k\epsilon K .l\epsilon L.i\epsilon N$

c_{ij}^g Repair completion of g th group

$C_{ij'}^h$ Completion of helicopter mission

s_{ij}^g Start time of repair

$s_{ij'}^h$ Start time of helicopter mission
x_{ijji} If repair of ij before ji

Model

$$f_1 = Min\left(Max\left\{\sum_{(i,j)\in E}\sum_k\sum_t I_j.\left(\frac{d_{ij}\ y_{ijkt}}{v_k}\right)\right\}\right) \tag{1}$$

$$f_2 = \min\sum_{j\in M}\sum_{t\in T}f_j\,x_{jt} + \sum_{k\in K}\sum_{(i,j)\in E}\sum_{t\in T}c_k\cdot d_{ij}\cdot y_{ijkt} + \sum_{(i,j)\in E}\sum_{t\in T-e_{ij}}w\cdot e_{ij}\cdot P_{ijt} \tag{2}$$

$$f_3 = \max\sum_{(i,j)\in E}\sum_{t\in T}\sum_{k\in K}r_{ij}\cdot y_{ijkt} \tag{3}$$

$$f_4 = \min\sum_{(i,j)\in E}c_{ij'}^h\cdot I_j \tag{4}$$

Subject to:

$$x_{it} \geq y_{ijkt}\ \ \forall i\in M\,,(i,j)\in E, k\in K, t\in T : i\neq j \tag{5}$$

$$x_{it} \geq z_{ikt}\ \ \forall i\in M\,,(i,j)\in E, k\in K, t\in T \tag{6}$$

$$z_{ikt} \geq y_{ijkt}\ \ \forall i\in V\,,(i,j)\in E, k\in K, t\in T : i\neq j \tag{7}$$

$$z_{ikt} \geq VF_{ikt}\ \ \forall i\in V, k\in K, t\in T \tag{8}$$

$$\sum_{k\in K}VF_{ikt} = 1, \forall k\in K, t\in T \tag{9}$$

$$\sum_{k\in K}y_{ijkt} \leq 1, \forall (i,j)\in E, t\in T : i\neq j \tag{10}$$

$$\sum_{j\in V}y_{ijkt} \leq 1, \forall i\in N, k\in K, t\in T : i\neq j \tag{11}$$

$$\sum_{i\in M}\sum_{j\in N}y_{ijkt} \leq 1, \forall k\in K, t\in T \tag{12}$$

$$\sum_{i\in M}\sum_{j\in N}\sum_{k\in K}q_{jlkt}\cdot z_{ikt} \leq Q_l, \forall l\in L, t\in T \tag{13}$$

$$deV_{jlt} = D_{jlt} - \left(\sum_{k\in K}q_{jlkt}\right) \geq 0, \forall j\in N, l\in L, t\in T \tag{14}$$

$$\sum_{j\in N}\sum_{l\in L}uv_l\cdot q_{jlkt} \leq L_k, \forall k\in K, t\in T \tag{15}$$

$$\sum_{j/(j,i)\in E} y_{jikt} - \sum_{j/(j,i)\in E} y_{ijkt}$$

$$= \begin{cases} 1 & VF_{ikt} = 1, i \in N, t \in T : i \neq j \\ -1 & z_{ikt} = 1, i \in M, \forall k \in K, t \in T : i \neq j \\ 0 & else \end{cases} \tag{16}$$

$$\sum_{i\in V}\sum_{k\in K} y_{ijkt} \geq 1, \forall j \in N, t \in T \tag{17}$$

$$r_{ijt} = 1 - h_{ij}\left(1 - \sum_{t'=1}^{t-e_{ij}} p_{ijt'}\right), \forall i, j \in V, t \in T \tag{18}$$

$$y_{ijkt} \leq 1 - \sum_{t'=t+e_{ij}+1}^{t} p_{ijt'}, \forall t \in T, k \in K, i, j \in V \tag{19}$$

$$\sum_{i\in V}\sum_{j\in V} p_{ijt} \leq G, \forall t \in T, i \neq j \tag{20}$$

$$\sum_{i\in M} y_{ijkt} = 0, \forall j \in M, k \in K, t \in T \tag{21}$$

$$u_{ikt} - u_{jkt} + n.y_{ijkt} \leq n - 1, \forall i, j \in N, k \in K, t \in T : i \neq j \tag{22}$$

$$x_{it} \geq y'_{ij'ht} \quad \forall i \in M, j' \in O, h \in H, t \in T \tag{23}$$

$$\sum_{h\in H} y'_{ij'ht} \leq 1 \quad \forall i \in M, j' \in O, t \in T \tag{24}$$

$$\sum_{i\in V} y'_{ij'ht} \leq 1 \quad \forall j' \in O, h \in H, t \in T \tag{25}$$

$$\sum_{i\in M}\sum_{j'\in O} y'_{ij'ht} \leq 1 \quad \forall h \in H, t \in T \tag{26}$$

$$\sum_{i\in V}\sum_{j'\in O}\sum_{h\in H} y'_{ij'ht} \leq N_h \quad \forall j' \in O, h \in H, t \in T \tag{27}$$

$$\sum_{i\in V}\sum_{j'\in O}\sum_{t\in T} y'_{ij'ht} = 1 \quad \forall j' \in O, h \in H, t \in T \tag{28}$$

$$\sum_{i\in M} y'_{ij'ht} = 0 \quad \forall j' \in O, h \in H, t \in T \tag{29}$$

$$\sum_{i\in M}\sum_{j'\in O}\sum_{h\in H} y'_{ij'ht} \leq G_h \quad \forall t \in T \tag{30}$$

$$c_{ij'}^h = (\frac{d_{ij}}{v_h} + B'_i + B_j) \cdot y'_{ij'ht} \tag{31}$$

$$q_{ilkt} \geq 0, \forall (i,j) \in E, l \in L, k \in K, t \in T \tag{32}$$

$$s_{ij}^g, s_{ij}^h, c_{ij}^h \geq 0, ; \ i \in M, j \in N, g \in G, h \in H \tag{33}$$

$$x_{it} \in (0,1), \forall i \in M, t \in T \tag{34}$$

$$y_{ijkt} \in (0,1), \forall (i,j) \in E, k \in K, t \in T \tag{35}$$

$$z_{ikt} \in (0,1), \forall i \in V, k \in K, t \in T \tag{36}$$

$$VF_{ikt} \in (0,1), \forall i \in N, k \in K, t \in T \tag{37}$$

$$u_{ikt} \in (0,1), \forall i \in N, k \in K, t \in T \tag{38}$$

$$x_{iji'j'} \in (0,1), \forall i, i' \in M, j, j' \in N, i \neq i', j \neq j' \tag{39}$$

$$y'_{ij'ht}, u_{ji'ht}, x'_{j't} \in (0,1) \tag{40}$$

In the first objective function, we minimize the maximum time of vehicle passing by choosing the disaster zone based on the defined priority. The maximum transit time means the lowest time to complete the service among all the disaster nodes. In the second objective function, total costs are minimized. The model simultaneously provides the number and location of distribution centers, the allocation of disaster zones to distribution centers and vehicle routes so that the total cost includes three components: (1) fixed cost of establishing distribution centers; (2) transportation costs of vehicles; (3) Cost of repairing damaged roads. In objective function three, the minimum route reliability is maximized: In an earthquake, many infrastructures (roads, bridges, tunnels) are damaged, and these damages may increase in aftershocks. In this study, to prevent secondary injuries, we determine the reliability of the route as the probability of saving workers to accelerate across demand points. In the fourth objective function, the goal is to minimize the time required to complete the routes in air relief, where j' ∈ O and O are the set of candidates for field hospital construction also, j' must be very close to j. It should be noted that the completion time of ground and airborne vehicles are considered in two separate objective functions (1 and 4). This is because the innovation in consideration of air vehicles can be more easily compared with the time of ground vehicles.

Equations (5) and (6) state that distribution centers can only obtain services. Equation (7) guarantees that any vehicle in any given period can travel through the connection (i, j) if and only if node i is in the path of each vehicle. Equation (8) specifies that the nodes at the end of the path of each vehicle in each period must deal with one vehicle. Equation (9) guarantees that each vehicle in each period must eventually be placed in the disaster area or distribution center. Equation (10) shows that in each period, only one vehicle is selected for each route. Equation (11) guarantees that each vehicle operates once in most cases for each critical area in each period. Equation (12) ensures that each vehicle is shipped at most from one distribution center at any period. Equation (13) ensures that the amount of relief delivered from all distribution centers to critical areas cannot exceed the amount of relevant relief available. Equation (14) shows that the amount of relief distributed for each node in each period does not exceed the required value of this node. Equation (15) is shown in each period, the total amount of relief delivered at critical points by any vehicle not exceeding its capacity. Equation (16) expresses the consecutive motion and guarantees the assumption of path opening (Please see Appendix A). Equation (17) ensures that each disaster area can be visited at least once in each period. The split delivery hypothesis in this constraint has well demonstrated that the reliability of each road should be reliable (e.g., it will be equal to 1) considering the percentage of damage in each period calculated and after road reconstruction. Hence, Formula (18) calculates the reliability of roads in each period based on the relevant assumptions. Equation (19) shows that the destroyed or damaged road should not be used to deliver relief during reconstruction. Equation (20) limits the number of destroyed roads that can be rebuilt in each period based on the number of existing work teams. Equation (21) ensures that distribution centers are not correlated, meaning that goods are not exchanged between centers. Equation (22) is the constraint for the elimination sub-tours. Constraint (23) states that if the distribution center is built in a place, the transmission will take place from there. Restriction (24) ensures that up to one helicopter is assigned to each route. Constraint (25) guarantees that each route goes at most once. Equation (26) implies that a region receives one time from an air rescue distributor. Equation (27) ensures that the helicopters' missions to build a hospital will not exceed the number of helicopters. Equations (28, 29) indicate that each helicopter is assigned to one route and cannot visit other distributions. Formula (30) guarantees the capacity limitation of the number of helicopters available in air support. Formula (31) provides the completion time of the air relief process. From constraints (32) to (40) apply to nonnegative, integer, zero, and one values for decision variables.

4 Solution Methods

In the rest of the research, we consider two different problems, then we will solve them and evaluate the results.

These problems are solved by GAMS (LP-metrics technique), and obviously, all objectives cannot be satisfied simultaneously because this is a multi-objective problem. This technique minimizes the gaps which appear between optimal results and multi-objective results [21]. Moreover, it should be noticed ANTIGONE solver is selected to solve our problems.

$$L_p = \left[\sum_{k=1}^{k} (\pi_k W_k - b_k|)^p \right]^{\frac{1}{p}} \quad \text{II}$$

In Formula II, p determines which family of LP-metric is used, and the weights of objectives are determined by πk [22].

5   Numerical Example

5.1   Computational Experiments

The GAMS approach is an exact approach that is used for small and medium-scale problems. ANTIGONE solver is used for solving three problems. These problems' data are determined below (Table 2).

The first problem's data is presented below:

Table 2. Initial information of the first problem

Sets	
Set of potential distributions	A, B
Set of demand points	1, 2, 3
Set of ground vehicles	K1, k2
Set of relief items	Med, Food
Set of helicopters	h1, h2, h3

This example assumes two distributions and three disaster areas. Fix costs of establishing distributions are considered A = 40, B = 30. The relief items' volumes are Medicine = 2, Food = 1. Transportation costs for ground vehicles are k1 = 5, K2 = 7 and the normal speed of them are Vk1 = 50, Vk2 = 60. Its vehicles' capacities are k1 = 300 and K2 = 400. The helicopters' speeds are considered Vh1 = 30, Vh2 = 35, Vh3 = 20, moreover, both loading and discharging times for helicopters are considered 2. Weight of demand points are 1 = 0.2, 2 = 0.3 and 3 = 0.5 (Tables 3, 4 and 5).

Table 3. Distance between distributions and demand points of the first problem

	1	2	3
A	100	100	80
B	150	120	110

Table 4. Road damage of the first problem

	1	2	3
A	0.2	0.1	0.4
B	0.3	0.4	0.3

Table 5. Repair time of route for the first problem

	1	2	3
A	1	2	1
B	2	1	1

The following tables show the data used to solve the model in the second instance (Tables 6 and 7):

Table 6. The demand for relief items for the first problem

	1	2	3
Medicine	5	7	8
Food	6	5	5

Table 7. Initial information of the second problem

Sets	
Set of potential distributions	A, B, C
Set of demand points	1, 2, 3, 4, 5, 6
Set of ground vehicles	k1, k2
Set of relief items	Med, Food
Set of helicopters	h1, h2, h3, h4, h5, h6

In the second instance, we considered three distribution centers and six demand areas. The fixed cost of establishing distributions is considered A = 40, B = 30, C = 36. The relief items' volumes are Medicine = 2, Food = 1. Transportation costs for ground vehicles are K1 = 5, K2 = 5 and the normal speed of them are Vk1 = 50, Vk2 = 60. Its vehicles' capacity of the first and second vehicles are 300 and 400, respectively. The helicopters' speeds are considered Vh1 = 30, Vh2 = 35, Vh3 = 20, Vh4 = 30, Vh5 = 20 and Vh6 = 30. Moreover, both loading and discharging time for helicopters are considered 2. Weight of demand points are 1 = 0.2, 2 = 0.3, 3 = 0.5, 4 = 0.4, 5 = 05 and 6 = 0.7 (Tables 8, 9, 10, and 11).

Table 8. Distance between distributions and demand points of the second problem

	1	2	3	4	5	6
A	100	100	80	100	180	80
B	150	120	110	120	200	60
C	100	90	200	139	130	100

Table 9. Road damage of the second problem

	1	2	3	4	5	6
A	0.2	0.1	0.4	0.3	0.2	0.4
B	0.3	0.4	0.3	0.4	0.1	0.4
C	0.5	0.7	0.4	0.1	0.7	0.4

Table 10. Repair time of route for the second problem

	1	2	3	4	5	6
A	1	2	1	2	1	2
B	2	1	1	2	3	2
C	2	3	1	1	3	2

Table 11. The demand for relief items for the second problem

	1	2	3	4	5	6
Medicine	5	7	8	3	7	7
Food	6	5	5	6	5	4

5.2 Model Validation

In this section, we verify the quality of the model and evaluate some analyses. Finally, we approve the validation of the model.

If all loading times are multiplied twice at the first problem, the completion time of the helicopter will be changed (old result + (number of helicopters*2)). As Fig. 1 illustrates in the first problem, number 2 means that each loading time is doubled, and number 4 shows that every loading bar is multiples four times and so on.

Figure 2 shows the repair cost that is added to several units (old number + 3*number of the route). As the given bar char presents in the first problem, number 3 shows repair cost is added three units and so on.

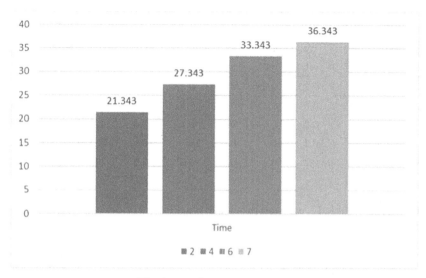

Fig. 1. Loading time validation

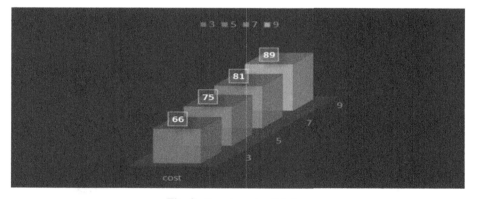

Fig. 2. Repair cost validation

6 Result and Discussion

6.1 Results

All initial parameters are determined and used in the GAMS. Different objective weights obtain various answers, but all objectives cannot satisfy, simultaneously.

All the first problem results are shown in Table 12:

Table 12. Small scale's results

Test problem number	Weight of objective function				Objective function value			
	w1	w2	w3	w4	f1	f2	f3	f4
1	0.25	0.25	0.25	0.25	12.42	25087	5.2	8.25
2	0.6	0.2	0.1	0.1	11.65	23365	5	9.04
3	0.2	0.6	0.1	0.1	12.6	17365	5	9.04
4	0.2	0.1	0.6	0.1	12.6	26550	5.6	9.04
5	0.2	0.1	0.1	0.6	12.6	26550	5	7.34

According to Table 12, each test problem has a different priority, and selecting each of them depends on the situation. In test problem number 1, all objectives have equal weight, so priority is not considered. The first and second objectives have some challenges to each other. As a result, neither of them is satisfied. The first objective selects vehicles with the highest speed while they are expensive, so the second objective chooses other vehicles. In other test problems, the results have changed.

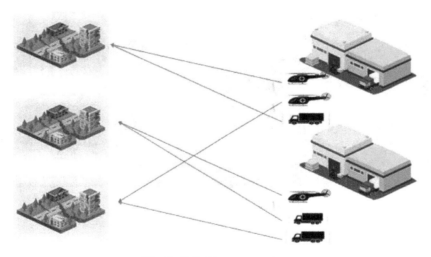

Fig. 3. Relief items operation

The given Fig. 3 shows how helicopters and trucks are allocated to demand points. Distribution A sends helicopter 3 to disaster area number 1 and helicopter 2 to disaster area number 3. Also, distribution B sends helicopter 1 to disaster area number 2. Ground vehicles are allocated like helicopters. Distribution A covers one area and distribution B covers two areas. GAMS determine this allocation.

All the second problem results are shown in Table 13:

Table 13. Medium scale's results

Test problem number	Weight of objective function				Objective function value			
	w1	w2	w3	w4	f1	f2	f3	f4
1	0.25	0.25	0.25	0.25	28.996	16981	15.7	12.04
2	0.6	0.2	0.1	0.1	27.196	17028	14.1	18.94
3	0.2	0.6	0.1	0.1	29.112	14821	14.1	18.94
4	0.2	0.1	0.6	0.1	29.112	19821	16.8	18.94
5	0.2	0.1	0.1	0.6	29.112	19821	14.1	9.87

As Table 13 illustrates, different test problems are evaluated. It depends on the situation in which managers select the results. Like the previous problem, all objectives are not satisfied. Test problem 1 considers all objective functions have equal priority and solves the model.

6.2 Sensitive Analysis

Now in this section key parameters are evaluated, but at the first, the research wants to show how much its innovation has an effective impact on the response time. As mentioned above, the chief objective of this research is minimizing the response time. As Fig. 4 shows, airborne vehicles have an efficient effect on response time. Obviously, if helicopters are not used for humanitarian operations, vital equipment will be delivered to demand points at an inappropriate time.

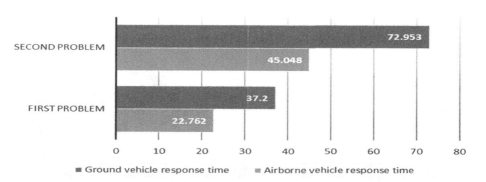

Fig. 4. Comparing airborne and ground vehicles

Figure 4 indicates that without helicopters at the first problem, this operation needs 37.2 h, and by considering helicopters, it needs 22.762 h. This time is reduced to about 15 h by airborne vehicles, also at the second problem, this reducing time is about 28 h. This completion time does not consider the demand points' weight. Thus, all the above results prove that its innovation has an efficient effect on humanitarian operations.

Three important parameters that have a deep effect on response time are loading time, discharge time, and helicopters' speed. If the summation of these parameters is more than the ground vehicles' time, airborne vehicles will not be used. The given chart in Fig. 5 shows the comparison between objective number 4 and number 1.

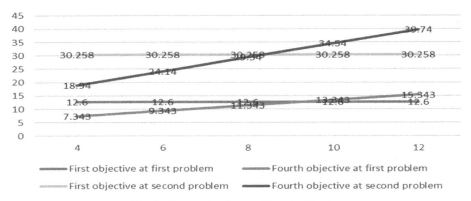

Fig. 5. Comparing first and fourth objective

In Fig. 5, the horizontal axle shows the sum of loading and discharge time for airborne vehicles. If total loading and discharge time have increased, ground vehicles will reach demand points sooner than airborne vehicles and it is not efficient because medicines have higher priority than the other relief items. Also, the chief objective of using helicopters is to minimize the response time and build a field hospital at an appropriate time. Therefore, using helicopters in this situation is not intellectual and economical due to helicopters are far more expensive than the trucks. Moreover, they reach demand points after trucks.

Now the road repair time is evaluated. As Fig. 6 presents, repair time has a deep effect on ground vehicles' response time, so if the repair times are more than a specific

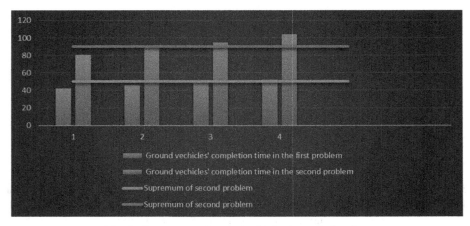

Fig. 6. Supremum of ground vehicles 'completion time

amount of time, the ground vehicles will not be used for operation since they cannot transfer relief items at the appropriate time. Figure 1 shows that every repair time is added by one and so on.

The manager of relief items operation can determine the specific number for ground vehicles' completion time. When their completion times exceed their supremum, the manager can use just airborne vehicles to the minimum completion time.

6.3 Discussion

We evaluate all results in two problems, and in this section, some challenges are examined. In the previous section, results prove that in some cases airborne vehicles are more efficient than ground vehicles and, in some cases, ground vehicles are more efficient, while in most cases cooperate both vehicles have the appropriate effect on humanitarian relief items operation. This approach can reduce the casualty and increase the response time to extend possible. But sometimes, airborne approaches are not used because of inappropriate weather; it is clear that helicopters cannot start their mission if weather is unsuitable for airborne operation. Moreover, it is so dangerous. So, the manager should focus on ground vehicles and repair groups to send relief items and medicine.

In some cases, the geographical location of disaster areas or demanding points is mountainous or hilly, so it is difficult to transfer relief items to demand points by ground vehicles or establishing field hospitals. In this case, the manager should decide on operations. Airborne vehicles are efficient in these areas and can transfer the injured people to appropriate situations and after the nurses and doctors start to cure them.

Ground and airborne vehicles are utilized to minimize the response time to extend possible. Although in some cases one of them is more efficient, the cooperation of them can help the humanitarian operation effectively, also manager role should not be ignored.

7 Conclusion

This research illustrates one humanitarian relief item operation which considers repair groups, different relief items, airborne and ground vehicles. It has four objectives for reducing response time and cost and increasing the reliability of the route. Reliability is one of the important evaluation criteria that is utilized for assessing different products [23]. Therefore, it is used to assess the route in this paper. Two problems (small and medium) are considered and solve by GAMS, and their results are presented in previous sections. The chief objective of this model is reducing response time to decrease casualties, and this aim is satisfied by considering airborne vehicles. They are used to transfer initial field hospitals' equipment and ample doctors and nurses; meanwhile, ground vehicles and repair groups transfer relief items. The models' results prove that helicopters have a special effect on the results and minimize response time very efficiently. Some challenges are available for this model, and managers should decide between them to choose the best approach for each of them.

Different weights are assigned to demand points by operation's information, and the operation starts to satisfy the demand point which has the highest weight or priority. This approach needs to correct information to determine their weight appropriately.

Further research for this study can be done in different sections. The limited budget for establishing distributions or using airborne or ground vehicles can be added. Limited ground or airborne vehicles will be an appropriate innovation that shows the importance of demand points better.

Appendix A

In this section, the linear model of Eq. 16 is presented. To solve the model by the GAMS (ANTIGONE solver), the linear model should be used.

$$\sum_{j/(j,i)\in E} y_{jikt} - \sum_{j/(j,i)\in E} y_{ijkt} \leq VF_{ikt}.M \quad , i \in N, t \in T : i \neq j$$

$$\sum_{j/(j,i)\in E} y_{jikt} - \sum_{j/(j,i)\in E} y_{ijkt} \geq -z_{ikt}.M \quad , i \in M, \forall k \in K, t \in T : i \neq j$$

Two above equations are presented in Eq. 16 in the model.

References

1. Chang, F.-S., et al.: Greedy-search-based multi-objective genetic algorithm for emergency logistics scheduling. Expert Syst. Appl. **41**(6), 2947–2956 (2014)
2. Edrissi, A., Nourinejad, M., Roorda, M.J.: Transportation network reliability in emergency response. Transp. Res. Part E: Logist. Transp. Rev. **80**, 56–73 (2015)
3. Leknes, H., et al.: Strategic ambulance location for heterogeneous regions. Eur. J. Oper. Res. **260**(1), 122–133 (2017)
4. Vahdani, B., Veysmoradi, D., Shekari, N., Mousavi, S.M.: Multi-objective, multi-period location-routing model to distribute relief after earthquake by considering emergency roadway repair. Neural Comput. Appl. **30**(3), 835–854 (2016). https://doi.org/10.1007/s00521-016-2696-7
5. Nedjati, A., Vizvari, B., Izbirak, G.: Post-earthquake response by small UAV helicopters. Nat. Hazards **80**(3), 1669–1688 (2015). https://doi.org/10.1007/s11069-015-2046-6
6. Lerner, E.B., et al.: Mass casualty triage: an evaluation of the science and refinement of a national guideline. Disaster Med. Public Health Prep. **5**(2), 129–137 (2011)
7. Rezaei-Malek, M., et al.: An approximation approach to a trade-off among efficiency, efficacy, and balance for relief pre-positioning in disaster management. Transp. Res. Part E Logist. Transp. Rev. **93**, 485–509 (2016)
8. Rezaei-Malek, M., Tavakkoli-Moghaddam, R.: Robust humanitarian relief logistics network planning. Uncertain Supply Chain Manag. **2**(2), 73–96 (2014)
9. Ghodratnama, A., Tavakkoli-Moghaddam, R., Azaron, A.: Robust and fuzzy goal programming optimization approaches for a novel multi-objective hub location-allocation problem: a supply chain overview. Appl. Soft Comput. **37**, 255–276 (2015)
10. Jemai, Z., Aboueljinane, L., Sahin, E.: Emergency ambulance deployment in Val-De-Marne department a simulation-based iterative approach. In: Proceedings of the 3rd International Conference on Simulation and Modeling Methodologies, Technologies and Applications, vol. 1, pp. 565–576 (2013)
11. Altay, N. and Green III, W.G.: OR/MS research in disaster operations management. Eur. J. Oper. Res. **175**(1), 475–493 (2006)

12. Travers, D.A., et al.: Five-level triage system more effective than three-level in tertiary emergency department. J. Emerg. Nurs. **28**(5), 395–400 (2002)
13. Kahn, C.A., et al.: Does START triage work? An outcomes assessment after a disaster. Ann. Emerg. Med. **54**(3), 424–430. e1 (2009)
14. Pidgeon, N., O'Leary, M.: Man-made disasters: why technology and organizations (sometimes) fail. Saf. Sci. **34**(1–3), 15–30 (2000)
15. Cutter, S.L., et al.: A place-based model for understanding community resilience to natural disasters. Glob. Environ. Chang. **18**(4), 598–606 (2008)
16. Berlin, G.N., Liebman, J.C.: Mathematical analysis of emergency ambulance location. Socioecon. Plann. Sci. **8**(6), 323–328 (1974)
17. Kim, S.H., Lee, Y.H.: Iterative optimization algorithm with parameter estimation for the ambulance location problem. Health Care Manag. Sci. **19**(4), 362–382 (2015). https://doi.org/10.1007/s10729-015-9332-4
18. Abazari, S.R., Aghsami, A., Rabbani, M.: Prepositioning and distributing relief items in humanitarian logistics with uncertain parameters. Socio-Econ. Plan. Sci. **74**, 100933 (2021)
19. Danesh Alagheh Band, T.S., Aghsami, A., Rabbani, M.: A post-disaster assessment routing multi-objective problem under uncertain parameters. Int. J. Eng. **33**(12), 2503–2508 (2020)
20. Momeni, B., Aghsami, A., Rabbani, M.: Designing humanitarian relief supply chains by considering the reliability of route, repair groups and monitoring route. Adv. Ind. Eng. **53**(4), 93–126 (2019)
21. Rabbani, M., Aghsami, A., Farahmand, S., Keyhanian, S.: Risk and revenue of a lessor's dynamic joint pricing and inventory planning with adjustment costs under differential inflation. Int. J. Procur. Manag. **11**(1), 1–35 (2018)
22. Ringuest, J.L.: Lp-metric sensitivity analysis for single and multi-attribute decision analysis. Eur. J. Oper. Res. **98**(3), 563–570 (1997)
23. Azadeh, A., Sheikhalishahi, M., Aghsami, A.: An integrated FTA-DFMEA approach for reliability analysis and product configuration considering warranty cost. Prod. Eng. Res. Devel. **9**(5–6), 635–646 (2015). https://doi.org/10.1007/s11740-015-0642-7

Author Index

Printed in the United States
by Baker & Taylor Publisher Services